Planetary Science: A Lunar Perspective

STUART ROSS TAYLOR

The Moon depicted in a painting by Donato Creti. Oil on canvas, 1711. [*Bedini, S. A. (1980) PLC 11: xiii.*] *Courtesy the Vatican Monumenti, Musei e Gallerie Pontificie. This illustrates the state of planetary observation at the beginning of the eighteenth century. Most of the succeeding advances in this field have occurred in the past decade.*

Planetary Science:
A Lunar Perspective

STUART ROSS TAYLOR

Lunar and Planetary Institute
Houston, Texas, U.S.A.

and
Research School of Earth Sciences
Australian National University
Canberra, Australia

LUNAR AND PLANETARY INSTITUTE
3303 NASA ROAD 1
HOUSTON, TEXAS 77058

SCI
QB
601
.738
1982

Cover: An artistic perception of a portion of Mare Ingenii (S. Adlis-Vass, artist).

Back cover: An artistic rendition of the view south across Mare Imbrium (S. Adlis-Vass, artist).

To
Noël
Susanna, Judith and Helen

Preface

The technical triumph of manned landings on the Moon and the
return of samples from the lunar surface has provided scientists with a
unique opportunity to advance our understanding of the nature, evolution
and origin of the solar system. The nine suites of samples recovered by the
Apollo and Luna missions have provided so many answers that it is now
difficult to recall our state of ignorance before July, 1969. The new knowledge is contained in a massive outpouring of scientific literature, conservatively estimated at 18,000 articles which have appeared in the past decade.

In a previous book, I attempted to survey and comprehend the state of
lunar science at the close of the active period of lunar exploration. That
work records the state of lunar science as of April, 1974, following the evaluation of the results from the final manned mission, Apollo 17, in
November, 1972. In that book, I was concerned with extracting the evolutionary history of the Moon from the great mass of data on the lunar samples. The story as recorded in *Lunar Science: A Post-Apollo View* has
survived reasonably well. Many new data have appeared, new models have
been proposed, and refinements of existing models have been carried out.
No substantial reason exists to attempt the essentially minor changes of a
second edition: instead I have adopted the maxim "what I have written, I
have written" [1], letting the work stand as a statement of lunar science at
that time, and I have essayed a new approach.

In 1974, only early results from other planets were available. Now a
considerable advance in our knowledge of the solar system has occurred
and I have attempted to extend the knowledge gained from the lunar data
to provide some insights into this wider sphere. Accordingly, I have used
the lunar story as a basis for understanding the origin and evolution of the
solar system. In the early days of the manned missions, the term "Rosetta
Stone" was often used. In a sense, the usage is correct. The study of the
lunar samples has proven to be a key to understanding, for example, much
about meteorites. The Moon remains the only body which we have visited
and sampled with a proper understanding about the relation of the samples
to the overall evolution of the planet. The Moon, as revealed from those
studies, is sufficiently different from the Earth, both to render terrestrial
analogies hazardous, and to provide another example of planetary differentiation processes. The tendency to regard the Earth as a norm, and terrestrial geochemical processes as a model for those on other planets, has not
survived the encounter with the lunar data. Many details, such as the contrasting behavior of chromium on the two planets, have provided salutary
warnings of the hazards of extrapolating from one planet and of the problems of the statistics of small numbers. The same warning applies to the
lunar samples. Accordingly, I have tried to utilize the lunar data along with

our long terrestrial experience to provide a synthesis. The subject matter and treatment is heavily biased toward the lunar data, since these have provided the major advance in our understanding of planetology.

Some of the information given in the previous book has changed little since 1974, and it has proven necessary to retraverse some familiar ground. For example, the lunar stratigraphic sequence was already well established, and only the subdivision of the Pre-Imbrian into Nectarian and Pre-Nectarian is new. Some of the same figures have been reused. The beautiful earth-based lunar photographs taken by the Lick Observatory have never been surpassed and it seemed pointless to replace them. They are still worthy of study, for our spacecraft photographs of many planets and satellites are at about the same resolution. The reconstructions of the lunar surface in previous ages by Davis and Wilhelms reappear for similar reasons and there are a few other similar instances throughout the book. In some disciplines, such as organic geochemistry, no significant new material has appeared, and the reader is referred to the previous text for enlightenment.

During the writing of *Lunar Science: A Post-Apollo View*, a very large number of questions about the Moon had not been resolved, and were the subject of intense debate. Thus the question of whether water was present on the Moon was under consideration; the nature of the regolith-forming processes awaited proper study of the core tubes. The questions of origin of the Cayley Formation and light plains deposits, and of whether volcanism existed in the highlands continued to persist. The composition and origin of the highland crust was a major problem. Possibly, it had been added to the Moon like a coat of plaster. Although the volcanic versus impact debate over the origin of the craters had been resolved before the landings, the formation of multi-ring basins was in dispute. The origin of mare basalts, whether from primitive or cumulate sources, was under active discussion, together with the question of the origin of the europium anomaly. There were still questions about model age interpretations. No Sm-Nd data were available. These were first reported in March, 1975, at the 6th Lunar Science Conference and immediately disposed of single-stage models for mare basalt genesis. Whether the Moon had been wholly melted, or not melted at all, was also a matter of dispute. The solar nebula was generally held to have been isotopically homogeneous with the data from the oxygen isotopes just beginning to suggest otherwise. Thus the intellectual climate in which the book was written was very different from that of today. It was at the time a fascinating intellectual exercise to judge among the many competing explanations, particularly when they lay outside my own specialty.

Further analyses of the lunar samples have resolved many of the problems listed above, but a number remain unsolved. Many questions suffer from lack of sufficient resolution in the data, a problem which may be referred to as the "Martian Canal Syndrome."

As in the previous book, the emphasis is on interpretation and synthesis. The Moon appears particularly well suited to this type of approach. It has a comparatively simple history, and responds well to large-scale overviews. The Moon is built on a large scale. The infinite complexity of some lunar samples such as the highland breccias is due to the repeated effects of large impacts, a comparatively simple concept, although the effects are complex.

I have avoided the temptation to include much about the Earth in this book, for various reasons: (a) the amount of information available is huge; (b) adequate treatment would demand so much space that it would be difficult to do justice to the subject and avoid swamping the data from the other planets and satellites; (c) the surficial geology, volcanic landforms, plate tectonic mechanisms and much else, including the atmosphere and hydrosphere, are unique to this planet and cannot be addressed adequately in the space available; and (d) our insights into the composition, structure and evolution of the core, mantle and crust are in many ways at an early stage. Much work remains to be carried out on the significance of the trace element and isotopic evolution of the mantle. In this context, the picture of lunar evolution which has emerged has provided new insights into the early history of the Earth. Occasionally I have been unable to resist commenting on various aspects, such as the development of the continental crust and the bulk composition of the terrestrial mantle (Chapter 8), which form an important constraint on our ideas of planetary evolution, and on the relationship between the Earth and the Moon. Finally, the origin of the Earth is addressed briefly in Chapter 9.

The problems in dealing with the information from the other planets and satellites are large. Those workers who have expected more insights on planets other than the Moon should recall that our knowledge of these bodies is comparable to that of the Moon in the pre-Apollo era. There is excellent photography, some geophysical data and a little chemistry from remote samplers. The analogy with the data base for the Moon in 1968–69 is nearly complete.

Our experience with the lunar samples was salutary and informed us that much of our pre-mission thinking was in error. This is a common situation in science. In a book such as this, concerned with the scientific results of the space program, we must be cautious not to push the data beyond the limits of resolution, and so fall into Martian canals or dust-filled maria.

The quality of the photographs returned from the Pioneer, Mariner, Voyager and Viking missions has provided one of the more aesthetic experiences of the age. Analogies with the heroic period of the early terrestrial navigators are appropriate. Perhaps the beauty and strangeness of the surfaces of the planets and satellites will prompt a new literary or poetic flowering. We must not let those excellent pictures deceive us into believing that we fully understand all that is visible or, by extrapolation, that we comprehend the nature of the planets or satellites. This phase of exploration is a preliminary, rather than a terminal phase. The appropriate analogy is with the pre-Apollo and pre-Surveyor stages of lunar exploration. The Ranger, Orbiter and Russian photographs of the lunar surface resolved only a few questions about the nature of the lunar surface. "Information was limited to the images and once again it was demonstrated that a surface cannot be characterized by its portrait . . . the heightened resolution of the pictures did not resolve the arguments. The moon remained inscrutable at all scales" [2]. "It is evident that pictures of strange material in a strange environment do not lead to unique deductions as to composition" [3]. These cautionary tales are to remind us that our knowledge of the other members of the solar system is similar to our pre-Apollo understanding of the Moon. This is a reason for continued exploration at an enhanced level, including sample return. Accordingly, in this book, which is biased toward cosmochemistry, I have emphasized the lunar data, despite many temptations to extrapolate beyond our present understanding of the other planets and satellites.

Meteorites provide unique and crucial evidence on early conditions in the solar nebula, and the lunar data have shed much light on previously obscure problems. It is, however, not possible in a book of this length to make more than passing reference to them, mainly in Chapters 8 and 9. Theories for the origin and evolution of the solar system must accommodate the new meteoritic evidence, which demands book-length treatment in its own right.

The very large body of literature has called for much selection and evaluation. I have attempted to provide a synthesis, rather than a catalogue, believing the former approach to be of greater service to the reader. Accordingly, I have had to make various choices among competing models such as on the questions of the existence of lunar cores, terminal cataclysms, magma oceans, KREEP volcanism, hot or cold nebulae, planetesimals or gaseous protoplanets, and many other fascinating but hazardous exercises. I have been fortified in these endeavors by discussions with the superb body of lunar scientists [4] who form an imposing tribute to the rational methods of scientific inquiry.

I have tried generally to reference the latest comprehensive statements on particular topics. I have also attempted to list references to all sources of fact, information, opinion and interpretation other than my own. It is essential in a book of this kind both that a proper perspective be presented and that appropriate formal referencing be attempted. Unsupported statements amount to expressions of dogma, acceptable in works on science fiction, astrology, religion, paranormal phenonoma, extra-sensory perception and the occult, but not, since the Renaissance, in scientific literature. Books which are commonly not subject to peer review processes may occasionally fall into such a credibility trap [5]. Accordingly, I have attempted both extensive referencing and a measure of peer review [4].

The number of references cited has led, as in the previous book, to the use of an abbreviated style of referencing. It is not practical in a work of this size to cite the fully expanded reference for each paper. Since many of the references are to the *Proceedings of the Lunar and Planetary Science Conferences*, an enormous amount of repetition would occur. One of the aims of this book is to serve as an introduction to the lunar literature, so that the interested reader can pursue any topic in depth and consult the original references. The numbered references at the end of each chapter contain (1) authors names (where there are more than two, the initial author followed by et al., is used); (2) year; (3) abbreviated journal or book title; (4) volume number; (5) initial page number or page number quoted. This provides enough information to enable the reference to be quickly located. Full referencing details including the journal titles are given in Appendix I. In Appendix II, a list of primary data sources is given. I have tried to avoid references to unrefereed papers, internal reports and other "grey" literature, but sometimes this has been unavoidable.

In a book of this size, it is difficult to include an extensive data base. As an example, the recent treatise on *Basaltic Volcanism on the Terrestrial Planets* runs to over 1300 pages on a topic which I have treated in one chapter. Some of the material which I prepared for use in that book has been incorporated in this text with minor amendments.

I have attempted not to dwell excessively on facts which are well known and widely accepted, but have devoted more space to the controversial aspects of the subject. This carries the risk of producing some imbalance in the book, but I have judged it more profitable to adopt this approach rather than to inflict tedious descriptions of well-known material which is readily available in other sources.

The literature coverage extends to November, 1981, including references to the *Multi-ring Basin Conference*, the *Apollo 16 Workshop*, the *Proceedings of Lunar and Planetary Science Conference*, 12B, and to *Basaltic Volcanism on the Terrestrial Planets*. All radiometric age dates have been recalculated in terms of the new decay constants [6].

Stuart Ross Taylor
Houston and Canberra
November, 1981

References and Notes

1. Pontius Pilate (The Gospel according to St. John, 19:22, King James Version).
2. Scott, R. F. (1977) *Earth Sci. Rev.* 13: 379.
3. Urey, H. C. (1966) *Science.* 153: 1420.
4. See Acknowledgments.
5. See for example Smith, J. V. (1980) *J. Geol.* 88: 250.
6. See Chapter 6, reference [105].

The Author

Stuart Ross Taylor, M.A., D.Sc. (Oxford), M.Sc. (N. Z.), Ph.D. (Indiana), F.A.A., is a Professorial Fellow at the Research School of Earth Sciences, the Australian National University. Dr. Taylor has carried out extensive investigations in lunar and terrestrial trace element geochemistry, including research work on analytical techniques, the composition and evolution of the continental crust, island-arc volcanic rocks, tektites, meteorite impact glasses, lunar samples and models for the geochemical evolution of the Moon. He has published 170 scientific papers and three books, including *Lunar Science: A Post-Apollo View* (Pergamon, 1975).

Dr. Taylor was a member of the Lunar Sample Preliminary Examination Team (LSPET) for Apollo 11 and 12 and carried out the first chemical analysis of a lunar sample. He is a Principal Investigator for the lunar sample program and a Fellow of the Meteoritical Society and of the Australian Academy of Science. He is an editor of *Geochimica et Cosmochimica Acta* and *Chemical Geology*.

Acknowledgments

This book was written over a period of fourteen months (October 1980–November 1981), principally at the Lunar and Planetary Institute, Houston. I am grateful to Dr. Roger Phillips, Institute Director, for enabling me to work for extended periods in Houston, providing access both to the library and data facilities and the Johnson Space Center, and to a wide spectrum of lunar scientists.

I owe, as do all workers in planetary research, an initial debt to Harold Urey and Ralph Baldwin for their early and perceptive insights, which set the intellectual stage for planetary exploration. A further tribute is due to the scientists of the lunar and planetary community on whose work this book is based. During the past decade, I have had fruitful discussions with innumerable individual scientists, and have formed many lasting friendships, so that it is difficult to single out those who have made a special contribution. I am grateful to them for the encouragement to write this book. This body of scientists, interdisciplinary and international, can deal competently with any current scientific observation in the solar system and is perhaps the greatest benefit to have come from the space program.

Books have long been considered as an authoritative source of information, but unless subject to peer review, may commonly reflect the prejudices of the author. Some bias is inevitable, but I have attempted to avoid this trap by prevailing upon several scientists to read chapters dealing with their own specialties. Dr. Don E. Wilhelms has read Chapter 2 on geology and Chapter 3 on cratering. Dr. Fred Hörz also read Chapter 3 and Chapter 4 on planetary surfaces. Professor Larry Taylor has read Chapter 5 on planetary crusts and Dr. David Walker, Chapter 6 on basaltic volcanism. Chapter 7, which deals with many geophysical aspects, has been read by Dr. Roger Phillips, Dr. Len Srnka and Dr. Michael McElhinny. Professor J. V. Smith read both Chapter 8 on planetary compositions and Chapter 9 on the origin and evolution of the solar system.

I am exceedingly grateful to these workers for their time and patience in toiling through rough draft material. Their comments have substantially improved the text, and saved me from various grievous errors. The responsibility for interpretations, shortcomings and errors remains my own.

The production of this book on a short time scale would not have been possible without the aid and help of the staff of the Lunar and Planetary Institute. Rosanna Ridings, Managing Editor, contributed greatly to the efficient editing and production of this book. She was ably assisted by Dory Brandt, Kelly Christianson, Reneé Edwards, Lanet Gaddy and Donna Theiss. The preparation of the figures was carried out with artistic skill by Donna Jalufka-Chady, Sharon Adlis-Vass, and David Powell, with S. Adlis-Vass acting as lead designer. Carl Grossman, assisted by Gayle Croft, was responsible for the typesetting of the book. Production of final pages was coordinated by Pamela Thompson. Ron Weber was responsible for supplying much of the photographic material. Jack Sevier read over the entire text. Fran Waranius placed the excellent and essential facilities of the library freely at my disposal. In Canberra, Gail Stewart typed many preliminary drafts with speed and precision.

Many colleagues have supplied illustrative material. The length of the list and the diversity of scientific disciplines represented is a tribute to the interdisciplinary nature of the subject. I am grateful for the help of the following individuals:

I. Adler	J. B. Murray
E. Anders	L. E. Nyquist
C. G. Andre	M. Norman
R. N. Clayton	J. J. Papike
J. W. Delano	R. J. Phillips
R. A. F. Grieve	C. M. Pieters
W. K. Hartmann	J. B. Pike
J. W. Head	S. Pullan
B. R. Hawke	L. Raedeke
F. Hörz	D. J. Roddy
R. M. Housley	C. T. Russell
J. S. Huebner	G. Ryder
R. L. Kovach	M. Sato
J. Longhi	R. S. Saunders
G. W. Lugmair	J. V. Smith
I. S. McCallum	S. C. Solomon
J. F. McCauley	D. Stöffler
D. S. McKay	R. G. Strom
G. A. McKay	L. A. Taylor
A. L. Metzger	G. J. Wasserburg
J. W. Morgan	

A particular acknowledgment is due to Dr. Robin Brett, who ensured my initial participation in lunar studies.

Acknowledgment is made to the following for permission to use copyrighted material:

Anglo-Australian Observatory (Fig. 9.1)

Cerro Tololo Inter-American Observatory (Fig. 9.2)

Lick Observatory (Figs. 2.1, 2.2, 2.4, 2.7, 6.9a)

Scientific American (Fig. 9.3)

The National Aeronautics and Space Administration (NASA) provided the initial sample material, collected with care and expertise, which has made possible the writing of this book. Additional support for the writing of this book was provided through NASA Contract No. NASW 3389, made possible by the kind efforts of Dr. William L. Quaide of NASA.

CONTENTS

Preface   i

Acknowledgments   ix

Chapter 1   The Exploration of the Solar System   1
  1.1   Early Influences   1
  1.2   Lunar Sampling   3
  1.3   The Moon and the Solar System   12
        References and Notes   15

Chapter 2   Geology and Stratigraphy   17
  2.1   The Face of the Moon   17
  2.2   Stratigraphy of the Lunar Surface   26
  2.3   Radiometric and Stratigraphic Lunar Time Scales   29
  2.4   Detailed Lunar Stratigraphy   32
        2.4.1   Pre-Imbrian   32
        2.4.2   Pre-Nectarian   33
        2.4.3   Nectarian   33
        2.4.4   The Imbrian System   36
        2.4.5   Eratosthenian System   39
        2.4.6   Copernican System   39
  2.5   Mercury   40
        2.5.1   Intercrater Plains   40
        2.5.2   The Heavily Cratered Terrain   44
        2.5.3   Smooth Plains   47
        2.5.4   Lobate Scarps   47
  2.6   Mars   50
        2.6.1   The Ancient Cratered Terrain   51
        2.6.2   Volcanic Plains and Mountains   51
  2.7   Venus   55
  2.8   Gallilean and Saturnian Satellites   57
        References and Notes   57

Chapter 3   Meteorite Impacts, Craters and Multi-Ring Basins   61
  3.1   The Great Bombardment   61
  3.2   The Volcanic/Meteorite Impact Debate   62
  3.3   The Mechanism of Crater Formation   64
  3.4   Simple Bowl-Shaped Impact Craters   68
  3.5   Large Wall-Terraced Craters, Central Peaks and Peak Rings   71

3.6   The Multi-Ring Basins   76
3.7   Depth of Excavation of Basins   87
3.8   Basin Diameter-Morphology Relationship   91
3.9   Impact Melts and Melt Sheets   92
3.10 Secondary Impact Craters   93
3.11 Dark-Halo Craters   96
3.12 Smooth-Rimmed Craters and Floor-Fractured Craters   97
3.13 Isostatic Compensation of Craters   99
3.14 Linné Crater and Transient Phenomena   100
3.15 Swirls   100
3.16 Meteorite Flux   100
3.17 The Cratering Record and the Lunar Cataclysm   104
3.18 Secondary Cratering and the Cayley Problem   109
      References and Notes   110

Chapter 4   Planetary Surfaces   **115**
4.1   The Absence of Bedrock   115
4.2   The Extreme Upper Surface   116
4.3   The Regolith   117
     4.3.1 Thickness and Mechanical Properties   118
     4.3.2 Structure   121
     4.3.3 Rate of Accumulation   123
     4.3.4 "Ages" of the Soils   123
     4.3.5 Soil Breccias and Shock Metamorphism
           of Lunar Soils   127
4.4   Glasses   128
     4.4.1 Agglutinates   128
     4.4.2 Impact Glasses   130
     4.4.3 Tektites   135
4.5   Chemistry and Petrology of the Regolith   139
     4.5.1 Metallic Iron in the Lunar Regolith   142
     4.5.2 Volatile Element Transport on the Lunar Surface   147
     4.5.3 The Meteoritic Component   151
4.6   Microcraters and Micrometeorites   151
4.7   Irradiation History of the Lunar Surface   155
     4.7.1 Solar Wind   155
     4.7.2 Solar Flares   155
     4.7.3 Galactic Cosmic Rays   157
4.8   Physical Effects of Radiation   157
4.9   Chemical and Isotopic Effects   159
     4.9.1 Carbon, Nitrogen and Sulfur   159
     4.9.2 The Rare Gases   161

4.9.3	The Argon-40 Anomaly	162
4.9.4	Cosmogenic Radionuclides	163
4.9.5	Thermal Neutron Flux	163
4.10	Exposure Ages and Erosion Rates	163
4.10.1	Nuclear Track Data	165
4.10.2	Rare Gas Data	166
4.11	Solar and Cosmic Ray History	167
4.12	The Lunar Atmosphere	169
4.13	Rare Gases and Planetary Atmospheres	169
4.14	Organic Geochemistry and Exobiology	171
	References and Notes	171
Chapter 5	**Planetary Crusts**	**177**
5.1	The Lunar Highland Crust	180
5.1.1	Thickness and Density	180
5.1.2	The Megaregolith	186
5.2	Breccias	187
5.2.1	Monomict Breccias	191
5.2.2	Dimict Breccias	192
5.2.3	Feldspathic Fragmental Breccias	194
5.2.4	Impact Melt Breccias	195
5.2.5	Granulitic Breccias	198
5.2.6	Basalts in Highland Breccias	200
5.3	Melt Rocks	201
5.4	Highland Crustal Components	201
5.4.1	Anorthosites	205
5.4.2	The Mg Suite	206
5.4.3	KREEP	208
5.4.4	KREEP Volcanism in the Lunar Highlands?	214
5.5	Volatile Components in the Highland Crust	216
5.6	The Ancient Meteorite Component	219
5.7	Pristine Rocks in the Lunar Highlands	221
5.8	The Orbital Chemical Data	223
5.9	The Chemical Composition of the Highland Crust	227
5.9.1	Element Correlations	228
5.9.2	Highland Crustal Abundances	230
5.10	Age and Isotopic Characteristics of the Highland Crust	233
5.10.1	The Oldest Ages	233
5.10.2	The 4.2 Aeon Ages	238
5.10.3	Basin Ages and the Lunar Cataclysm	240
5.11	Evolution of the Highland Crust	242
5.11.1	The Magma Ocean	242

5.11.2 How Long did the Early Highland Crust
   Take to Evolve?   244
5.11.3 Crystallization History of the Lunar Crust   246
5.12 The Crust of the Earth   253
5.13 Other Planetary Crusts   256
  References and Notes   258

Chapter 6 Basaltic Volcanism   **263**
6.1 Floods of Basaltic Lava   263
 6.1.1 Thickness of Mare Fill   265
 6.1.2 Age of the Oldest Mare Surface   266
 6.1.3 The Lunar Lava Flows   269
 6.1.4 Dark Mantle Deposits   272
 6.1.5 Domes and Cones   274
 6.1.6 Ridges   276
 6.1.7 Sinuous Rilles   278
 6.1.8 Straight and Curved Rilles   282
6.2 Mare Basalt Rock Types   282
 6.2.1 Mineralogy   284
 6.2.2 Reduced Nature of Mare Basalts   292
 6.2.3 Volcanic Glasses   297
6.3 Chemistry of Mare Basalts   300
 6.3.1 Major Elements   300
 6.3.2 Large Cations   301
 6.3.3 Rare-Earth Elements and the
    Europium Anomaly   308
 6.3.4 High Valency Cations   309
 6.3.5 The Ferromagnesian Elements   311
 6.3.6 Sulfur and the Chalcophile Elements 313
 6.3.7 The Siderophile Elements   315
 6.3.8 Oxygen and Carbon   316
6.4 Ages and Isotopic Systematics of the Mare Basalts 317
 6.4.1 The Commencement of Mare Volcanism 317
 6.4.2 Radiometric Ages for the Mare Basalts 318
 6.4.3 Isotopic Indexes of Mantle Heterogeneity and
    Basalt Source Ages   320
6.5 Origin of the Mare Basalts   321
6.6 Cooling and Crystallization of the Lavas   331
 6.6.1 Primary Magmas   331
 6.6.2 Fractional Crystallization   333
6.7 The Record Elsewhere   335
 6.7.1 Mercury—Are Basalts Present?   335

6.7.2	Mars	335
6.7.3	Venus	336
6.7.4	Io	337
6.7.5	Vesta	338
	References and Notes	338

Chapter 7   Planetary Interiors — **343**

7.1	The Advantages of Geophysics	343
7.2	Radii, Densities and Moments of Inertia	343
7.3	Lunar Center of Mass/Center of Figure Offset	344
7.4	Gravity	345
7.4.1	Lunar Gravity	345
7.4.2	The Mascons	350
7.4.3	Mars	352
7.4.4	Venus	354
7.5	Seismology	354
7.5.1	The Lunar Record	354
7.5.2	The Martian Record	356
7.6	Internal Structures	356
7.6.1	The Moon	356
7.6.2	A Lunar Core?	358
7.6.3	Mars, Venus and Mercury	359
7.7	Heat Flow and Internal Temperatures	361
7.7.1	Heat-Flow Data	361
7.7.2	Electrical Conductivity	361
7.7.3	The Lunar Temperature Profile	362
7.8	Magnetic Properties	364
7.8.1	Lunar Magnetism	364
7.8.2	Remanent Magnetism	364
7.8.3	Lunar Magnetic Field Paleointensities	366
7.8.4	Crustal Magnetic Anomalies	367
7.8.5	Origin of the Lunar Magnetic Field	368
7.8.6	Planetary Magnetic Fields	370
	References and Notes	372

Chapter 8   The Chemical Composition of the Planets — **375**

8.1	The New Solar System	375
8.2	The Type 1 Carbonaceous Chondrites (C1)	380
8.3	The Earth	381
8.3.1	Primitive Mantle (Mantle Plus Crust)	381
8.3.2	The Crust	389
8.4	The Moon	390

8.4.1	The Highland Crust	390
8.4.2	The Mare Basalts	393
8.4.3	Bulk Moon Compositions	395
8.4.4	Earth-Moon Comparisons	396
8.5	The Planets	402
8.6	Meteorite Parent Bodies and Satellites	404
	References and Notes	405

Chapter 9   The Origin and Evolution of the Moon and Planets   **409**
9.1	The Beginning	409
9.2	Initial Conditions in the Solar Nebula	411
9.3	Primary Accretion Models and "Condensation" in the Nebula	415
9.4	Variations Among Refractory, Volatile, Chalcophile and Siderophile Elements	417
9.5	Early Solar Conditions	418
9.6	Secondary Accretion Process	420
9.7	Formation of the Moon	423
9.8	The Formation of the Earth	429
	References and Notes	431

Chapter 10   The Significance of Lunar and Planetary Exploration   **435**
10.1	The Lunar Experience	435
10.2	Future Missions	436
10.3	Man's Responsibility in the Universe	438
	References and Notes	439

Appendix I
Reference Abbreviations   **443**
Appendix II
Lunar Primary Data Sources   **446**
Appendix III
Diameter and Location of Lunar Craters Mentioned in the Text   **450**
Appendix IV
Lunar Multi-ring Basins   **454**
Appendix V
Terrestrial Impact Craters   **455**
Appendix VI
Appendix to Table 8.1: Sources of Data   **458**

Appendix VII
    Element Classification, Condensation Sequences   460
Appendix VIII
    Chondritic Rare Earth   462
Appendix IX
    The Lunar Sample Numbering System   463
Appendix X
    Planetary and Asteroidal Data   464
Appendix XI
    Satellite Data   465
Glossary   467
Index   475

Chapter 1

THE EXPLORATION OF THE SOLAR SYSTEM

1.1 Early Influences

In 1419, Henry the Navigator (1394–1460) founded, at Sagres on the southwestern tip of Portugal, what we would now call an Institute of Maritime Research. This date conveniently marks the commencement of the heroic age of oceanic exploration and of our understanding of the geography of this planet. Under Henry's sponsorship, the Portuguese captains discovered the Canary and Cape Verde Islands and sailed as far south as Sierra Leone, dispelling medieval terrors that the edge of the world lay just south of Cape Bojador (latitude 26° N), which accordingly marked the southern limit of safe navigation [1].

This exploration was made possible not only through the administrative skills of Henry but also because of technical advances in ship design. These advances led to the construction of truly ocean-going vessels such as the caravel, and to improvements in navigational devices, of which the magnetic compass was the most important.

We are now at an analogous stage in history. The advances in technology, which have resulted in rockets, spacecraft, computers and rapid data transmission have, in two or three decades, enabled an unparalleled exploration of the solar system. This has opened perspectives so new that we are still endeavoring to assimilate and comprehend the information. The present state of planetary exploration is shown in Fig. 1.1. This figure emphasizes the fact that we have orders of magnitude less information for each successive planet or satellite, knowing most about the Moon, but almost nothing about Pluto. This book reflects our current state of knowledge which is heavily influenced by the extensive lunar data. We are more fortunate now than in our studies of the planets in pre-Apollo time, when all our direct experiences and analogies were confined to the Earth.

1.1 The current status of exploration of the solar system. The amount of information increases by orders of magnitude for each step upwards on the vertical scale.

For example, our experience in dating surfaces by crater counting techniques, although well understood by some workers, was so insecurely based or accepted that estimates varying by orders of magnitude appeared in the literature as late as 1969. Such uncertainties persist as we voyage toward the outer reaches of the solar system, where the meteorite flux rates become less well understood [2].

The scientific exploration of the solar system represents the culmination of a process whose roots go back to the earliest stages of human thought and development. The strange motions of the planets, wandering among the fixed stars, the monthly waxing and waning of the Moon, the cycles of the seasons, the occasional occurrences of eclipses, and the apparition of comets provided an incentive to record and understand all those celestial events.

In this context the Moon, as the closest and most obviously variable heavenly body, has played a dominating role [3]. Tantalizingly out of reach to poets and princesses alike, its features are sufficiently intriguing to stimulate not only myth-making and the production of calendars, but also the construction of telescopes and spacecraft.

1.2   Lunar Sampling

Six Apollo missions returned a total of 382 kg of rocks and soil from the Moon. Three Russian unmanned landers brought back 250 gm (see Table 1.1 for details of the lunar landings and Table 1.2 for a listing of successful planetary missions). Various questions arise from these visits: (a) Was the sampling adequate? (b) How much can we tell about the Moon from nine suites of samples? (c) Were the manned landings necessary [4,5]?

The sampling sites for the lunar missions are shown in Fig. 1.2. The Apollo 11 mission collected 22 kg rather hurriedly within about 30 m of the Landing Module (LM). This first landing on that distant and alien shore was brief. The accessible surface, the regolith, contained rocks excavated by meteorite impacts of varying depths from the local mare basalts; bedrock in the terrestrial sense lay several meters deeper. The ubiquitous debris blanket—the regolith—however, mirrors with reasonable faithfulness the local bedrock so that extended field work in many different sites was a useful and productive exercise. The development of a lunar vehicle, the Rover, enabled the astronauts to traverse distances up to 20 km on Apollo missions 15, 16 and 17. Accordingly, detailed collecting of specialized samples from differing terrains became possible, particularly since precise navigation (Fig. 1.3) enabled landings to be carried out in narrow valleys (e.g., Apollo 15 at Hadley-Apennines and Apollo 17 in the Taurus-Littrow region) adjacent to mare-highland boundaries. These achievements enabled the collecting of samples which reasonably can be related to various mappable formations.

Table 1.1 Lunar exploration by spacecraft.

Successful Pre-Apollo Lunar Landings

Spacecraft	Date	Landing site	Data returned
Ranger 7	August, 1964	Mare Cognitum	Photographs
Ranger 8	February, 1965	Mare Tranquillitatis	Photographs
Ranger 9	March, 1965	Crater Alphonsus	Photographs
Luna 9	February, 1966	Western Oceanus Procellarum	Photographs
Surveyor I	June, 1966	Oceanus Procellarum, north of Flamsteed	Photographs
Luna 13	December, 1966	Western Oceanus Procellarum	Photographs; soil physics
Surveyor III	April, 1967	Oceanus Procellarum (Apollo 12 site)	Photographs; soil physics
Surveyor V	September, 1967	Mare Tranquillitatis (25 km from Apollo 11 site)	Photographs; soil physics; chemical analyses
Surveyor VI	November, 1967	Sinus Medii	Photographs; soil physics; chemical analyses
Surveyor VII	January, 1968	Ejecta blanket of Crater Tycho (North Rim)	Photographs; soil physics; chemical analyses

Apollo Lunar Landings

Mission	Landing Site	Latitude	Longitude	EVA duration (hours)	Traverse distance (km)	Date	Sample Return (kg)
11	Mare Tranquillitatis	0°67'N	23°49'E	2.24	—	July 20, 1969	21.7
12	Oceanus Procellarum	3°12'S	23°23'W	7.59	1.35	Nov. 19, 1969	34.4
14	Fra Mauro	3°40'S	17°28'E	9.23	3.45	Jan. 31, 1971	42.9
15	Hadley-Apennines	26°06'N	3°39'E	18.33	27.9	July 30, 1971	76.8
16	Descartes	8°60'S	15°31'E	20.12	27	April 21, 1972	94.7
17	Taurus-Littrow	20°10'N	30°46'E	22	30	Dec. 11, 1972	110.5

Russian Lunar Sample Missions

Mission	Landing site	Latitude	Longitude	Date	Sample Return (grams)
Luna 16	Mare Fecunditatis	0°41'S	56°18'E	Sept. 1970	100
Luna 20	Apollonius highlands	3°32'N	56°33'E	Feb. 1972	30
Luna 24	Mare Crisium	12°45'N	60°12'E	Aug. 1976	170

Russian Lunar Traverse Vehicles

Vehicle	Landing site	Date	Traverse Length
Lunokhod 1 (Luna 17)	Western Mare Imbrium	Nov. 1970	20 km
Lunokhod 2 (Luna 21)	Le Monnier Crater, Eastern Mare Serenitatis (180 km north of Apollo 17 site)	Jan. 1973	30 km

Table 1.2 Planetary exploration.[†]

Mission	Launch Date	Target	Encounter Date
Pioneer 5	March 1960	Interplanetary	—
Mariner 2	August 1960	Venus flyby	Dec. 1962
Mariner 4	Nov. 1964	Mars flyby	July 1965
Pioneer 6	Dec. 1965	Interplanetary	—
Pioneer 7	August 1966	Interplanetary	—
Venera 4	June 1967	Venus landing	Oct. 1967
Mariner 5	June 1967	Venus flyby	Oct. 1967
Pioneer 8	Dec. 1967	Interplanetary	—
Pioneer 9	Nov. 1968	Interplanetary	—
Venera 5	Jan. 1969	Venus landing	May 1969
Venera 6	Jan. 1969	Venus landing	May 1969
Mariner 6	Feb. 1969	Mars flyby	August 1969
Mariner 7	March 1969	Mars flyby	August 1969
Venera 7	August 1970	Venus landing	Dec. 1970
Mars 3	May 1971	Mars landing	Dec. 1971
Mariner 9	May 1971	Mars orbit	Nov. 1971
Pioneer 10	March 1972	Jupiter flyby	Nov. 1973
Venera 8	March 1972	Venus landing	July 1972
Pioneer 11	April 1973	Jupiter	Nov. 1974
		Saturn	Sept. 1979
Mars 5	July 1973	Mars orbit	March 1974
Mariner 10	Nov. 1973	Venus flyby	Feb. 1974
		Mercury flyby	March, Sept. 1974 March 1975
Helios	Dec. 1974	Sun approach	—
Venera 10	June 1975	Venus landing	Oct. 1975
Viking 1	August 1975	Mars landing	July 1976
Viking 2	Sept. 1975	Mars landing	Sept. 1976
Helios 2	Jan. 1976	Sun approach	—
Voyager 2	August 1977	Jupiter	July 1979
		Saturn	August 1981
		Uranus	Jan. 1986
Voyager 1	Sept. 1977	Jupiter	March 1979
		Saturn	Nov. 1980
Pioneer Venus 1	May 1978	Orbiter	Dec. 1978
Pioneer Venus 2	August 1978	multiprobe	Dec. 1978
Venera 11	Sept. 1978	Venus flyby and probe	Dec. 1978
Venera 12	Sept. 1978	Venus flyby and probe	Dec. 1978

[†]Mariner, Pioneer, Viking and Voyager were US missions. Helios was a joint US/Federal Republic of Germany mission, and Mars and Venera were USSR missions.
Note: Only successful missions are listed.

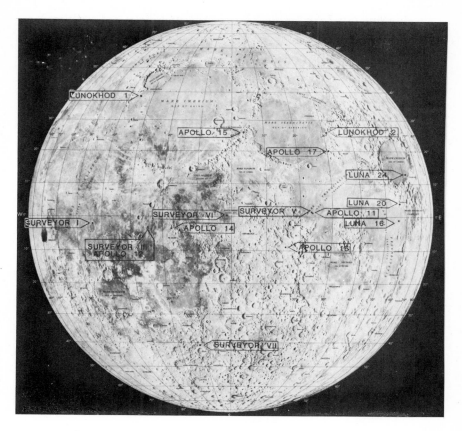

1.2   Surveyor, Apollo, Luna and Lunokhod landing sites on the Moon.

The sampling was thus adequate for us to obtain a first order appreciation of the nature of the lunar surface and of the varying stratigraphic relationships. The limited number of missions has raised a number of detailed stratigraphic problems whose resolution can only be achieved by further missions. Such problems include the evolution of the early highland crust and the detailed sequence of the gigantic basin collisions—pressing intellectual questions of much significance for the early history of the solar system.

In this book, the claim is made that the data from the nine lunar missions provide a key to unlock both lunar evolutionary history and to shed light on the restricted and circumscribed information from the other planets and satellites. This claim would not be valid for the Earth where no combination of three spot samples and six from areas of a few square kilometers could, without hindsight, have led to a synthesis of terrestrial geological processes.

1.3   Astronaut Pete Conrad at the Surveyor III lander, during the Apollo 12 mission. The Lunar Landing Module is in the background.

The overall geology of the Moon is much simpler, with a basic two-fold division into the dark maria and the highlands. Sampling of the basalts has yielded about 20 varieties, but these are related by reasonably well-understood variations in chemistry. An important conclusion is that they are not all uniform, but indicate some heterogeneity, again within our comprehension, deep in the lunar interior.

The samples from the highlands were so smashed up by the early bombardment of the Moon that traces of the original crust are exceedingly difficult to identify. Nevertheless, the chemical composition survives to tell its tale of these events, close in time to the formation of the solar system. The moon-wide nature of these events lends special significance to the correlation of the surface sampling both with the photogeological mapping and remote-sensing data and with the orbital geochemical values obtained from XRF and gamma-ray experiments. The integration of all this information enables us to relate the surface sampling to the broad lunar perspective established by stratigraphic procedures. We are thus able to construct models, with the aid of the vital ages established on the returned samples, which are tantalizingly close to final answers. The question of a lunar core, the resolution of the magnetic puzzles, the detailed evolution of the highland crust, the origin of KREEP, and some other problems, await only a minimal addition of data and samples from future lunar missions.

Although manned exploration of the Moon sometimes has been considered superfluous, it was crucial to a proper understanding of the Moon. The reason lies in the nature of the lunar surface (and of the surfaces of other bodies which lack atmospheres). An automated sample return, in the current state of technology, obtains a small drill core of soil hopefully with a few rock fragments. Only our experience with the lunar samples enables us to extract correct information from such a sample. If our lunar sampling had been restricted to such material it would have been difficult and perhaps impossible to discern the true story. The lunar soil is a complex mixture, formed by the prolonged meteoritic bombardment of the lunar surface over a period of three to four aeons. The rock samples collected by the astronauts enable us to investigate the individual components of the mixture (Fig. 1.4).

The most critical observation was that of the age of the material. The basaltic rocks from the initial landing site had clearly established crystallization ages of 3.6–3.8 aeons. The complicated soil mixtures indicated *model* Rb-Sr ages of around 4.5 aeons, close to the accepted age for the formation of the solar system. This paradox was understood to result from a combination of a primordial differentiation of the basaltic source regions at about 4.4 aeons, and a small separation of Rb from Sr during the formation of the basaltic magma at the younger epoch. Thus the soils recorded the earlier event [6]. The redistribution of volatile Rb during meteorite impacts led to some apparent ages in excess of 5 aeons, while the addition of exotic components such as KREEP, rich in Rb or anorthosite with primitive $^{87}Sr/^{86}Sr$ ratios, all contributed to confuse the story. If the total lunar sample had been only a few grams, it might have been impossible to disentangle the true age sequences. An alternative scenario, difficult to disprove, would have compressed the entire lunar evolutionary story (formation of the highland crust, meteoritic bombardment and basaltic eruptions) into a time span of 100–200 million years producing a catastrophic picture of early planetary evolution, as misleading to planetology as the phlogiston theory was to chemistry.

Now, with the skill and experience obtained from lunar sample studies, it is possible to extract information from a few grams of soil and rock fragments

1.4   Astronaut Harrison H. (Jack) Schmitt collecting samples during the Apollo 17 mission.

as was demonstrated by the studies of the samples from Luna 16, 20 and 24. But these skills were not easily acquired. Faced with a minute amount of sample, two problems arise: (1) Which problems shall be attacked, and (2) which laboratories shall carry out the analyses? It was by no means clear in 1969 that the age and isotopic results, closely followed by the trace element chemistry, would provide the most significant information. The allocation of over 2 kg of sample for biomedical testing indicates a differing set of priorities, a scenario repeated in the Viking missions to Mars.

It was also not clear in 1969 which scientific teams possessed or would develop the highest skills, for who "can look into the seeds of time, and say which grain will grow and which will not" [7]. The scientific community is reluctant to accept one result from one laboratory as the ultimate truth. One of the great benefits from the large amount of sample available from the Apollo missions, and of the enlightened policy of distribution established by NASA, has been the formation of a new science of planetology, comprising a scientific community rich in expertise, self-checking and self-regulating which contains many individuals, unknown in 1969, who have made significant contributions to our understanding. It has also been found that high quality scientific work can be carried out rapidly without loss of precision, accuracy or understanding [8]. It is to be hoped that the administrative successors to Henry the Navigator, will ensure the survival of this unique asset, in a society which shows some signs of relapsing into medieval patterns of thought.

Accordingly, the manned missions saved us from probable errors of interpretation and, with hindsight and experience, we now are capable of extracting significant information from a small sample return. The Viking experience on Mars, however, warns us that a soil sample from that planet is unlikely to contain useful rock fragments, but is more likely to resemble a wind-blown desert sand. Accordingly, some device for breaking off pieces of the abundant rocks—coupled with surface mobility extending to kilometer ranges at least—is required for a Martian sample return. Possibly, we may see an advancement in mass spectrometric techniques that will enable us to obtain reliable ages by remote sampling, but the experience even with a relatively straightforward technique such as X-ray fluorescence in obtaining chemical information from Mars illustrates the difficulties. The biological experiments indicate the problems in interpreting unusual or unexpected data in a mini-laboratory on a distant planetary surface [9].

A further question, which can be addressed with hindsight, is whether the Apollo sites and sampling techniques could have been different. The experience gained in the early missions was in practice rapidly incorporated into successive visits. The walking traverses of Apollo 11 and 12 were supplemented by a hand-drawn cart resembling a golf buggy, on Apollo 14, and by the roving vehicles on the final three missions. Each carefully selected site provided unique samples. In retrospect, more attention to magnetic and heat-flow measurements earlier in the missions would have provided useful information, but the major gaps would have been filled by the three cancelled missions. Most damage was done by the premature termination of the landing program and the decision to turn off the ALSEP experiments on September 30, 1977, when many instruments were still recording data [10]. The seismic data from one large impact on the far side on July 17, 1972, provided not only unique information about the lunar interior, but also the expectation of further such events. The most useful immediate information can now be gained from a polar orbiter, providing a moon-wide picture of the surficial distributions of the radioactive elements, the variations in Al/Si ratios in the highlands, the mapping of the differing mare basalt types, and the moon-wide variations in surface magnetism.

1.3   The Moon and the Solar System

In a celebrated comment, Newton said that if he saw further, it was because he stood on the shoulders of giants [11]. The Moon provides us with an analogous platform from which to comprehend the other planets and satellites.

The first, and possibly the most critical advantage, is that it provides us with a well-established stratigraphic sequence, to which an absolute chronology may be fixed by the radiometric dating of the returned samples. Such information, discussed in the next chapter, enables us to apply similar reasoning to the less accessible surfaces of other planets and satellites. This concept is of particular importance because of the ubiquitous evidence of extensive early cratering throughout the solar system. The cratering question has had a long and varied history, hampered by our experience of living on the surface of a planet from which most of the record has been erased. The efficiency of terrestrial erosion indeed made it difficult for the scientific community to recognize and accept impact processes. As T. H. Huxley remarked, "it is the fate of new truths to begin as heresies." Even now, vestiges of alternative internally generated processes appear [12], although the mineralogical evidence for instantaneous shock pressures exceeding 500 kbar at impact sites has removed internal volcanic explanations from consideration [13].

As discussed in Chapter 3, the Moon provides us with sequences of crater forms only dimly perceived on Earth. The great lunar craters have always excited interest. The recognition of the existence of a larger class of multi-ring basins, with diameters reaching thousands of kilometers, was a product of detailed lunar mapping and has provided critical evidence for the existence of large objects up to several hundred million years after the formation of the planets. This early bombardment record is interpreted to provide evidence in support of the planetesimal hypothesis for planetary growth. One lesson which has become apparent from the studies of the giant multi-ring basins, and of the large size of Martian canyons and volcanoes, is that much of our comprehension of geological processes based on terrestrial experience has been on too small a scale. Indeed, Sir William Hamilton perceived this truth in 1773 when he commented, after many years of observations of Mt. Vesuvius, that "we are apt to judge of the great operations of Nature on too confined a plan" [14]. Much of terrestrial geology, examined in road cut, drill core or thin section does encourage the development of expertise in the minutae of geology. In this context, the plate tectonic revolution was wrought by ocean-going geophysicists, perhaps accustomed to wider horizons, than by land-based stratigraphers and paleontologists.

The Moon has provided vital information on the nature of surfaces developed on rocky planets in the absence of atmospheres. Early ideas that the mare basins contain kilometer thicknesses of fine dust were dispersed by the

Surveyor evidence of a firm cohesive surface. The debate over the presence of water on the lunar surface was resolved only after the Apollo sample return. In this context, the mineralogical evidence in the large rock samples returned by the astronauts provided decisive evidence of a dry Moon in a way that the fine-grained, often glassy soils could not [15]. Among many other features of the lunar surface discussed in Chapter 4 is the possibility of establishing the long term history of the sun.

The nature and origin of differentiated crusts on planets was illuminated by the lunar missions. Early geochemical thinking had considered the Moon to be a primitive object, captured into Earth orbit and resembling the carbonaceous chondrites in composition (the Martian satellites, Phobos and Deimos, are probably examples of such objects). It will become apparent to readers of this book that the Moon has provided us with much more information than if it had been a large carbonaceous chondrite. The highly differentiated lunar crust was a surprise to most lunar workers and stimulated thinking in general about early planetary models. Clearly, if one had to produce a strongly chemically fractionated crust close on the heels of accretion of the planet [16], then considerable deviations from formerly accepted models of planetary formation were called for. Decisive evidence of the operation of crystal-liquid fractionation, rather than of gas-solid condensation processes, as described in Chapter 5, indicated early moon-wide melting processes. The feldspathic crust of the Moon, generated by flotation during crystallization, stands in great contrast to either the oceanic or continental crusts of the Earth, generated by varying episodes of partial melting from the mantle. Although the lunar highland crust bears a superficial resemblance to the terrestrial continental masses, the distinction in origin reminds us that each planet may be unique. The Mercurian crust may be closest to that of the Moon, but the differing densities and bulk compositions must engender caution until we have more geochemical and petrological data. Mars and Venus present different aspects of crustal genesis, so far as we can judge from the available evidence. The surfaces of the satellites of Jupiter and Saturn, lately revealed for our curious inspection, provided so many surprises that "the sense of novelty would probably not have been greater had we explored a different solar system" [17]. The tendency of solar system bodies to develop crusts distinct from their bulk composition by processes possibly unique in detail for each body provides a major stimulus to develop theories of planetary evolution.

Basaltic eruptions have long been familiar on the Earth [18], although their full extent was only realized with the discovery of the mid-oceanic ridges and of the basaltic composition of the oceanic crust. The lunar maria constitute a second example of the widespread occurrence of lavas generated by partial melting deep within planetary mantles. The Moon provided examples which indicated that terrestrial petrological experience was not all-embracing.

The surprising differences in titanium enrichments and europium depletions from familiar terrestrial lavas provided evidence for differing evolutionary histories for lunar and terrestrial mantles. The isotopic systematics told of extensive early differentiation of the Moon, while the trace elements revealed the complementary nature of the highland crust and the deep source regions of the mare basalts. The early assumptions that the lunar interior, in so far as it is sampled by the basaltic lavas, might be primitive, and so yield the bulk composition of the planet, gave way to models of zoned mantles of varying mineralogy (Chapter 6). These scenarios contrast strongly with our models of the terrestrial mantle. Accordingly, we must expect surprises from Martian lavas, even though Olympus Mons has a profile resembling that of Mauna Loa. The composition of basalts and possibly even of granites on Venus, for which the Venera gamma-ray data for K, U, and Th hold promise, is likely to provide unique information on the internal constitution of that planet. Basaltic volcanism on Mercury remains an enigma.

The state of planetary interiors, as discussed in Chapter 7, illustrates just how many data are needed to make unique interpretations from the geophysical data. We lack adequate resolution from the lunar seismic experiments [10] to decide whether the Moon has a core, and to pass judgement on the reality of discontinuities within the lunar mantle. The heat-flow data suffers from having only two measurements, although the frustrations of geochemists have been tempered somewhat by the realization that the bulk uranium content of a planet is not a simple function of the heat flux. The magnetic evidence has proven perplexing, but an understanding is slowly being reached with the development of techniques for preserving the magnetic memory of the sample (carried by fine-grained iron) in a wet oxidizing terrestrial atmosphere.

It is sometimes considered surprising that geochemists are bold enough to construct tables of planetary composition from a few basic parameters. As discussed in Chapter 8, various interlocking sets of constraints from isotopic and element ratios, coupled with the observation that planetary compositions differ in their contents of refractory, volatile and siderophile elements, enables a large degree of internal self-consistency to be achieved in these estimates. When integrated with geophysical parameters such as density, moment of inertia, magnetic properties and mantle structures revealed by seismology, significant statements can be made about bulk planetary compositions to an extent not possible before the lunar missions. The data from the meteorites, in all their complexity, are relevant to our understanding of much of early solar system history. The study of the lunar samples has shed much light on meteoritic problems, formerly so intractable that a distinguished geochemist, in 1965, pronounced the chemical evidence in the meteorites to be unreadable [19].

Chapter 9 addresses the basic intellectual question of the origin of the planets in the light of the evidence assessed in this book. A sober reading of the literature on this topic over the past three decades since the appearance of *The Planets* by Harold Urey might daunt the most accomplished reviewer, but progress in realistic scenarios and reduction in the numbers of free parameters is occurring rapidly. Although it is conventional to lament the complexity of modern knowledge and the difficulty of obtaining an overview, it should be recalled that the Renaissance scientists, often envied for working in a supposedly simpler situation, had to comprehend the complexities of medieval thought, if only to dismiss such topics as alchemy and astrology from rational consideration.

References and Notes

1. The Institute at Sagres was destroyed in 1587 in a raid led by Francis Drake, designed to disrupt preparations for the attack by the Spanish Armada. [Mattingly, G. H. (1959) *The Defeat of the Spanish Armada*, Jonathan Cape, London.]

2. See, for example, the controversy over the ages of Martian features. [Neukum, G., and Hillier, K. (1981) *JGR*. 86: 3097.]

3. The influence of the Moon on primitive art is illustrated with many beautiful photographs in Bedini, S. A., et al. (1973) *Moon*, Abrams, N.Y.

4. The traverses and details of sample collecting are described in the following sources:
    Apollo 11:   LSPET (1969) *Science*. 165: 1211; NASA SP 214 (1969); NASA SP 238 (1971); USGS Map I-619 (1970); Beaty, D. W., and Albee, A. L. (1980) *PLC 11*: 23.
    Apollo 12:   LSPET (1970) *Science*. 167: 1325; NASA SP 235 (1970); USGS Map I-627 (1971).
    Apollo 14:   USGS Apollo Geology Team (1971) *Science*. 173: 716; NASA SP 272 (1971); USGS Map I-708 (1970).
    Apollo 15:   USGS Apollo Geology Team (1972) *Science*. 175: 407; NASA SP 289 (1972); USGS Map I-723 (1971).
    Apollo 16:   USGS Apollo Geology Team (1973) *Science*. 179: 62; NASA SP 315 (1972); USGS Map I-748 (1972); USGS Prof. Paper 1048 (1981).
    Apollo 17:   USGS Apollo Geology Team Report (1973) *Science*. 182: 672; NASA SP 330 (1973); USGS Map I-800 (1972).

5. The lunar sample numbering system is described in Appendix IX.

6. Wetherill, G. W. (1971) Of Time and the Moon, *Science*. 173: 383.

7. Shakespeare, W. (1606) *Macbeth*, Act 1, Scene III (comment by Banquo to the three witches on the blasted heath).

8. Creative work can be accomplished in brief time scales, contrary to popular wisdom. Thus Handel wrote the Messiah between August 22 and September 14, 1741. Mozart produced his three final symphonies (No. 39 in E flat, K 543 ; No. 40 in G minor, K 550; and No. 41 in C minor, K 551) within a period of two months (early June–August 10, 1788). The G minor symphony has been considered by at least one critic to provide sufficient justification for the existence of Homo sapiens [Einstein, A. (1957) *Mozart: His Character, His Work*, 3rd ed., Cassell, London].

9. See Cooper, H. S. F. (1980) *The Search for Life on Mars*, Holt, Rinehart, and Winston, N.Y., for a readable account of these problems. See also Soffen, G. A. (1981)

Chapter 9 in *The New Solar System* (eds. Beatty, J. K., et al.), Sky Publishing, Cambridge, Mass.

10. Bates, J. R., et al. (1979) ALSEP Termination Report. NASA Ref. Pub. 1036. This publication provides a description of the Apollo Lunar Surface Experiments Package (ALSEP) for Apollo missions 11–17, of their operational history, and of the significant scientific results.

11. Letter to Robert Hooke, Feb. 5, 1675.

12. See introduction in Roddy, D. J., et al., eds. (1977) *Impact and Explosion Cratering*, Pergamon Press.

13. Such overpressures cannot be built up at shallow depths in the crust where the confining pressure at 40 km is only 10 kbar.

14. Hamilton, W. (1773) *Observations on Mt. Vesuvius, Mt. Etna and other Volcanoes*, T. Cadell, London, p. 161. This is one of the first modern works on volcanology. This distinguished naturalist is, alas, better known to history as the husband of Emma, Lady Hamilton.

15. A sample of terrestrial desert sand could be so used to infer the absence of water on Earth.

16. The Moon is commonly referred to as a planet in this text. The large size of the satellite relative to the primary justifies consideration as a double planet system. In addition, as suggested by one worker, it makes for simpler sentences.

17. Smith, B. A., et al. (1979) *Science*. 204: 951.

18. The igneous nature of basalt was demonstrated by James Hall (1805); See Lofgren G. E., in Hargraves, R. B. (1980) *Physics of Magmatic Processes*, Princeton, Chap. 11.

19. A recent review by J. V. Smith [(1982) Heterogeneous accretion of meteorites and planets especially the Earth and Moon. *J. Geol.*, in press] provides an excellent, if brief, statement of the significance of the meteoritic evidence, and much else.

Chapter 2

GEOLOGY AND STRATIGRAPHY

2.1 The Face of the Moon

The Moon is close enough to the Earth to enable its major surface features to be discerned. The effect of this fact on human intellectual development has provided a major stimulus to subjects as diverse as cosmology and the construction of telescopes. Accordingly, it is appropriate to begin this chapter with a brief discussion of the lunar surface. This provides the best stepping stone to the complexities of the inner solar system.

Following Galileo's observation in 1610 that the Moon did not have a smooth spherical surface, but was rough and mountainous, speculation about the origin of the lunar landforms continued for three and a half centuries. Our appreciation of the forces responsible for shaping the surface of our own planet has taken about the same time to comprehend. Many of the major lunar features can be ascertained in Figs. 2.1 and 2.2. Figure 2.1 is a full-moon photograph, on which the most prominent features are the young rayed craters, such as Tycho, Copernicus, and Kepler. These rayed craters are the results of the most recent events on the Moon, and the debris resulting from the great collisions that formed them overlies the older formations. The details of the surface morphology are seen in Fig. 2.2, a composite photograph. The division into light-colored heavily cratered highlands and smooth dark maria is clearly apparent.

Many stratigraphic truths may be obtained from a study of these photographs. The best earth-based telescopic views of the Moon are equivalent in resolution (about 1 km) to many of those obtained by spacecraft photography of other planets and satellites. Thus, it is of value to study the lunar surface features and stratigraphy, since the surface of the Moon serves as a yardstick against which geological and stratigraphical interpretations of photographs of more distant objects may be tested.

17

2.1 This view of the full Moon is at about the same resolution as many spacecraft photos of other planets and satellites. The rays from Tycho dominate the southern portion of the photograph. Mare Imbrium is prominent in the northwest quadrant, immediately north of the bright ray crater Copernicus. Mare Serenitatis lies to the east of Imbrium. (Lick Observatory photograph.)

The basic stratigraphy of the Moon was reasonably well understood prior to the initial manned lunar landing in 1969. In the *Geology of the Moon: A Stratigraphic View*, Mutch [1] presented essentially the pre-Apollo view of lunar stratigraphy. With the exception of the interpretation of the Cayley Formation, the broad stratigraphic picture which he presented has survived reasonably well. Most changes have occurred due to our increased understanding of the dominating effect of meteorite impact processes (and the downgrading of volcanic processes) as responsible for shaping the morphology of the highlands.

2.2 This composite full-moon photograph shows the contrast between the heavily cratered light highlands and the smooth dark maria. Note the radial structure southeast from the Apennine Mountains bordering Mare Imbrium, which provided the first clue to the impact origin of the Imbrium basin. (Lick Observatory photograph.)

A fundamental step in understanding lunar geology was the notion that the widespread ridges radial to Mare Imbrium, and the circular structure itself, were the result of a gigantic collision. This interpretation was first advanced by Gilbert [2] in 1893, and was later discovered independently several times. Another crucial step was the recognition that the smaller craters were mostly due to meteorite impact and not to volcanic processes. This latter view is still not entirely extinct [3], and remnants of this view survived long enough to influence Apollo site selection [4]. The Orbiter spacecraft photography first revealed to full view the Orientale basin (Fig. 2.3), although its

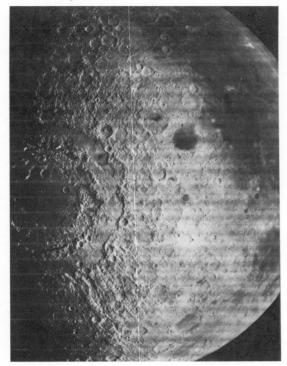

2.3a Mare Orientale, the classic example of a multi-ring basin. The diameter of the outer mountain ring (Montes Cordillera) is 900 km. The structures radial to the basin are well developed. The small mare basalt area to the northeast is Grimaldi. The western edge of Oceanus Procellarum fills the northeast horizon (NASA Orbiter IV 181 M).

2.3b An oblique view of the Orientale multi-ring basin (NASA Orbiter IV 193 M).

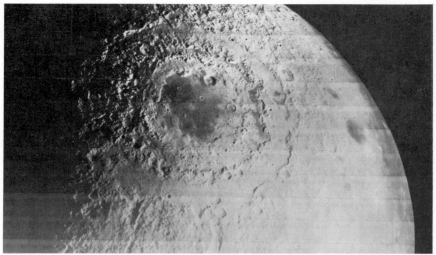

2.3c The central mare basalt fill and the outer rings of Mare Orientale (NASA Orbiter IV 194 M).

presence had been inferred earlier by Hartmann and Kuiper [5]. This nearly perfectly preserved far-side feature provided the best evidence for the impact of very large objects and clarified the nature of the circular basins, which was less obvious on the earth-facing side. These concepts of an early lunar history dominated by cratering and giant basin forming events opened the way to construct a stratigraphy for the lunar surface based upon the observed super-position relationships.

The oldest regions observable are the lunar highlands, saturated with large craters (50–100 km diameter) (Fig. 2.4). Coincident with the cratering was the excavation of the giant ringed basins. The youngest examples, Imbrium and Orientale, have affected wide areas of the lunar surface, but many more are clear (Fig. 2.2). Near-side and far-side mapping of these has revealed the existence of 30 basins with diameters in excess of 300 km. Fourteen others probably exist [6] (Table 3.1, Appendix IV). The basins are apparently distributed evenly on both near and far sides of the Moon. The apparent anisotropy of the Moon is caused by the paucity of mare basalt flooding on the far side, not by a different cratering history.

It is possible to reconstruct the appearance of the Moon at the close of the great bombardment, at about 3.9 aeons [7] as shown in Fig. 2.5 [8] by removing the younger formations.

Following closely, and indeed overlapping the terminal or earlier stages of the intense cratering, came the floods of basaltic lava, which are so prominent on the visible face. These events on the Moon have terrestrial analogies, such as the flood basalts of the Columbia River Plateau, the Deccan Traps,

2.4 The heavily cratered far-side highlands of the Moon. Note the paucity of mare basalts. Mare Crisium is the circular mare on the northwest horizon. The other patches of basalt are Mare Smythii and Mare Marginis (Apollo 16 metric frame 3023).

and others [9]. The lava floods continued for at least 800 million years. The latest flows are possibly as young as 2.6 aeons, close to the time of the Archean-Proterozoic transition on the Earth (Chapter 6). A reconstruction of the surface of the Moon at the terminal stages of the mare basalt flooding (about 3.0 aeons) is shown in Fig. 2.6 [8].

An important stratigraphic conclusion was drawn from the study of earth-based photographs; namely, that the major lava flooding of the basins occurred well after the ringed basins were formed. The southeast region of Mare Imbrium (Fig. 2.7), which includes the Apollo 15 landing site, is revealing. The Apennine ridge, the rim of the Imbrium basin, runs diagonally from the southwest corner (near the crater Eratosthenes). Archimedes, the large prominent crater in the north central portion, was formed after the excavation of the basin but before the lava flooded both it and the Imbrium basin, since the crater ejecta is buried by basalt. Other craters (Eratosthenes, Autolycus, and Aristillus, in the northeast quadrant), as well as many small craters, formed after the lava floods. Figure 2.2 reveals other post-basin, pre-fill features (e.g., Plato, Sinus Iridum). Eight craters with diameters in the range 11–266 km were formed within the Imbrium basin before the major flooding of the mare basalts.

From a study of Figs. 2.2 and 2.6, a further stratigraphic truth is available. There is no debris from the excavation of the giant ringed basins lying on

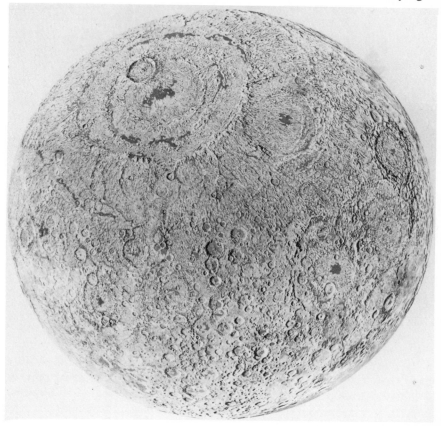

2.5 A reconstruction of the face of the Moon at 3.9 aeons following the Imbrium basin collision, showing the highland surface prior to the onset of the mare basalt flooding. Sinus Iridum in the northwest corner is post-Imbrium basin formation. Archimedes and Plato have not yet formed. (Diagram courtesy of D. E. Wilhelms, reproduced from *Icarus*. 15: 368, Academic Press.)

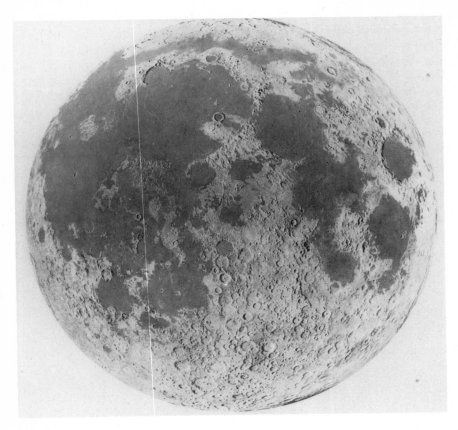

2.6   A reconstruction of the face of the Moon at 3.2 aeons. This shows the surface at the end of the period of mare basalt flooding. Note that craters such as Archimedes, Plato, and Sinus Iridum are post-Imbrium basin formation, but predate the mare basalt fill. Note that the Moon at this remote epoch already looks familiar. (Courtesy D. E. Wilhelms, reproduced from *Icarus* 15: 368, Academic Press.)

the lava plains. No Imbrium ejecta is observed on the smooth plains of Mare Serenitatis. Conversely, no material thrown out during the excavation of Serenitatis is seen over the surface of the Imbrium basin lavas. This observation [10] also means that the eruption of the mare basalts came well after the excavation of the basins which they occupy. These are clearly of differing ages, shown by the various degrees of destruction of their rims. This indicates that any impact melting caused by the collisions that formed the ringed basins was not responsible for the major floods of mare basalt that fill them. This deduction was abundantly confirmed by the petrographic, geochemical and age dating studies on the samples returned by the Apollo missions.

2.7 An earth-based photo of the southeast sector of Mare Imbrium. The Apennine Mountains form the prominent northeast-trending ridge which forms the outer ring of the Imbrium basin. The large lava-filled crater is Archimedes. This is post-basin formation but pre-lava fill. The two craters northeast of Archimedes are Autolycus and, to the north, Aristillus. These postdate the lava fill. The large crater in the southwest corner is Eratosthenes. (Lick Observatory photograph.)

2.8 Oblique view of Copernicus, 93 km in diameter, showing central peak and slump terrain [the rim of the Imbrium basin forms the horizon (NASA AS 17-151-23260)].

Superimposed on lava plains and highlands are craters, some older and subdued such as Eratosthenes, and others younger with bright rays [e.g., Kepler, Copernicus (Fig. 2.8), and Tycho]. One estimate for the age of formation of Copernicus is 900 million years, but this conclusion depends on the interpretation that the light grey layer in the regolith at the Apollo 12 site is ray material from Copernicus [11].

Tycho is estimated to be about 100 million years old. It is the youngest major cratering event. The age estimate depends on the interpretation that the landslide at the Apollo 17 site, whose surface is dated by cosmic ray exposure ages, was caused by the impact of ejecta from Tycho (Section 4.10.2). Crater counting statistics (Chapter 3) indicate a range in ages from 200 to 1000 million years with a best estimate of 700 million years, considerably older than the previous figure. If Tycho formed during the Mesozoic era on Earth (65–225 million years ago), it provides evidence that large objects were in earth-crossing orbit at that time. Accordingly the notion of a similar impact on the Earth at the Cretaceous-Tertiary boundary receives additional credibility [12].

Tectonic features in the terrestrial sense are rare on the Moon. Mostly they appear to be associated with vertical motion, producing small grabens, for example. Most of these features are probably related to crustal loading by the filling of the impact basins by mare basalts (see Chapters 6, 7). No vestige of a plate tectonic regime appears on the lunar surface.

Thus, the visible face of the Moon has changed little throughout the past 3 billion years, during which time complex geological events have occurred on the surface of the Earth. A space traveler visiting the Earth 3–4 aeons ago would have seen the Moon (Fig. 2.1) rather like it is today. The red glow of a mare basalt flood could have been visible during a particularly well-timed visit. The spectacular but nearly instantaneous production of the Imbrium or Orientale basins or of a large impact crater would require finer timing to witness. Meanwhile, the Earth began the slow and infinitely complex synthesis of organic compounds that resulted ultimately in the appearance of Homo sapiens and the reconstruction of lunar and terrestrial history.

2.2 Stratigraphy of the Lunar Surface

Shoemaker and Hackman [13] proposed a lunar stratigraphic sequence based on the stratigraphy observed around the Imbrium basin. They divided this sequence into five recognizable systems (Table 2.1). This classification formed the basis for further mapping, but complexities arose. The most important concerned the subdivision of the Imbrian system. An early, or Apenninian series, referred to the excavation of the basin itself. Included in this was the ejecta blanket of the Imbrium basin, named the Fra Mauro

Table 2.1 Original stratigraphic subdivision of the Imbrium basin sequence.[†]

System	Events
Copernican	Young craters, rays, bright albedo
Eratosthenian	Older craters, no rays, low albedo
Procellarum	Mare lava flooding
Imbrian	Basin excavation
Pre-Imbrian	Highland crust cratering

[†]From Shoemaker and Hackman [13].

Formation. A later Archimedian series included the period of formation of the post-basin pre-fill craters (Archimedes, Plato, Sinus Iridum). After this came the Procellarum group, which suffered many changes in status in the stratigraphic hierarchy [14] before finally vanishing, when the wide spread in ages of mare basalt flooding was recognized.

Further complexities became apparent as the mapping was extended into regions of the Moon remote from the Imbrium basin. The rugged Fra Mauro ejecta blanket close to the Imbrium basin (Fig. 2.9) graded insensibly into smoother deposits. These smooth plains were widespread over the lunar highlands. Age relationships became confusing since these deposits appeared to be of differing ages in different regions (Figs. 2.10 and 2.11). In many areas they were the latest premare fill material and so presumably younger than the Imbrium basin ejecta. This dilemma was partially resolved by assigning them to a new formation, the Cayley Formation [14], of which we shall hear more.

The recognition of the importance and effects of the great ringed basins led to the concept that much of the highlands contains a complex overlap of

2.9 A low sun angle oblique view of the Fra Mauro ejecta blanket. The spacecraft boom points to the center of the crater Fra Mauro. The Apollo 14 site is northwest of the Fra Mauro crater rim (NASA 16-1420).

2.10 Light plains deposit in the 150-km diameter crater Ptolemaeus (Apollo 16 metric frame 990).

ejecta blankets. This view has been modified by the recognition that the secondary cratering due to the great collisions extends beyond the deposition of primary crater ejecta (Section 3.18). Many other workers, perplexed by the complexity of highlands stratigraphy, appealed to volcanic ash flows, of *nuée ardente* type, to account for the smooth highland plains. Such processes could be invoked to provide small or large deposits of variable ages but raised more questions than they answered. The petrological consequences of a volcanic process differing markedly from the observable mare basaltic volcanism [15]

2.11 Light plains fill in the northwest quadrant of Mendeleev Crater (diameter 330 km). The higher density of craters suggests an older age for these light plains compared with those in Fig. 2.10 (Apollo 16 metric frame 2078). [Mattingly, T. K., and El-Baz, F. (1973) *PLC* 4: 52.]

were rarely addressed. As the other basins were studied, the familiar problem of correlation arose to bedevil the lunar stratigraphers attempting to extend the Imbrium basin formations to the younger (Orientale) and older (Humorum, Serenitatis, and Nectaris) basins. It was soon realized that the Imbrium event was only one of many such, although the full extent of the bombardment (Chapter 3) was only slowly realized.

Separate formations were established related to major basins. The Orientale basin ejecta blanket was subdivided into Hevelius formation (smooth facies) and Cordilleran formation (hummocky facies). For other basins the ejecta were mapped as Vitello formation (Humorum basin) and Janssen formation (Nectaris basin), establishing the basic stratigraphy of the lunar surface [14].

This phase of lunar investigation has been completed by the publication of the six 1:5,000,000 series of lunar maps by the U. S. Geological Survey [16]. These provide detailed accounts of lunar stratigraphy, as well as useful statements on many details of the lunar surface. A definitive account of lunar geology and stratigraphy has been written by Wilhelms [17] to which the reader is referred for book-length treatment of this subject.

The most recent major change to the lunar stratigraphic column has been the subdivision of the Pre-Imbrian. This has been divided into two new systems based on the recognition of the importance of the formation of the Nectaris basin as a convenient marker in Pre-Imbrian time [18]. Thus, the Nectarian system extends from the base of the Janssen Formation (the ejecta blanket of the Nectaris basin) up to the base of the Fra Mauro Formation, which was formed during the Imbrium collision.

The rocks older than Nectarian are referred to as Pre-Nectarian. Although these two systems replace the Pre-Imbrian in many areas of the Moon, the latter term is very firmly entrenched in the lunar literature. In this review the term "Pre-Imbrian" is used as equivalent to "Nectarian and Pre-Nectarian." This subdivision of lunar time becomes important in considering the question of the possible lunar "cataclysm" or "terminal cataclysm."

The final stratigraphic scheme adopted is shown in Table 2.2. In this table, a division has been made between the maria, craters, highlands and circum-basin materials as the principal regions for which it is useful to establish relative stratigraphic sequences. A terrestrial analogy is the facies variation among differing tectonic regimes.

2.3 Radiometric and Stratigraphic Lunar Time Scales

Before the Apollo missions, many attempts were made to establish a lunar chronology using the only available dating method, that of crater counting. This method can yield valuable relative age sequences, if properly

Table 2.2 Lunar stratigraphic sequence.

System	Maria	Craters	Circumbasin materials	Highlands
Copernican		Young ray craters with high albedo (e.g., Copernicus, Tycho)		
Eratosthenian	Younger basalt flows in Imbrium	Older subdued craters No bright rays, low albedo e.g., Eratosthenes		
	Marius Hills domes			
Imbrian	Older basalts	Plato, Archimedes	Hevelius Formation (Orientale)	
			Fra Mauro Formation Alpes Formation Montes Apennius	Cayley Formation Descartes Formation
Pre-Imbrian				
Nectarian		Alphonsus	Vitello Formation (Humorum) Janssen Formation (Nectaris)	
		Clavius		
Pre-Nectarian		Ptolemaeus Hommel		Heavily cratered terrain

employed, and if care is taken not to include secondary craters in the count (Sections 3.17, 3.18 and 5.10). Even secondary craters are useful, since they assist in establishing the relative chronology of the great impact basins. The pre-Apollo understanding in this respect might be best described as chaotic. Primary craters, provided the meteorite flux is known, can provide an absolute time scale. This is the only method applicable to the other bodies in the solar system until sample returns are possible. The method has and can provide unique information, and has led to much understanding of the chronology of the inner solar system. The success of the method is due to calibration with the ages obtained from the returned lunar samples, so that

Table 2.3 Comparison between stratigraphic and radiometric ages for the Moon.[†]

Stratigraphic Age	Event	Radiometric Age
Copernican	South Ray Crater (Apollo 16 site)	2 m.y.
	Shorty Crater (Apollo 17 site)	19 m.y.
	Cone Crater (Apollo 14 site)	24 m.y.
	North Ray Crater (Apollo 16 site)	50 m.y.
	Camelot Crater (Apollo 17 site)	90 m.y.
	Tycho Crater	107 m.y.
	Copernicus Crater	900 m.y.
Eratosthenian		
Imbrian	Apollo 12 basalts	3.08–3.29 Ae
	Apollo 15 basalts	3.25–3.34 Ae
	Luna 24 basalts	3.26–3.30 Ae
	Luna 16 basalts	3.35–3.41 Ae
	Apollo 11 high-K basalts	3.57 Ae
	Apollo 17 high-Ti basalts	3.59–3.77 Ae
	Apollo 11 low-K basalts	3.69–3.86 Ae
	Imbrium basin (Fra Mauro)	3.82 Ae
Nectarian	High-Al basalts (Fra Mauro Apollo 14)	3.85–3.96 Ae
	Serenitatis basin	3.86 Ae
	Nectaris basin	3.92–4.2 Ae
Pre Nectarian		
	Primary lunar differentiation	~4.4 Ae

[†]Data from sources discussed throughout the text. See Sections 3.17, 4.3.4, 4.10, 5.10 and 6.4 for details.

cautionary tales from the lunar experience are in order. Hartmann [19] estimated the age of the lunar maria at 3.5 aeons, a remarkably accurate estimate, and consistent with earlier values of Baldwin [15]. As the Apollo sample return came closer, the ages based on crater counting began to fall, the extreme examples being 3–6 *million* years for parts of Oceanus Procellarum [20] and with many estimates giving ages of a few hundred million years for the maria [21]. These estimates, 1–2 orders of magnitude too low, were in error principally on account of the confusion of secondary with primary craters, and errors in estimating meteorite flux rates.

The first age determinations were made by the Preliminary Examination Team on the Apollo 11 samples, using potassium-argon (K-Ar) dating. They found ages ranging up to 3.7 aeons, later confirmed by other techniques, for the mare basalts [22].

The significance of these ages was immediately apparent, and led them to comment that "Perhaps the most exciting and profound observation made in the preliminary examination is the great age of the igneous rocks from this lunar region—it is clear that the crystallization of some Apollo 11 rocks may date back to times earlier than the oldest rocks found on Earth." The age information from the returned samples alone makes the Apollo sample return of unique scientific value. The correlation between the stratigraphic and radiometric time scales is shown in Table 2.3. Two facts should be noted. First, events on the Moon are ancient by terrestrial standards. Second, the period of intense cratering predates 3.8 aeons. This provides a time-frame for similar intense cratering observed throughout the inner solar system.

2.4   Detailed Lunar Stratigraphy

Frequent reference is made throughout the book to the various stratigraphic terms used to subdivide the lunar record and, accordingly, some brief descriptions of these units are given here to set the stage for further discussion. For full details, the reader must consult Mutch [1], and Wilhelms [17].

2.4.1   Pre-Imbrian

The identification of the impact origin of the Imbrium basin, and the dominating effect which it produced on the near side of the Moon, make it a useful time marker. the Orientale basin is demonstrably younger, with secondary impact craters from that event being superposed on Imbrium secondaries [23]. The formation of the Imbrium basin occurred nearly at the close of the great cratering of the Moon (Section 3.17). Only ten impacts that produced craters or basins greater than 150 km in diameter occurred subsequent to the Imbrium collision, compared to 73 identified in that class before Imbrian time [24]. Because of crater and basin saturation effects, the actual number must be even greater.

As noted earlier an increased understanding of the early history of the Moon has led to subdivision of Pre-Imbrian time into Nectarian and Pre-Nectarian. Pre-Imbrian time remains useful in the same context as the concept of Pre-Cambrian time on Earth, Nectarian and Pre-Nectarian being analogous to the Proterozoic and Archean respectively but of course being much older and not equivalent in absolute age.

2.4.2 Pre-Nectarian

This system is equivalent to the older part of the Pre-Imbrian. The Pre-Nectarian system contains twice as many basins as does the overlying Nectarian, although Pre-Nectarian craters are less numerous. This is ascribed to destruction of older craters by new craters or basin collisions [18].

The distribution of the Pre-Nectarian units is given in Fig. 2.12, which shows the distribution of large ringed basins. The implications both for meteoritic flux rate and of the possibility of a spike in the cratering record (the lunar cataclysm) lend importance to the establishment of a relative chronology for the Pre-Nectarian period. The Pre-Nectarian highlands are generally divided into "cratered" and "heavily cratered" terrain (e.g., Fig. 2.12). The less heavily cratered terrain has probably been covered with basin ejecta. The oldest basins have been nearly totally destroyed (e.g., Al Khwarizimi). How many older basins once existed? These questions are addressed in the sections on meteorite flux (3.16, 3.17) and on planetary crusts (Section 5.10), but it is probable that we see only the latest part of the record.

2.4.3 Nectarian

This system forms an important and useful time-stratigraphic unit, since the Janssen Formation, representing ejecta from the Nectarian basin, is widespread in the southeast regions of the Moon, remote from the effects of the Imbrium and Orientale collisions. The type area for the Janssen Formation is in the crater Janssen, although a somewhat wider region (Lat. 40–48° S, Long. 37.5–65° E) is defined as the type area which includes the crater Janssen (180 km diameter) as well as the prominent valley Vallis Rheita, long a source of curious speculation, but almost certainly due to low-angle secondary ejecta from the Nectarian basin collision (e.g., Baldwin [15]). The Nectarian system is much more heavily populated with impact structures than is the Imbrian system, containing about five times more basins and craters with diameters larger than 150 km. The age of formation of the Nectarian basin thus becomes of crucial importance in our understanding of cratering flux rates and of cataclysms. The base of the Nectarian system (the Janssen Formation) is extensive and forms a useful marker in Pre-Imbrian time. The view is taken here, and throughout the book, that the cratering flux shows a steep but steady decline down to 3.8 aeons and that the evidence of a late spike is not demanded by either the age or geological data.

The Descartes Formation may well belong to the Nectarian system, if it is finally identified as Nectarian basin ejecta. This is an important question for

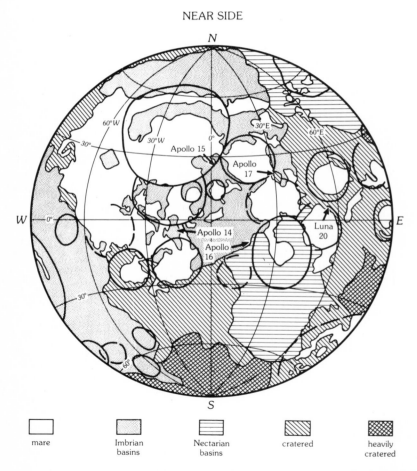

2.12 Province map of the lunar highlands, showing the distribution of 43 large basins with diameters greater than 220 km. Mare lava fill is shown in white. The terrain affected by the young "Imbrian" basin collisions (e.g., Imbrium, Orientale) is distin-

dating the basin sequence, since samples could be represented among the Apollo 16 collection. This question is still unresolved. The Kant plateau, to the east of the Apollo 16 site, is probably the primary Nectaris ejecta, but the Descartes Formation alternatively may be primary Imbrium ejecta piled up against the Kant plateau. Other alternatives are possible (see Sections 3.18 and 5.10 and the discussion in the following section).

FAR SIDE

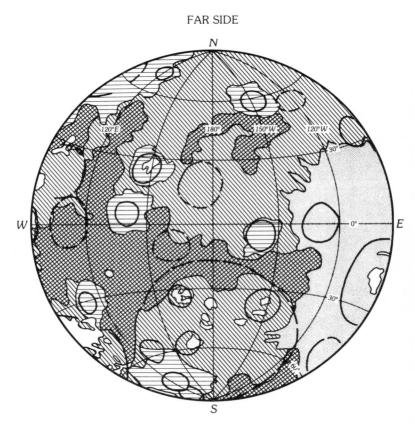

guished from that due to the older Nectarian basins. The other parts of the highlands are divided into *cratered* and *heavily cratered* terrain. Those highland regions with low numbers of craters have been covered by basin ejecta. The distribution of basins is random. Some old basins have been nearly obliterated. Presumably older basins have been totally destroyed. The highland surface appears to be saturated with basins as well as impact craters, so that we see only the final stages of the bombardment history.

This interpretation favors a continued impact history extending back in time toward the origin of the highland crust, rather than a sudden spike of great collisions at about 4.0 aeons. [Howard, K. A., et al. (1974) *Rev. Geophys. Space Phys.* 12: 322.]

2.4.4   The Imbrian System

The base of this great system is defined as the basal unit of the Fra Mauro Formation. Since the formation of both the Imbrium and Orientale basins occurs within the time frame encompassed by this system, it is one of the most important stratigraphic systems on the Moon, marking the decline of the great bombardment. This datum may be widespread throughout the solar system.

Fra Mauro Formation

Two facies are recognized, "hummocky" and "smooth" respectively. The type areas are close to the Fra Mauro crater (Fig. 2.9). The Fra Mauro Formation is generally interpreted as an ejecta blanket from the Imbrium basin-forming event, with qualifications on the amount of primary and locally derived material present. The thickness of the Fra Mauro Formation is somewhat model dependent, but estimates ranging from 900 m (at the outer basin rim) to 550 m (at 900 km south) have been made [1]. This definition of the formation includes all deposits radial to Imbrium; for example, the Haemus Mountains.

Cayley Formation

The rugged Fra Mauro "hummocky" facies grades insensibly into the "smooth" facies. Such smooth plains are widespread in the lunar highlands. Crater counting revealed differing ages in different areas (Figs. 2.10, 2.11). As noted earlier, this problem was partly resolved by the time-honored method of creating a new stratigraphic formation. Two major controversies have swirled about the Cayley Formation (and the associated hilly Descartes Formation). Initially, a widespread view was that both the Cayley and the Descartes Formations were igneous in origin, with the Cayley representing ash-flow or ignimbritic deposits while the more rugged Descartes terrain was analogous to terrestrial rhyolitic domes. This interpretation had a widespread influence on site selection for the Apollo 16 landing as well as other implications as far afield as the origin of tektites. The Apollo 16 sample return effectively demonstrated that both formations were comprised of impact breccias and melt rocks of impact origin, and that no volcanic rocks were present (Sections 5.2, 5.3). The second controversy concerned the extent of primary and secondary components in the ejecta blankets. The major question turns upon local versus distant origin for the smooth plains deposits. One major piece of evidence in support of a local origin is the variability in the Al/Si ratios observed from the orbital chemical data (Sections 3.9, 5.8). This observation would not be expected to follow from a debris sheet of uniform composition originating from Imbrium, but is consistent with deriving some of the ejecta locally, ploughed up by secondaries from Imbrium. Terrestrial examples have demonstrated this effect. The study of the Ries (West Germany) ejecta blanket, where it crosses a geological boundary south of the crater, indicates a high percentage of locally derived debris [25]. On the Moon, the situation may not be so simply resolved. The upper portions of the highland crust are comprised of a complex of overlapping ejecta blankets from the basin collisions. Excavation of this material by later impacts, accompanied by ploughing up ejecta by secondaries may produce much chemical heterogeneity superimposed on primary differences. A further question turns on the relationships of the Fra Mauro, Cayley and Descartes Formations. More distant Cayley plains units are not likely to be related to the Imbrium event, but a large percentage of primary Imbrium ejecta is expected at Fra Mauro (Apollo 14 site). The problems of accounting for smooth plains deposits far removed from the Imbrium basin have perhaps overshadowed the importance of primary basin ejecta. The question of the relative amounts of basin primary and secondary ejecta at the Descartes site is also critical for the interpretation of the Descartes (Apollo 16) ages. The controversy between the two opposing schools of thought is well set out in references [26, 27].

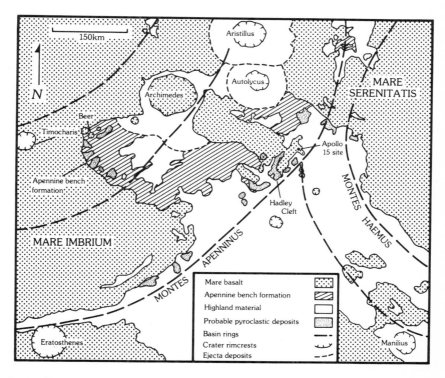

2.13 Sketch map of the Apennine Bench Formation, showing the extensive distribution west of the Apollo 15 site. (Courtesy B. R. Hawke.)

Apennine Bench Formation

This enigmatic formation was apparently not sampled during the Apollo 15 mission; therefore, its origin remains in dispute. It is exposed south of the crater Archimedes, within the Imbrium basin and immediately northwest of the Apennine mountain scarp (Figs. 2.7, 2.13). Various origins have been proposed and the crucial question is whether it was formed during the Imbrium collision, or post-dates it as a volcanic fill. This question will be discussed in connection with the orbital data for Th and K and the possible occurrence of KREEP basalt postdating the collision (Section 5.4.4). The evidence as assessed here is that it is equivalent to the hummocky material between the Rook and Cordillera mountains of the Orientale basin (Montes Rook Formation) (Fig. 3.16). This so-called knobby facies is interpreted as a fine-grained debris sheet or an impact melt sheet ponded against the Cordillera scarp. A corollary is that this mountain ridge must have been in place before the melt or debris sheet arrived (within a matter of minutes after the Orientale impact). Debate on this point as on many others in this book will continue until the resumption of manned spaceflights to the Moon.

Cordillera and Hevelius Formations

Just outside the Cordillera Mountain ring of the Orientale basin occurs a hummocky facies of the ejecta sheet (Fig. 3.16). This grades outwards into a smooth facies, named the Hevelius Formation (Fig. 2.14). These relationships are a little clearer than those observed in the Fra Mauro Formation, and have aided in the interpretation of the latter. The Hevelius Formation must have flowed over the surface, since in many places it is piled up against obstacles [28] (e.g., Southeast wall of Crater Inghirami). These units are discussed in more detail in the chapters on cratering and planetary crusts. The observable effects of the Orientale collision extend out for many hundreds of kilometers from the outer ring of the Montes Cordillera.

Mare Materials

The older basalts are conventionally included in the Imbrian system and represent the youngest member of that system. Probably the Marius Hills volcanic cones and domes belong in the Imbrian system (D. E. Wilhelms, pers. comm., 1981). Recent mapping practice has been to distinguish mare materials as a separate formation, because of the wide spread in their ages, rather than to include them in the formal lunar stratigraphic schemes. This is because the rate of filling of the large basins is not known, so that only the ages of the uppermost units in the basins are established, either by dating of returned samples, or by photogeologic techniques. Individual basin filling extended probably over a long time, overlapping perhaps with some of the basin-forming events.

2.14 An Orbiter photo of the Hevelius Formation, 700 km southeast of the center of the Orientale basin. Note the strong flow pattern toward the southeast (NASA Orbiter IV 167-H3).

2.4.5   Eratosthenian System

The type locality for this system is the crater Eratosthenes (Fig. 2.7) (58 km diameter) northeast of Copernicus. All post-mare craters whose ejecta blankets are subdued, and where the rays are no longer visible or are very faint under high sun angle illumination, are included. The latest episodes of volcanism are also generally included in this system.

2.4.6   Copernican System

This is the youngest of the systems employed for lunar mapping. It includes all young, fresh, bright-rayed craters. The type locality is Copernicus (93 km diameter) (Fig. 2.8).

The detailed discussion of absolute time assignments to these geologically established systems will be taken up later in the book. Much controversy surrounds this topic, in particular the age of the formation of the Nectaris basin, which is critical for much of our interpretation of early lunar history.

2.5 Mercury

The geology of Mercury shows strong similarities with that of the Moon (Fig. 2.15). Thus, it can be argued that a similar sequence of events must have occurred on both bodies. This has important implications for extrapolating the well-dated sequence on the Moon to other bodies. The surface of Mercury is therefore highly significant in the interpretation of early solar system history, contrary to frequently expressed opinions that Mercury is "just like the Moon."

The accretion of Mercury must have been closely followed by differentiation. Two principal observations support this conclusion: (a) the high density of Mercury (5.44 gm/cm^3), and (b) the presence of silicate material and a lunar-like topography at the surface. These observations lead to the conclusion that the planet has a high iron content, which must be segregated into a core about 0.75–0.80 of Mercurian radius, overlain by a silicate crust. The heavy cratering of the crust must have been early, by analogy with the Moon, and the crust must have been thick enough and cold enough to preserve the record of this bombardment well before 4.0 aeons.

Detailed descriptions of the planet may be found in the report on the Mariner 10 mission [29] and in the proceedings of two conferences on Mercury [30, 31]. The photographic coverage from the Mariner 10 mission is restricted to less than one-half the planet. Reliable image resolution over this area is about 4–5 km, with the best resolution about 1 km, comparable to earth-based telescopic views of the Moon. A very small number of high resolution (100 m) views are available. Accordingly, the interpretation of the geology and stratigraphy of Mercury depends strongly on lunar analogies.

There are three dominant landscape forms on Mercury: (a) intercrater plains; (b) heavily cratered terrain; and (c) smooth plains.

2.5.1 Intercrater Plains

The intercrater plains (Fig. 2.16) may represent an "ancient primordial surface." The principal distinction from the smooth plains is that they are much more heavily cratered. The smooth plains (Fig. 2.17) may be compared with the Cayley Formation on the Moon. The intercrater plains occupy about one-third of the Mercurian surface which was visible to Mariner 10 [32]. As noted above, they differ from the later "smooth plains" principally because of the large number of craters in the range of 5–16 km diameter. There is no difference in the albedos of the intercrater plains and the smooth plains (albedo = 0.15 ± 0.02 for both). There is little contrast between the albedos of the heavily cratered terrain and the plains units, in contrast to the Moon. Accordingly, difficulties in interpretation arise. Are these plains units volcanic, or do they represent deposits from large craters and basins?

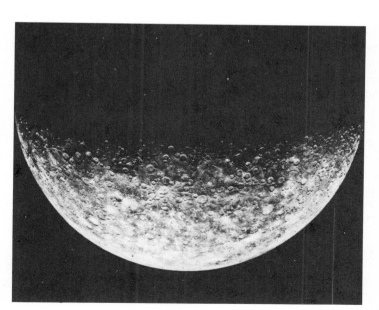

2.15a   At left is a mosaic of 18 pictures taken when Mariner 10 was 200,000 km away from Mercury. In this view, which resembles the highland regions of the Moon, the largest craters are about 200 km in diameter. Two-thirds of the surface seen here is in Mercury's southern hemisphere. The 18-picture mosaic at right was taken after Mariner passed Mercury. The distance was 210,000 km. The north pole is at the top and the equator about two-thirds down. The large 1300-km (800-mile) diameter basin is Caloris.

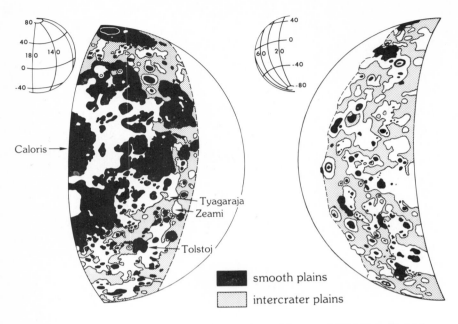

2.15b   The distribution of the major plains units on Mercury. [After *Basaltic Volcanism* (1981) Fig. 5.5.3.]

This question appears to be inherently unresolvable from the present photographic data. The absence of contrast in albedo is not consistent with a basaltic composition for the Mercurian plains [33]. The analogy with the Cayley problem on the Moon prior to the Apollo 16 landing is nearly complete. Arguments in favor of a volcanic explanation for the plains units (both "intercrater" and "smooth" should have the same origin) are primarily negative. Thus the intercrater plains are extensive, and there is an apparent paucity of multi-ring basins which might have supplied ejecta. Such ejecta have a more restricted ballistic range than on the Moon, due to the higher gravity of Mercury. The lunar terrain which most closely resembles that of the Mercurian intercrater plains is the so-called Pre-Imbrian "pitted terrain," southwest of the Nectaris basin (35–65°S; 10–30°E) (Fig. 2.18). The distinction from Cayley-type plains is mainly in a higher density of craters. These lunar plains have been suggested to represent an early phase of volcanism [34] since it is difficult to assign them to particular multi-ringed basins, if they are basin ejecta. This is the same problem which is encountered on Mercury. However, the resemblance between the Cayley plains and these "pitted terrain" plains must be interpreted cautiously; this author finds no compelling evidence to interpret them as other than the reult of impact-produced debris from basin

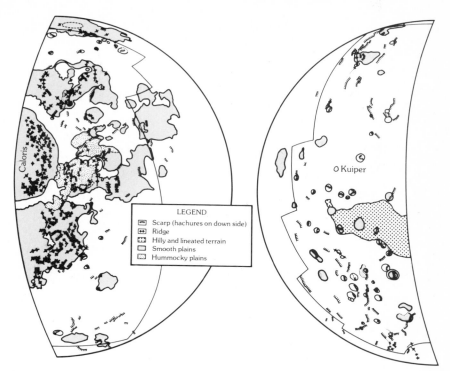

2.15c Tectonic map of Mercury, showing lobate scarps and ridges. Caloris basin at left. [After *Basaltic Volcanism* (1981) Fig. 6.5.11.]

formation. The lack of identifiable sources may be simply due to the destruction of old basins by new ones, a view consistent with an extended period of basin formation rather than a lunar cataclysm.

Another problem which bedevils the interpretation of the intercrater plains of Mercury is that their stratigraphic position is uncertain. Significant areas may be younger than the heavily cratered terrain [35]. Many of these questions can only be resolved by future missions. The view taken here is that both the intercrater plains and the smooth plains represent Cayley-type deposits resulting from basin-forming impacts [33], although there is a considerable body of opinion that they represent lava plains [36]. Few visible morphological indicators of volcanism can be recognized on Mercury. Perhaps the most persuasive evidence for volcanism on Mercury is the presence of some dark albedo areas within craters (e.g., Tyagaraja) [37, 38].

Finally, there is the problem that all analogies are with the Moon, but that body has a distinctly lower bulk density and hence a different composition, as well as a different interior structure, so that the mantle of Mercury is

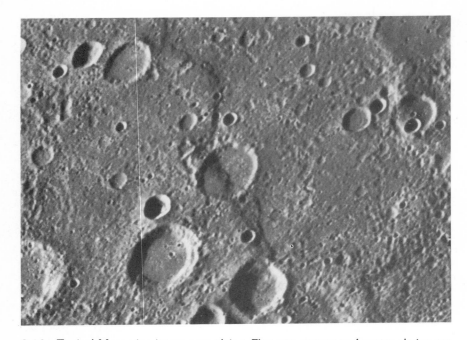

2.16 Typical Mercurian intercrater plains. Elongate craters and crater chains are present on the intercrater plains and are probably secondary impact craters from basins and craters of the heavily cratered terrain. The lobate scarp (Santa Maria Rupes) is probably a thrust fault which post-dates the intercrater plains and transected craters. The picture is 400 km across; north is at the top (Mariner 10, FDS 27448) [34].

probably very different than that of the Moon. Lavas erupted from it may not resemble lunar basalts in albedo. There is accordingly a complex situation with respect to interpretations of the Mercurian photographs. The Moon provides the only viable analogy, but due to planetary density differences and probable mantle differences, Mercurian volcanism may be sufficiently different to make photogeological interpretations difficult.

In summary, the evidence for extensive basaltic style volcanism on Mercury is judged to be slender and the plains are here interpreted as of impact origin.

2.5.2 The Heavily Cratered Terrain

The origin of the heavily cratered terrain (Fig. 2.19) is less enigmatic than that of the plains. The craters are similar in general morphology to those on

2.17 Hilly and lineated terrain on Mercury. Smooth plains occur in the large degraded crater. The linear valley extending northwest is Arecibo Vallis. Center of area is at 28°S, 22°W. Region shown is 450 km across (Mariner 10 Frame 27370) [32].

the Moon, and all the features such as ejecta blankets, terraces, central peaks and peak rings, secondary crater fields, etc., are present. Differences due to gravity, possible target strength, average impact velocity, and the source of the impacting objects lead to minor differences (see Chapter 3 for an extended discussion).

Crater diameters range from thirty to several hundred kilometers and appear in general to be very similar to those of the lunar highlands, with the differences in detail noted above. The Caloris basin may be conveniently noted here. This great feature recalls Mare Orientale on the Moon, but is somewhat larger (1300 km diameter). It is older than most of the smooth plains deposits [39].

By analogy with the Moon, the heavily cratered terrain is dated as earlier than about 4.0 aeons. Crater counting techniques (Chapter 3) give best estimates of 3.8 aeons for the Caloris basin and 3.9–4.1 aeons for the heavily cratered south polar uplands. These ages provide evidence that the massive

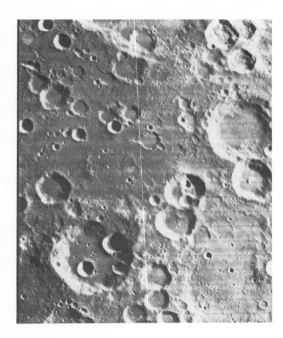

2.18 Lunar pitted plains north-east of the crater Mutus (78-km diameter), which is the large crater enclosing three smaller craters in the southwest corner (NASA Orbiter IV 82 H3).

2.19 Heavily cratered surface on Mercury, with a 2-km high scarp. Width of picture is 500 km (P75-61 JPL 654).

2.20   Mariner 10 photograph of the northern limb of Mercury showing a prominent east facing scarp, extending from the limb near the center of the photograph southwards for hundreds of kilometers. The linear dimension along the bottom of the photograph is 580 km. The picture was taken at a distance of 77,800 km from Mercury (NASA Photograph P.75-61-JPL-654-5-75).

cratering episodes recorded on the Mercurian surface persisted for several hundred million years following accretion of the planet as is the case for the Moon.

2.5.3   Smooth Plains

The smooth plains (Fig. 2.17) have been discussed earlier. They are younger than the heavily cratered terrain, and closely resemble the Cayley plains on the Moon. Their large area comprises the best evidence for a volcanic origin, as well as the apparent lack of source basins for an origin as basin ejecta. All the arguments recall the controversy over the Cayley Formation and only future missions can resolve these questions.

2.5.4   Lobate Scarps

Lobate scarps (Fig. 2.20) are unique to Mercury. They vary in length from 20 to 500 km and in height from a few hundred meters to about three

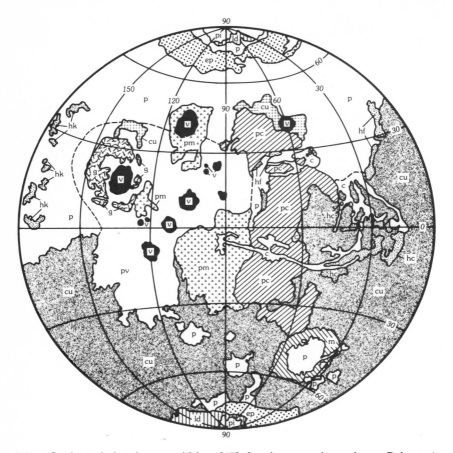

2.21 Geological sketch map of Mars [47]. Lambert equal-area base. *Polar units* include *pi* (permanent ice), *ld* (layered deposits), and *ep* (etched plains). *Volcanic units* include *v* (volcanic constructs), *pv* (volcanic plains), *pm* (moderately cratered plains), and *pc* (cratered plains). *Modified units* include *hc* (hummocky terrain,

kilometers. They are reverse thrust faults, formed due to compressive stresses, and appear to have a rather uniform distribution over the photographed portion of the planet. The decrease in estimates of the surface area associated with these scarps ranges from about 6×10^4 to 13×10^4 km^2, which corresponds to a decrease in Mercurian radius by about 1–2 km [40, 41].

The lobate scarps appear to have formed relatively early in Mercurian history. They occur mainly on the intercrater plains and on the older parts of the heavily cratered terrain. They are demonstrably of tectonic origin and none require a volcanic explanation (e.g., as flow fronts [42]).

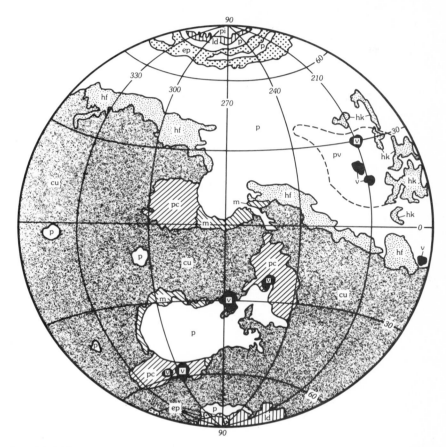

chaotic), *hf* (hummocky terrain, fretted), *hk* (hummocky terrain, knobby), *c* (channel deposits), *p* (plains, undivided), and *g* (grooved terrain). *Ancient units* include *cu* (cratered terrain, undivided) and *m* (mountainous terrain).

The origin of the lobate scarps is most generally ascribed to contraction of the 600-km-thick silicate mantle around a cooling and shrinking iron core. A suggested origin by tidal despinning [42] has not been supported by other workers [43]. From the analogy of the heavily cratered terrain with that of the lunar highland surface, the scarps must have formed prior to 4.0 aeons and hence preserve a record that the radius of Mercury is essentially unchanged for that immense period of time. The wider significance is that the lack of evidence for any expansion on Mercury (and the Moon) places serious limits on models both for Earth expansion and variation in G with time [44].

Another unique feature of the Mercurian surface is the hilly and lineated terrain, antipodal to the Caloris basin. It is suggested that this is due to seismic shaking [45], but it may be due to intersecting crater chains formed by basin ejecta [46]. The lineated terrain, reminiscent of the Imbrium sculpture on the Moon, consists of lines of hills and valleys up to 200–300 km long and is probably formed in the same manner.

2.6   Mars

The principal distinction between Mars, and the Moon and Mercury, is the presence of a tenuous atmosphere, abundant evidence of volcanic activity, and the action of wind and water on the surface. Photographic coverage is now good: 20% of the planet is covered at 50 meter resolution, 80% at 200 meter resolution and 100% at 1 km resolution.

Extensive accounts of the geology of Mars have appeared in many sources and it is not possible to provide more than a brief summary of the salient points of Martian geology in this section. Those readers who wish for a

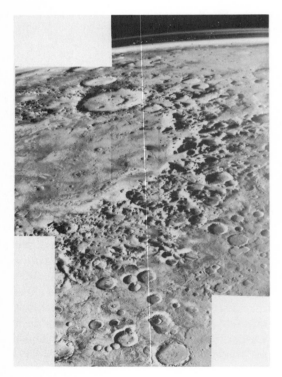

2.22 The Arygre Basin on Mars showing old heavily cratered terrain (NASA S.76-27774).

fuller treatment should consult *The Geology of Mars* [47], *The Surface of Mars* [48], the *Viking Mission Report* [49] and the *Proceedings* of the Second and Third Mars Colloquia [50].

The Martian surface is divided into several differing geological provinces, described below (Fig. 2.21). Discussion of some aspects of Martian impact cratering appears in Chapter 3, while comments on surface features appear in Chapter 4 on planetary surfaces.

2.6.1   The Ancient Cratered Terrain

There is a major subdivision inclined at about 28° to the equator (Fig. 2.21). To the north of this boundary are plains and volcanic ridges. To the south lies the oldest exposed surface on Mars, containing a large number of craters, typically in the size range 20–100 km diameter, with generally smooth floors. Intercrater plains are common. The craters are extensively degraded.

Three large impact basins have been recognized on Mars. The largest is Hellas (42° S, 293° W), 2000 × 1500 km, surrounded by an irregular mountainous rim which extends from 50 to 400 km in width. Argyre is the freshest multi-ring basin, 900 km diameter, surrounded by a mountainous rim extending out to 1400 km diameter. The center of the basin is at 50° S, 43° W (Fig. 2.22). Possibly four or five rings are present [51]. The third large basin is Isidis, about 1100 km in diameter, with an additional 300 km wide mountainous rim. About twenty smaller two-ring basins have been recognized [51]. The mountainous rims are interpreted as the degraded remnants of uplifted rim material, basin ejecta, mountain rings, etc., but most of the primary features have apparently been destroyed.

This cratered terrain occupies about one-half the planet. The best estimate for its age is 4.1 aeons (range 4.0–4.3) [52]. Many effects have contributed to the degradation of craters and basins on Mars. These include wind erosion, impacts, volcanic mantling, water erosion and thermal effects due to daily temperature differences of 130° C.

2.6.2   Volcanic Plains and Mountains

Extensive plains units lie to the north of the boundary. These are generally interpreted to be of volcanic origin. The chief evidence for this is the presence of lobate scarps, which are probably flow fronts. Although these plains are often compared to the lunar maria, there are many distinctions. The Martian plains are light, not dark, and they are comparatively high-standing (perhaps equivalent to the terrestrial continental flood basalts). Volcanic constructs are frequent, while they are rare on the lunar maria. Several divisions may be made, based on the relative crater ages. Thus, while the old heavily cratered terrain is dated at about 4.0 aeons, the densely cratered plains (examples include Lunae Planum, Tempe Fossae and Syrtis Major Planitia) range in age from 3.6 to 3.2 aeons (overlapping with the lunar maria ages) [52]. The moderately cratered plains do not appear to have flow fronts. Younger plains (e.g., Tharsis) have ages ranging from about 1.6 down to 0.3 aeons [52].

Volcanic constructional activity appears to have been of equally long duration on Mars. Thus, the Tyrrhena Patera volcanoes, in Hesperium Planum, are dated at 3.1 aeons, the Elysium volcanics at 1.9 aeons, the three shield volcanoes of Ascraeus Mons, Pavonis Mons and Arsia Mons at about 1.7 aeons, while Olympus Mons is dated at about 300 million years (range 100–700 million years) [52]. Olympus Mons is 550 km across and rises 21 km above its base [53]. The central caldera is 70 km in diameter. In contrast, the total volume of the island of Hawaii, made up of several shield volcanoes, is only 10% of that of Olympus Mons (Fig. 2.23).

The three large shields of Ascraeus Mons, Pavonis Mons and Arsia Mons are about 350–450 km across, and stand about 18 km above the Tharsis Plains. In addition to the Tharsis region, Elysium is a prominent volcanic area, with Elysium Mons forming a large shield volcano, 170 km in diameter, 14 km high with a single central caldera 12 km in diameter. Other positive relief features of probable volcanic origin are called paterae (inverted saucer shape) and tholii (domes). Alba Patera is of enormous size (1500–2000 km in diameter) but only 3 km high (Fig. 2.24). The flanks slope at fractions of a degree. A complex caldera 150 km across occupies the center, surrounded by long radial ridges which may be lava tubes, up to 350 km long. The chief

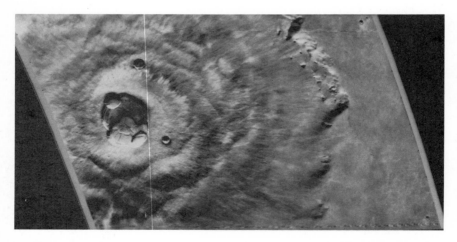

2.23a A view of Olympus Mons extending from the summit caldera to the 4 km high scarp at its base. Note that the scarp is covered by later lava flows at the bottom of the picture (641A52 JPL).

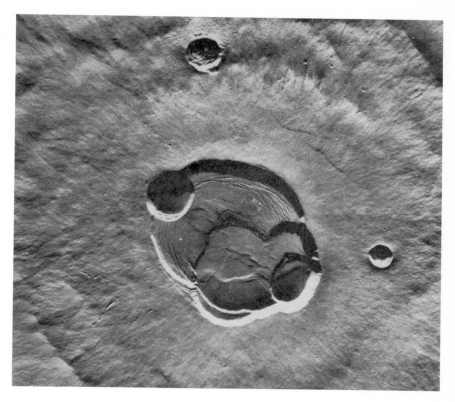

2.23b   A close-up view of the complex caldera of Olympus Mons, 70 km across. Two impact craters are superimposed on the flanks of the volcano (890A68 JPL).

interest in Alba Patera is that the lava flows, although resembling terrestrial lava flows in morphology, are about ten times larger. From the morphology, they appear to be basaltic. Presumably, extremely rapid effusion of lava is the probable basic cause of this extraordinary landform. Clearly the components of the lavas, viscosity, flow rates, yield strengths, and the surface conditions on Mars are all important parameters. Although some progress can be made from photogeology, remote sensing and from landers, only a sample return to Earth will produce definitive answers to these many Martian questions.

The same situation holds with the plains units on Mars. These comprise 60% of the surface of the planet. The origin of these plains provides many puzzles. As noted earlier, those with lobate flow fronts, or those clearly linked to volcanoes, are volcanic in origin. Many plains probably have a thin veneer of wind-blown sediment. The old plains units (cratered plateau plains), which occupy most of the southern hemisphere, are the most ancient plains on Mars,

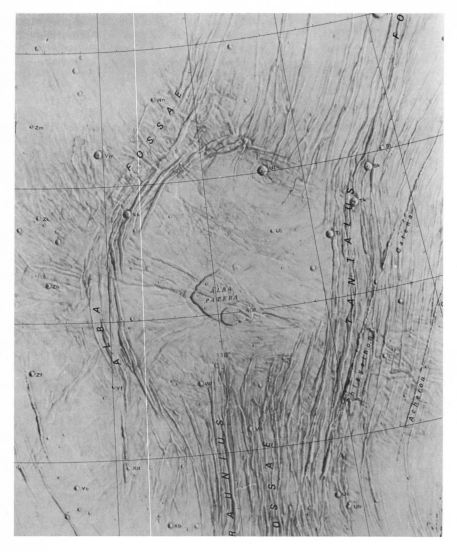

2.24   Alba Patera, the largest volcano in areal extent on Mars, is only 3 km high at its summit. The caldera complex at the center is 150 km in diameter. Numerous north-south trending grabens cut the volcanic edifice and form a fracture ring 600 km in diameter around the summit. (USGS shaded relief Map I-963.)

and their origin as volcanic units is as uncertain as that of the intercrater plains on Mercury. The flat floors of the major impact basins Hellas, Argyre and Isidis have wrinkle ridges and some flow fronts. They may be analogous to mare basin fill on the Moon. Areas such as Lunae Planum, Syria Planum, Sinai Planum and Hesperia Planum contain flow fronts and therefore are probably volcanic. Other plains, associated with central volcanoes, (e.g., NE of Olympus Mons) clearly seem to be of volcanic origin. The widespread plains in northern latitudes are again more enigmatic. They may be lava plains covered with wind-blown sediment as appears to be the case at the Viking 2 site on Utopia Planitia. The Viking 1 site in Chryse Planitia appears to be more complex, with lava emplacement, mantling with wind-blown sediment, and erosion. Primary evidence for volcanic origins for these northern plains appears to be lacking and they may have been much modified by past glaciations. The vesicular rocks observed at the landing sites appear to be volcanic and the chemical analyses suggest basaltic-style parents for the fine material obtained by the sample scoop. The analytical data are presented in Table 6.13 and some petrological implications are discussed in Section 6.8. The picture remains tantalizing and elusive. Most of the analogies used in interpreting the photogeological evidence are from the Moon. The Moon, however, is very different from Mars. Mars has an atmosphere and shows evidence of past water erosion, which provides the most reasonable mechanism for the formation of the famous channels early in martian history (although ice, wind and lavas have all been suggested as possible mechanisms). Mars also shows evidence of wind erosion, a bulk density different from that of the Moon, and differing volcanic styles, etc. The relative ages for Martian features employed here are based on those in *Basaltic Volcanism on the Terrestrial Planets* [52]. A differing interpretation of the Martian cratering statistics [54] leads to the view that most of the surface features on Mars are ancient (pre-3.5 aeons) with only the large Tharsis shield volcanoes being active at younger ages [54]. This debate recalls the controversies over lunar ages before the Apollo missions. Only returned samples will satisfy our curiosity (Fig. 4.4).

2.7 Venus

The surface of Venus covers an area larger than the combined areas of Mars, Mercury and the Moon. Accordingly, it occupies an important position for our understanding of planetary surfaces. Mapping through the cloud cover by radar altimetry on Pioneer Venus Orbiter has covered 93% of the planet to a vertical accuracy of 200 m [55, 56]. About 60% of the surface is very flat, within 500 m of the average lowland elevation. Only about 5% of the surface of Venus is elevated above 2 km (Fig. 2.25).

There are two large continental areas which compare in size to Australia and Antarctica. The largest is Aphrodite Terra which is located near the equator and is elevated between 2 and 5 km. The other, Ishtar Terra, is located

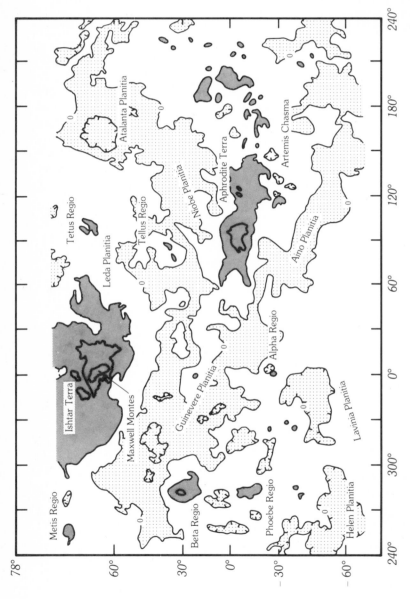

2.25 Topographic map of Venus. The major elevated features are Ishtar Terra, Aphrodite Terra and Beta Regio. Contour interval is 2 km.

between 50° and 80° N latitude and contains the Maxwell Mountains which rise to a height of 11 km. Two large isolated mountains, possibly shield volcanoes, occur at 25° N and 25° S latitude. The largest is 6 km high and 1000 km across. It has nearly the same volume as Olympus Mons, although it is not so high.

The topography on Venus is thus distinct from that on Earth, but few conclusions can be drawn at present. The horizontal resolution of the Pioneer Venus radar is about 100 km. One is reminded that Earth-based telescopic views of Mars have about that resolution. There appear to be some large valleys and possibly the relics of old impact craters. The presence or absence of an old cratered terrain on Venus is a necessary piece of information for evolutionary models of Venus. Further data are needed to avoid developing the Martian Canal Syndrome (see Sections 5.13, 6.7).

2.8 Gallilean and Saturnian Satellites

A list of these satellites along with their basic properties of radii, densities, etc., is given in Table 7.1, and various features are discussed in the chapters on cratering, volcanism and planetary crusts. An order of magnitude increase in our knowledge of these objects has occurred within the past two years with the Voyager missions. These have provided evidence of planetary expansion on Ganymede, sulfur volcanism on Io, ice tectonics on Europa, and cratering on Ganymede, Callisto, Mimas, Dione, Tethys, Rhea and others. The basic references for the Jovian satellites are given in references 57–59 and for the Saturnian satellites in reference 60.

References and Notes

1. Mutch, T. A. (1972) *Geology of the Moon: A Stratigraphic View* (2nd Ed.) Princeton Univ. Press.
2. Gilbert, G. K. (1893) *Bull. Phil. Soc. Wash.* 12: 241. According to Gilbert, Mädler in 1837 was one of the first lunar mappers to recognize this pattern.
3. Leonardi, P. (1976) *Volcanoes and Impact Craters on the Moon and Mars*, Elsevier, 463 pp. This work must surely represent the final statement of the volcanic hypothesis for the origin of the major lunar craters.
4. Muehlberger, W. R., et al. (1980) *Lunar Highlands Crust*, p. 1.
5. Hartmann, W. K., and Kuiper, G. P. (1962) *Comm. Univ. Ariz. Lunar Planet. Lab.*, 1, (12).
6. Wilhelms, D. E. (1980) Multi-ring Basins Abstracts, 115.
7. The term aeon, to denote one billion or 10^9 years, was proposed by Urey. The word is appropriate to designate the passage of a thousand million years, and conjures a vision of the vast abyss of time into which we look. It is greatly to be preferred to the non-euphonius term giga-year (Gy or Ga), a typical product of a committee.
8. Wilhelms, D. E., and Davis, D. E. (1971) *Icarus.* 15: 368.

9.   *Basaltic Volcanism* (1981) Section 1.2.3.
10. Urey, H. C. (1962) in *Phys. Astron. Moon*, Academic Press, N.Y., p. 484.
11. Eberhardt, P., et al. (1973) *Moon.* 8: 104. Dating of Copernicus by crater counting techniques give a somewhat older mean age of 1.5 aeons (Chap. 3).
12. Alvarez, L. W., et al. (1980) *Science.* 208: 1095. The geochemical evidence for a large impact is highly suggestive, but there is as yet no consensus on the detailed paleontological significance of this event. See for example *Cretaceous-Tertiary Boundary Events* (1979) (eds., Christensen, W. K., and Birkelund, T.), Univ. Copenhagen, Vols. 1, 2. It is clear that much further work is required to ascertain the exact nature and timing of the faunal extinctions at the Cretaceous-Tertiary Boundary. (See Large Body Impact Conference Abstracts, October 1981.)
13. Shoemaker, E. M., and Hackman, R. J. (1962) in *The Moon*, Academic Press, N.Y., p. 289.
14. Wilhelms, D. E. (1970) USGS Prof. Paper 599 F.
15. The nature of the lunar mare fill was correctly identified as basaltic lava by Baldwin, R. B. (1949) *The Face of the Moon*, Univ. Chicago Press; (1963) *The Measure of the Moon*, Univ. Chicago Press; Kuiper, G. (1966) in *The Nature of the Lunar Surface* (eds., Hess, W. N., et al.), John Hopkins Press, Baltimore; Whitaker, E. A. (1966) ibid., and others.
16. 1:5,000,000 USGS Misc. Geol. Invest. Maps:
     1-703, Near Side: Wilhelms, D. E., and McCauley, J. F. (1971);
     1-948, East Side: Wilhelms, D. E., and El-Baz, F. (1977);
     1-1034, West Side: Scott, D. H., et al. (1977);
     1-1047, Central Far Side: Stuart-Alexander, D. E. (1978);
     1-1062, North Side: Lucchitta, B. (1978);
     1-1162, South Side: Wilhelms, D. E., et al. (1979).
17. Wilhelms, D. E. (1982) *The Geologic History of the Moon.* NASA SP. In press.
18. Stuart-Alexander, D. E., and Wilhelms, D. E. (1975) *J. Res.* USGS 3: 53.
19. Hartmann, W. K. (1965) *Icarus.* 4: 164; (1966) *Icarus.* 5: 406.
20. Gault, D. E. (1970) *Radio Sci.* 5: 289.
21. Baldwin, R. B. (1969) *Icarus.* 11: 320, gave 640 million years; Gault, D. E. (1970) *Radio Sci.* 5: 289, 50–300 million years, except for the very young ages noted above.
22. LSPET (1969) *Science.* 165: 1211.
23. Wilhelms, D. E., pers. comm., 1981. See Orbiter frame 4H-189.
24. Baldwin, R. B. (1974) *Icarus.* 23: 97. Note that the age sequence of basin events given in this paper has been revised by Wilhelms [17]. In particular, the Serenitatis basin is now younger than Nectaris.
25. Hörz, F., and Banholzer, G. S. (1980) *Lunar Highlands Crust*, p. 211.
26. Oberbeck, V. R., et al. (1975) *Moon.* 12: 19.
     Oberbeck, V. R. (1975) *Rev. Geophys. Space Phys.* 13: 337.
     Hawke, B. R., and Head, J. W. (1978) *PLC 9:* 3285.
     Morrison, R. H., and Oberbeck, V. R. (1975) *PLC 6:* 2503.
27. Chao, E. C. T. (1975) *J. Res.* USGS 3: 379.
     Hodges, C. A., et al. (1973) *PLC 4:* 1.
     Ulrich, G. E., et al. (1981) USGS Prof. Paper 1048.
     Wilhelms, D. E., et al. (1979) LPS X: 1251.
     Wilhelms, D. E. (1981) Apollo 16 Workshop, 150.
28. Scott, D. H., et al. (1977) USGS Map I-1034.
29. Mariner 10 Report (1975) *JGR.* 80: 2341–2514.
30. First International Colloquium Report (1976) *Icarus.* 28: 429–609.
31. Report of LPI Conference on Mercury (1977) *PEPI.* 15: 113–314.

32. Trask, N. J., and Guest, J. E. (1975) *JGR*. 80: 2461.
33. Wilhelms, D. E. (1976) *Icarus*. 28: 551.
34. Strom, R. G. (1977) *PEPI*. 15: 156.
35. Gault, D. E., et al. (1977) *Ann. Rev. Astron. Astrophys*. 15: 97.
36. Advocates for volcanic origins include: Schultz, P. H. (1977) *PEPI*. 15: 202; Cintala, M. J., et al. (1977) *PLC 8*: 3409; Trask, N. J., and Strom, R. G. (1977) *Icarus*. 28: 559; Trask, N. J., and Guest, J. E. (1975) *JGR*. 80: 2461.
37. Hapke, B., et al. (1975) *JGR*. 80: 2431.
38. Schultz, P. H. (1977) *PEPI*. 15: 202.
39. Gault, D. E., et al. (1975) *JGR*. 80: 2444; McCauley, J. F. (1977) *PEPI*. 15: 220; McCauley, J. F., et al. (1981) *Icarus*. In press.
40. Strom, R. G., et al. (1975) *JGR*. 80: 2478.
41. Dzurisin, D. (1978) *JGR*. 83: 4883.
42. Dzurisin, D. (1978) *JGR*. 83: 4902.
43. E.g., *Basaltic Volcanism* (1981) Section 9.5.4.
44. McElhinny, M. W., et al. (1978) *Nature*. 271: 316.
45. Schultz, P. H., and Gault, D. E. (1975) *Moon*. 12: 159.
46. Wilhelms, D. E. (1976) *Icarus*. 28: 551.
47. Mutch, T. A., et al. (1976) *The Geology of Mars*, Princeton Univ. Press, 389 pp.
48. Carr, M. H. (1982) *The Surface of Mars*, Yale Univ. Press.
49. Viking Mission Report (1977) *JGR*. 84 (8).
50. Report on Second Mars Colloquium (1979) *JGR*. 84: 7909–8519; Report on Third Mars Colloquium (1982) *JGR*. In press.
51. Wilhelms, D. E. (1973) *JGR*. 78: 4084.
52. *Basaltic Volcanism* (1981) Chap. 8, Table 8.6.1. See also Section 5.6.2.
53. Davies, M. E. (1974) *Icarus*. 21: 230.
54. Neukum, G., and Hiller, K. (1981) *JGR*. 86: 3097.
55. Pettingill, G. H., et al. (1980) *JGR*. 85: 8261.
56. Masursky, H., et al. (1980) *JGR*. 85: 8232.
57. Smith, B. A., et al. (1979) *Science*. 206: 927 (Voyager 2 imaging results).
58. The Satellites of Jupiter (1981) *Icarus*. In press.
59. *The Satellites of Jupiter* (1980) ed. D. Morrison, Univ. Ariz. Press.
60. Saturn Satellites (1981) *Science*. In press.

Chapter 3

METEORITE IMPACTS, CRATERS AND MULTI-RING BASINS

3.1 The Great Bombardment

One of the principal new results of space exploration has been the demonstration that impact cratering was endemic throughout the solar system. Whether one is contemplating the cratered face of Mercury, the traces of the bombardment surviving on Venus or the Earth, the heavily cratered highland surface of the Moon or the southern hemisphere of Mars, the battered potato-like surfaces of Phobos and Deimos, the crater-saturated surface of Callisto, or the large impact crater on Mimas, the evidence is clear that a massive flux of large objects extended throughout the solar system. Although we have no direct evidence of cratering beyond Saturn, indirect evidence such as the inclination of Uranus at 90° to the plane of the ecliptic is suggestive of massive impact. Indeed, the random inclinations of the planets with respect to the plane of the ecliptic is reasonably explained as a result of massive late collisions during accretion. The evidence from samples returned by the Apollo missions provides us with a date for the steep decline in the cratering flux at 3.8–3.9 aeons. Unless special circumstances pertain to the lunar case (which were once argued, but now appear to be increasingly unlikely in view of the widespread evidence of similar heavily cratered surfaces), a general conclusion is that the bombardment continued until about 3.8 aeons, but did not persist beyond that period. The craters provide important insights into early solar system processes. The oldest observed surfaces are saturated with craters, providing evidence of the last stages of planetary accretion. The cratering rate curves for all bodies studied so far show a general resemblance, including a very high initial flux that declines rapidly between 4 and 3.5 aeons to a value close to the presently observed cratering rate. There does not seem to be any definitive evidence for a "spike" in the cratering rate at about 3.9 aeons, as has often been suggested for the Moon.

The terrestrial geological record begins at 3.7–3.8 aeons, tantalizingly close to the fall-off in the cratering rate. Earlier crusts may have been destroyed by the bombardment but there is no isotopic or chemical signature yet detected in the oldest rocks to suggest their presence, in contrast to the early crustal formation on the Moon.

Cratering hypotheses to explain the origin of the presently observed major features of the Earth have been suggested by many workers [1]. In general, these hypotheses have not survived detailed testing. Even the oldest terrestrial rocks occur at a period when the meteorite flux has declined to a low value, so that ringed basin-forming impacts are unlikely to have occurred at times younger than 3.8 aeons. Our greater understanding of terrestrial geology and of the origins of continents and ocean basins have rendered direct analogies with lunar highlands and maria of little value (see Sections 5.12 and 9.8). This chapter deals mainly with the macroscopic effects of cratering. The related topics of micrometeorites, microcraters, regolith formation, and soil breccias are discussed in Chapter 4 on planetary surfaces. The addition of meteoritic material to the lunar surface is also considered in that chapter, as well as in Chapter 5 on planetary crusts. The effects of cratering on lunar samples, breccia and melt rock formation are also dealt with principally in Chapter 5.

The implications of cratering history for the evolution of the highland crust are considered in Chapter 5, as well as a further discussion of the lunar cataclysm and an assessment of the age data for the highland rocks. The broader implications for planetary evolution and for the planetesimal hypothesis are discussed in Chapter 9. A list of the sizes and locations of all lunar craters mentioned in the text is given in Appendix III.

3.2   The Volcanic/Meteorite Impact Debate

The nature and origin of the great lunar craters was the subject of interest and speculation for over three centuries. The debate eventually became polarized between the volcanic and meteorite impact hypotheses. The volcanic hypothesis was championed by Dana [2], Spurr [3], Green [4], and others [5, 6]. Impact origins were defended principally by Gilbert [7], Baldwin [8] and Urey [9].

In his discussion, Gilbert [7] expended much effort on the question of why lunar craters were circular, an obvious difficulty for the impact hypothesis as it was understood at the time, in view of the fact that meteorites hit the Moon at all angles. The question was not resolved until familiarity with terrestrial explosion craters and the mechanics of impact made it clear that circular craters result from the explosive nature of impacts at high velocities, regardless of angle of impact [10].

However, various minor categories of lunar craters show some assymmetry. These include (a) elongate craters produced by the impact of secondary ejecta at low velocities, and (b) circular craters whose ejecta blankets have missing sections (e.g., Proclus, on the western edge of Mare Crisium). This ejecta pattern has been duplicated experimentally by impacts at oblique angles [11]. An alternative explanation favors the shielding effect of fault scarps [12].

Notwithstanding these special cases, the overall evidence of an impact origin for lunar craters was clear to Gilbert [7]. The argument is an interesting example of apparent constraints due to difficulties in extrapolation from familiar terrestrial conditions which occur in many scientific fields. Later analyses by Shoemaker [10] and experimental studies [13, 14], together with excellent lunar photography and familiarity with the characteristics of terrestrial explosion craters, enabled Baldwin [15] to comment in 1965 that the "136-year-old battle is over." This view has become the consensus. The first mention of the impact theory appears to have been by von Bieberstein in 1802 [16], making the debate 163 years old in 1965, although the 1829 work by Gruithuisen [16] is usually cited as the original source.

Definitive features favoring impact rather than volcanic origins include:

(a) Ejecta blankets. Simple collapse, as in terrestrial calderas, does not produce widespread deposits of thick ejecta blanket.

(b) The rayed ejecta pattern resembles that of explosion craters produced experimentally.

(c) Fresh impact craters always display pronounced elevated rims.

(d) The depth-diameter ratios fit those of terrestrial explosion and meteoritic impact craters.

(e) The energy required to excavate the largest craters and throw out the ejecta far exceeds that available from lunar seismic or volcanic processes.

(f) The production of breccias, shock-metamorphosed material and impact-melted glasses is not observed in terrestrial volcanic craters.

(g) The shock pressures, which often exceed several hundred kilobars, are not easily accounted for in volcanic eruptions. The confining pressure of the total thickness of the continental crust is only 10 kbar. Impact events produce instantaneous pressures as much as one thousand times greater than this.

Vestiges of the volcanic hypothesis for crater origin continued to haunt the lunar scene and influenced Apollo landing site selections in the lunar highlands. By that time the proponents of highland lunar volcanism had diverted their attention from the craters themselves to the nature of the smooth plains deposits covering their floors [17]. The debate continued even during the Apollo missions [18], a fact which has prompted this lengthy account. By now, one hopes that further discussion can be left to the historians of science. A recent summary by Pike [19] provides a fitting epitaph to the debate, in which he notes (p. 17) that "Although minute details of the crater or caldera proper are discussed at great length in such comparisons, the form and extent of the overall structure are almost always ignored. The latter aspects are of critical significance. . . . Two landforms so clearly different in gross surface geometry as terrestrial calderas and main-sequence lunar craters hardly can be proved to be analogous on the basis of minor morphologic characteristics."

3.3   The Mechanism of Crater Formation

Impact formed structures range in size from 0.01 microns, recorded as tiny "zap pits" in lunar samples (Fig. 3.1), to multi-ring basins over 1000 km in diameter (Fig. 3.2). As a result of intensive study over the past 20 years, we now have an excellent understanding of the broad features of impact phenomena [20–31], although problems still remain with scaling to the larger structures.

Formation times of impact craters are very short, with absolute time scales depending on the size of the ultimate crater, which in turn, is related to

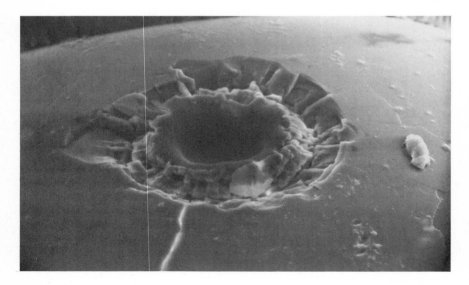

3.1   Zap pit. An SEM photo of a high velocity impact pit on a Apollo 11 glass sphere. The central pit is 30 microns in diameter. (Courtesy D. S. McKay, NASA S70-18264.)

the total kinetic energy of the impactor. Despite such short time scales, four physically distinct stages are usually recognized during the formation of a typical impact crater (Fig. 3.3):

(1) Meteorite collision, penetration and transfer of the projectile's kinetic energy into the target in the form of a shock wave.

(2) Rarefaction of this shock wave at the target's free surfaces, decompressing the volume traversed by the compressive shock wave.

(3) Acceleration of materials by the rarefaction wave and actual excavation of the crater cavity.

(4) Modification of transient crater cavity primarily due to gravitational forces and relaxation of compressed target materials.

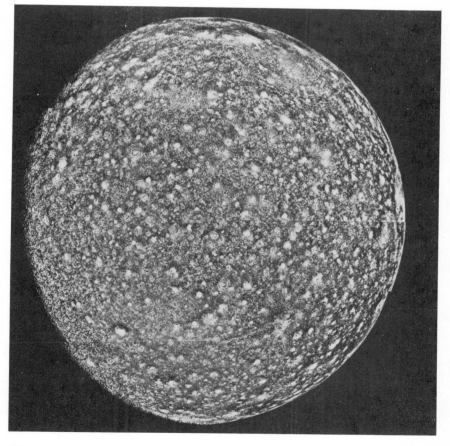

3.2 Callisto, from a distance of 390,000 km, showing the heavily cratered surface with a uniform distribution of craters. The Asgard basin is near the upper right limb.

CRATER GROWTH   FLOW FIELD

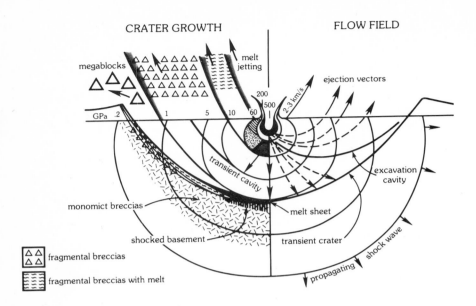

3.3   Schematic cross-section through a growing impact crater before crater modification takes place. Formation of various types of breccias are depicted on the left side.
Various regimes of the melt zone are shown. Shaded lines above the target surface
indicate ejecta plumes at different times. The compressed projectile is given in black.
The right-hand side shows the flow field of the particle motion. The solid line defines
the excavation cavity. (Courtesy D. Stöffler, *Workshop on Apollo 16.* LPI Tech.
Report No. 81-01.)

Depending on crater size, phase (1) may last fractions of a second to
seconds, as typical shock wave velocities are 10–13 km/sec. Phases 2 and 3
overlap in space and time and last typically 2 to 3 orders of magnitude longer,
with bulk ejection velocities generally being in about the 100 m/sec range. The
modification stage may last for comparable time scales. A 10-km diameter
crater is thus completed within minutes, although modest gravitational
adjustment may occur significantly later.

The principal energy transfer occurs via the compressive shock wave;
approximately 50% is transferred into thermal energy, and 50% remains as
kinetic energy in the form of ballistic crater ejecta and displacement of target
materials at depth. Peak pressures in the shock wave may be 5000 kbar at
cosmic impact velocities (15–25 km/sec) and sufficient to completely vaporize
the projectile and some target material. Volumes engulfed by an isobar >700
kbar are generally molten. Materials that experienced 400–600 kbar may be
molten in part if framework silicates are present, e.g., quartz and feldspar.

These minerals form characteristic solid-state glasses, e.g., maskelynite, at pressures between 250 and 400 kbar; rocks experiencing peak pressures < 250 kbar are mechanically deformed to variable degrees and framework silicates will display characteristic planar deformation features at pressures >100 kbar; materials subjected to < 100 kbar are typically just severely disrupted and fractured, although such phenomenon as polysynthetic twinning in, e.g., ilmenite may occur.

The total amount of melt is less than 10% of the excavated volume [31]; most of the ejecta are relatively unshocked (< 100 kbar). Although the melts have experienced the highest peak pressures, much of the melt may remain in the crater, particularly in large structures, where it accumulates in thick melt ponds and sheets [25]; this is explained by specific particle motions during the excavation stage: the highly shocked materials are predominantly driven downward into the growing cavity and the majority of the less shocked species receive a proportionally larger radial velocity component.

The rim deposits first contain unshocked material from near the surface. These are overlain by deep-seated more highly shocked materials,

3.4 Oblique view of crater Linné in Northern Mare Serenitatis. Rim crest diameter is 2450 m. Note ejecta blocks on rim, steep walls, dune-like features on the flanks, and secondary craters at 1–3 crater radii from rim crest [19] (Apollo 15 pan photo 9353).

which produces an inverted stratigraphy. The melted material lines the sides and floor of the cavity and some flows out over the rim.

During the final modification stage, fallback of ejecta into the cavity, collapse of the walls of the transient cavity basement, rebound and uplift of the center, together with various effects due to seismic shaking, occur very quickly. Melt sheets which are draped over some of these features provide evidence that such collapses occur before the melt sheets become viscous (i.e., within a few minutes).

These models satisfactorily account for simple bowl-shaped craters, but are difficult to extrapolate to multi-ringed basin formation, just as the simple Bohr hydrogen atom model cannot be extrapolated readily to describe the structure of the uranium atom. In addition, explosion cratering on Earth may not be the best model for craters formed by meteorite impact on planets lacking atmospheres.

A fictional account of a scenario involving meteorite impacts with the Earth has been provided by Fodor and Taylor [32]. A vivid eyewitness account of an impact is given on p. 62–65 and an excellent description of a freshly formed meteorite crater (4.2 km diameter, 750 m deep, rim height 40 m above the Kansas plain), a little larger than Arizona's Meteor Crater, appears on p. 104–108. Those who wish to toil through descriptions of numerous erotic episodes will find much useful information on meteorites and on the phenomena of impact in this book.

3.4 Simple Bowl-Shaped Impact Craters

The classical example of such a crater is the Arizona Meteor Crater [33] (1100 m diameter, depth 150 m, rim height 47 m), although due to joint patterns in the bedrock the surface plan is more square than circular. One estimate for the impacting body is that it was an iron meteorite about 30 m in diameter (1.7×10^8 kg) traveling at 15 km/sec. The total energy was 1.88×10^{16} J (4.5 megaton equivalent). About 150 m of mixed breccia underlies the floor, covered by 10 m of airfall breccia. The lunar crater Linné (2450 m diameter) provides an excellent example of a fresh bowl-shaped lunar crater (Figs. 3.4, 3.5) [34].

Stratification of the target rocks, volatile content, the inherent strength of the materials and planetary gravity are the predominating factors in crater morphology. What is clear on the Earth, and to a lesser extent on the Moon and Mars, is that the form of the craters depends on the terrain. Simple bowl-shaped craters on the Moon are stable only to about 15–20 km in diameter. The major change takes place at an average diameter of 17.5 km. Eleven changes in parameters take place, marking the transitions from simple to complex craters (Figs. 3.6, 3.7). These changes are a complex function of

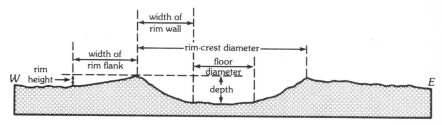

3.5 Crater terminology, illustrated by a cross-section of Proclus (28 km diameter). Adapted from [19].

the Moon's gravity, rock strength, and stresses produced by impact. Simple lunar upland craters are an average of 12% deeper than those of the same diameter on the maria. This is consistent with excavation in a deep megaregolith.

Mercury appears to have a very high target strength, stronger than that of the lunar surface. The Mercurian smooth plains and cratered terrain appear to

3.6 Transitions in crater morphology. The small crater at 11 o'clock from the left-hand crater is Diophantus C (5 km diameter), which resembles Linné. The adjacent 18-km diameter crater is Diophantus. The 25-km diameter crater to the right is Delisle, a complex impact crater in which slump terraces are not yet properly developed [19] (A.15 mapping camera 2738).

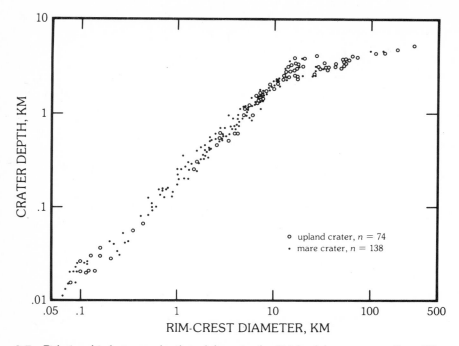

3.7 Relationship between depth and diameter for 212 fresh lunar craters (from Pike [19]). The distribution shows an inflection at about 15 km, which marks the transition from simple to complex craters on the Moon.

have the same target characteristics, suggesting that both are covered by similar material, in contrast to the difference between the lunar maria and highlands. This observation indicates the problems involved in comparing planets of differing bulk density and probably differing chemistries. However, it would be premature to use this evidence to infer that there is no megaregolith on Mercury, or that the Mercurian plains are of basaltic composition [35]. Ganymede and Callisto appear to have low target strength rocks, consistent with a water-ice crust.

The transition from simple bowl-shaped craters to complex types occurs at crater diameters of about 3 km on the Earth, about 6 km on Mars, 16 km on Mercury, and probably at 13 km on Ganymede and 15 km on Callisto. Both gravity and target characteristics appear to be involved (Fig. 3.8). Thus, simple-to-complex crater changes occur at much lower diameters in sedimentary rock than in crystalline rocks on the Earth [36, 37]. The density difference in the projectiles is also an important constraint. Iron meteorites will penetrate more deeply, whereas comet heads and carbonaceous chondrites will explode at shallower depths.

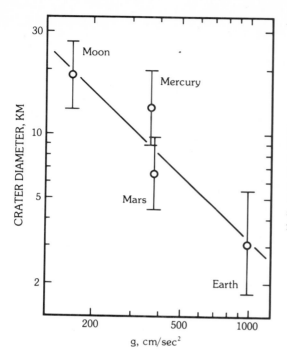

3.8 The inverse relationship between gravity and the crater diameter marking the simple to complex transition for impact craters on the Earth, Moon, Mars and Mercury. Bars are standard deviations. (Adapted from [37]. Courtesy R. J. Pike.)

The earlier concept that fresh craters in the size range 16–48 km were markedly more polygonal than smaller or larger craters [38] does not appear to be borne out by more recent data [34, p. 40]. A maximum in rim-crest circularity appears at about 10 km diameter. In comparison with mare craters, rim crests are less regular on highlands craters. The shapes of Proclus, King and Tsiolkovsky have been affected by impact into preexisting irregular topography. The rim crest relief increases rather regularly with crater diameter [34].

3.5 Large Wall-Terraced Craters, Central Peaks and Peak Rings

As noted earlier, modifications to the simple bowl shape for lunar craters begin to appear at diameters greater than about 15 km [34]. The larger craters have complex terraced walls, central peaks and flat floors. The wall terraces are well exhibited by the crater Theophilus (Figs. 3.9 and 3.10), which in addition displays a well-marked linear terrace where the crater wall has intersected the rim of the older crater, Cyrillus. Initial cavity depths may be

3.9 Oblique view of the lunar crater Theophilus, 100 km in diameter, showing well-developed slump terraces and a central peak complex. The straight terrace on the west wall is due to the intersection of Theophilus with the older degraded 98-km diameter crater Cyrillus (NASA Orbiter IIIM78).

estimated by restoring the slumped wall blocks to their original positions [39, 40]. The greater abundance of central peaks and wall terraces in mare craters is consistent with the greater strength of mare basalts compared to that of the highland megaregolith [34, 35]. The interest in central peaks in craters (Fig. 3.9) has been accentuated by the spectacular examples provided by the Voyager spacecraft photos of the Saturnian satellites, particularly of the 135–km diameter crater on Mimas and of others on Dione and Tethys (Fig. 3.11).

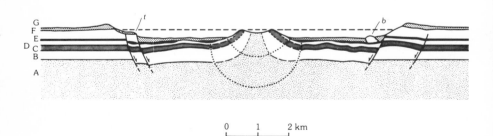

0   1   2 km

3.10 A schematic diagram of a large complex crater with slumped walls and a central peak showing uplift. The lowest unit, A, has been displaced upward by about 1 km. A and B dip vertically in the central peak. Unit G is ejecta on the crater rim, and breccia within the crater. Impact melt is not shown. The large semi-circle outlines the zone of total disruption. The smaller semi-circle marks the penetration limit of the impacting body. (Adapted from [37]. Courtesy R. J. Pike.)

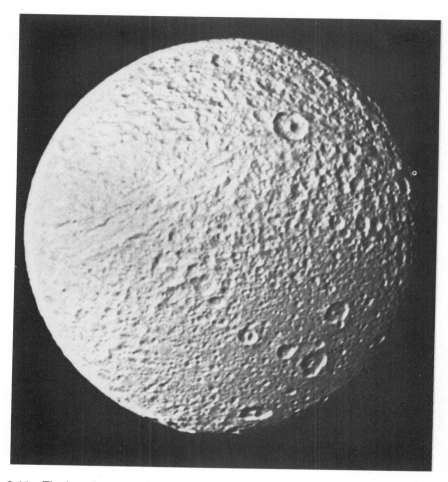

3.11 The heavily cratered surface of Tethys, showing craters down to about 5 km diameter. Note central peaks. There is a division between a heavily cratered region (top right) and a less heavily cratered area (bottom right). Tethys is 1050 km in diameter (JPL P.24065).

Central peaks appear in lunar craters at diameters greater than about 10–20 km, and all fresh lunar craters with diameters greater than 35 km have central peaks [42]. Two alternative explanations have been offered. The first is that the central peaks are thrust up by centripetal collapse of the rim [41]; the other is that they are caused by rebound. Both processes may be important [31].

The evidence from large terrestrial impact sites such as Gosses Bluff [43] and Sierra Madera, Texas [44] indicates very substantial uplift from stratigraphic horizons below those which could have been affected by rim slumping. The central peaks also seem to develop early in the cratering event [45] and so do not appear to form by isostatic readjustment (see Chapter 7). Since there is little discernible difference between craters in maria (10 m thick regolith) and highlands (2.5 km thick megaregolith), the theory that formation of central peaks requires a less cohesive surface layer overlying a more cohesive substrate must be reevaluated [46].

Although summit craters on central peaks of lunar impact craters have occasionally been reported by terrestrial observers, they are mostly illusory

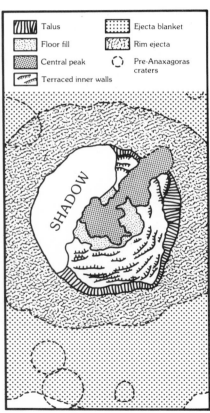

3.12 The lunar crater Anaxagoras (51 km diameter), showing an asymmetric central peak. The geological sketch map indicates that much of the central peak material overlies the terraced crater walls. The interpretation is that the central peak collapsed sideways. (Adapted from [47]. Courtesy J. B. Murray.)

3.13 The transition between central peaks and peak rings is well illustrated in this Orbiter photo (IV-8 M). The large central basin is Schrödinger (320 km diameter) which has a well-developed peak ring. The smaller crater Antoniadi (135 km diameter), at 4 o'clock from Schrödinger, has both a small central peak and a peak ring. The smaller crater just southwest of Antoniadi has a central peak only.

[34, p. 30]. A real example is the crater Regiomontanus A, a 6-km-diameter crater on the central peak of Regiomontanus (124 km diameter). It seems to be an impact crater fortuitously superimposed on the central peak. Based partly on this sort of observation, the formation of central peaks was long construed as evidence for the volcanic origin of lunar craters; however, their dissimilarity in form, compared to terrestrial volcanoes, long ago removed the analogy from serious consideration. Perhaps the most notable feature of the summit craters is their irregularity; it is rare to find symmetrical peaks. This lack of regularity constrasts with the well-developed circularity of impact craters (see [34] for details of the circularity index). The central peaks of Copernicus are well known, and show a rather jumbled topography. Other examples which show even more irregularity include Tsiolkovsky (180 km), King (77 km), Humboldt (207 km) and Anaxagoras (51 km) (Fig. 3.12).

Antoniadi (140 km) (Fig. 3.13) and Compton (165 km) represent a transitional type in that they have both central peaks and an inner peak ring. Asymmetrical ridges in Tsiolkovsky and Humboldt nearly link the central peak to the crater walls. Anaxagoras (Fig. 3.12) exhibits an extreme example of this structure, which is attributed to the sideways collapse of the central peak [47].

There is an interesting relationship between central peak size and crater diameter. The ratio of central peak diameter to crater diameter plotted as a function of crater diameter reaches a maximum value of about 0.3 for craters of about 50 km diameter and falls off at crater diameters larger than about 70 km, although there is much scatter in the data. Peak rings occur at diameters greater than about 100 km. The central peak size decreases when rings appear, indicating a connection between the formation of central peaks and peak rings [47], which will be discussed in the next section. Maxwell [48] notes that mare wrinkle ridge rings (Section 6.16) lie too close to basin edges to represent buried peak rings. The spacing of peak rings is not the same as that of the wrinkle ridges.

The proposal that the central peak of Copernicus may be composed of olivine [49], suggested by remote-sensing data, would indicate an origin by upward movement of deep-seated material in at least two stages. The original depth of material now present in the central peak of Copernicus was about 10 km. This observation, if correct, raises interesting questions about lunar crustal structures (Chapter 5), and the nature of mascons (Chapter 7). Possibly the Copernican impact was centered on a previously uplifted mantle plug. The ultrabasic composition of the central peak of Copernicus appears to be unique: other central peaks show typical crustal signatures (Carle Pieters, pers. comm., 1981).

3.6   The Multi-Ring Basins

These are among the most spectacular landforms in the solar system, rivalled only by Olympus Mons and Vallis Marineris on Mars. Examples extend from Mercury (Caloris basin, Fig. 3.14) to Callisto (Valhalla basin, Fig. 3.15). On the Moon, 30 basins with diameters exceeding 300 km are definitely recognized, and an additional 14 are probably present (Appendix IV).

Most discussions on the origin of multi-ring basins use the superbly preserved Mare Orientale (Fig. 2.3) as the type example [50]. This basin is most clearly defined by the Montes Cordillera, which forms an impressive mountain ring with a diameter of 920 km. Typical relief is about 3 km. An inner mountain ring with a diameter of 620 km is formed by the Montes Rook. There is an Inner Rook mountain ring with a diameter of 480 km and an Inner Basin Scarp with a diameter of 320 km which borders the inner mare basalt fill (see Figs. 2.3, 3.16).

As with most multi-ring basins, there is disagreement over the location of the original crater rim. According to Head [51], the outer Rook Mountain ring (620 km in diameter) forms the rim, while the inner Cordillera ring is considered to be a fault scarp and the region between Montes Cordillera and Rook is a megaterrace. The view adopted here, for reasons given later, is that the Montes Cordillera represent the true initial rim of the Orientale basin.

3.14 Photomosaic of Mercury, showing the Caloris basin, 1300 km in diameter, surrounded by an extensive ejecta blanket.

The deposits related to this basin are referred to as the Orientale Group and comprise the Hevelius Formation, which extends radially from the Montes Cordillera, the Montes Rook Formation, between the Cordillera and Rook Scarps, and the Maunder Formation, occupying the region between the central mare fill and the Rook Scarp. These deposits provide useful information on the origin of the basin. A significant point is that the lunar surface inside the Cordillera Scarp is very different from that outside, where a normal lunar cratered surface exists. Baldwin [52] notes that the ages of the inner basin and rings are the same, although they are much modified by the deposition of the Hevelius Formation. That is, the surface of the Orientale basin contained within the Cordillera Scarp is of the same age. A similar observation holds for the inner portion of the Nectaris basin up to the Altai Scarp. The Altai Scarp intersects the large crater Rothmann G. The portion of the crater rim inside the basin is missing. Baldwin [52] ascribes this to "fluidization," but there are many difficulties with that hypothesis. The mountain rings do not resemble frozen waves. No previous structures appear within

3.15   The multi-ring Valhalla basin on Callisto centered at 10° lat. 55° long.

the Apennine Mountain Ring at Imbrium, although the structure is much obscured by later mare basalt filling. The enigmatic Apennine Bench Formation (Section 2.4.4) is here interpreted as equivalent to the knobby facies of the Montes Rook Formation. Thus, it probably represents an impact melt or debris sheet ponded against the outer scarp. A controversy exists on both the diameter and depth of the original basin (Section 3.17). Here one is concerned principally with the diameter of the basin. The depth is model dependent, but the evidence appears to favor larger rather than smaller cavities. No traces of the pre-basin highland surface remain within the Cordillera, Altai or Apennine Rings.

Fault scarps circumferential to basins are occasionally observed. The most celebrated example is the set of three curved rilles, each about 2 km wide, concentric to Mare Humorum at a distance of about 250 km from the center of that basin (Fig. 3.17). These grabens, which follow ring faults, postdate the mare fill, do not resemble megaterraces and are interpreted here as lying within the original basin rim.

Several theories of basin formation have been proposed. A principal difficulty in dealing with such phenomena is that of scale. Cratering theory can account satisfactorily for simple bowl-shaped craters up to about 1 km in diameter, but the complexities of larger craters are less readily modelled. This tends to favor notions of the hand-waving variety. In such cases, it is necessary to examine carefully all the evidence available from geological relationships to constrain hypotheses.

The most popular hypothesis has been the slumping or "megaterrace" model [51, 53–56]. Initially, a deep central bowl-shaped transient crater is formed. Later subsidence along ring faults produces the outer rings. On this

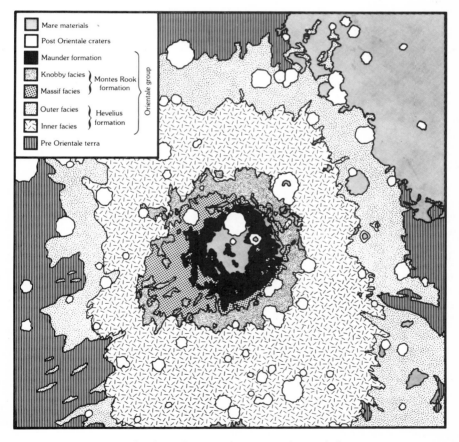

3.16 Geological sketch map of the Orientale basin.

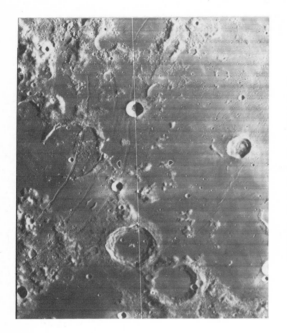

3.17 Three sets of curved rilles, about 2 km wide, concentric to Mare Humorum. The large crater, bottom right, is Campanus, 48 km diameter, flooded with mare basalt and containing a sinuous rille. The distance from the curved rilles to the center of Mare Humorum is about 250 km. The ruined crater intersected by the inner rille is Hippalus (58 km diameter) (NASA Orbiter IV-132-H).

3.18 The Imbrium basin, 1500 km in diameter. Copernicus (93 km diameter) is the bright ray crater at the bottom of the picture. Sinus Iridum is the crescentic feature on the northwest rim. Plato (101 km diameter) is the flooded crater on the northeast rim. The Apollo 15 site was southeast of Archimedes, the large crater flooded with lava in the southeast quadrant inside the basin. (Rectified photo, Courtesy D. E. Wilhelms.)

hypothesis, the original diameter of Orientale is given by the Montes Rook, rather than by the Montes Cordillera. Despite the popularity of this theory, it suffers from some serious defects. The evidence favors instantaneous production of the rings during the basin-forming collision. All preexisting topography is affected out to the main outer ring. The rings are generally quite concentric, although some exceptions occur, most notably at Imbrium, where there are gaps in the outer ring and where the Montes Caucasus are not concentric with the Montes Alpes. The fact that Imbrium is not a perfect multi-ring basin [61] may be due to the formation of the basin on a topography of older multi-ring basins. Thus, it has been proposed that the Montes Alpes slid into the Imbrium basin following the impact. This model has the advantage of restoring the outer ring of Imbrium to near-circularity [61]. Probably an earlier Procellarum basin, 3200 km in diameter, occupied the area covered by the present smaller Imbrium basin [57]. It is concluded that such topographic effects can account for the irregularities observed at Imbrium. A smaller-scale example is seen on the west rim of Theophilus, where it intersects the older rim of Cyrillus (Fig. 3.9).

The "Nested Crater" model was proposed by Wilhelms et al. [58]. In this model, the rings represent the rims of concentric craters reflecting layers of differing strength in the lunar interior. Thus, for the Orientale basin, the Montes Rook ring is equated with the 25 km seismic discontinuity, while the

3.19 The Korolev multi-ring basin, 440 km in diameter, of Nectarian age (NASA Orbiter I 038 M).

Inner Basin Scarp represents the discontinuity at the base of the crust. The Inner Rook ring is considered to be a slumped megaterrace.

Dence [59] proposed a very deep transient crater (230 km) for Imbrium (Fig. 3.18). Instantaneous collapse into this cavity produced the outer rings. The absence of samples in the Apollo collection derived from great depths (no identifiable mantle samples have been described) makes this hypothesis less attractive, although there is evidence from the Ries crater which supports the megaterrace hypothesis. Very little material, however, is derived from deeper than 0.6 km at the Ries [31, 60].

The megaterrace model has some difficulties in explaining the regularity of the Cordillera scarp. A slumping model might be predicted to produce a much more irregular and segmented scarp than is actually observed. A particularly good example of the cratering process is provided by the Schrödinger basin (Fig. 3.13). This basin, which contains a peak ring, is about 320 km in

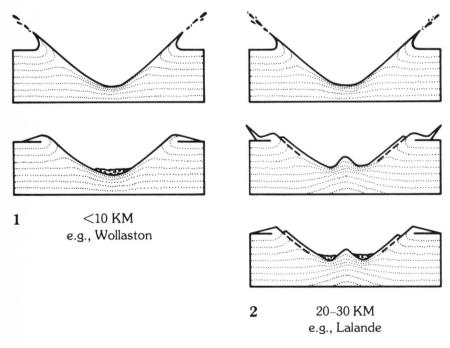

1   <10 KM
e.g., Wollaston

2   20–30 KM
e.g., Lalande

3.20 Cross-sections showing stages in the formation of impact structures in four different size ranges. The lower section in each case shows the final situation; the small circles represent crater fill. Dashed lines represent fractures. Dotted lines indicate subsurface deformation, and are not necessarily intended to represent differing layers in the substratum. See text for discussion [47]. (Courtesy J. B. Murray.)

diameter. The crater rim shows the usual terrace development as observed in Theophilus, for example (Fig. 3.9). There are no mountain rings outside the Schrödinger basin. It seems clear that the initial crater cavity for multi-ring basins must be significantly larger than Schrödinger. The ejecta blanket from Orientale is several times larger than that of Schrödinger. Accordingly, the viewpoint is taken here that the size of the initial basin is best represented by the outer mountain rings (Cordillera, Caucasus-Apennines, Altai). There does not seem to be a regular development of megaterraces as the size of the

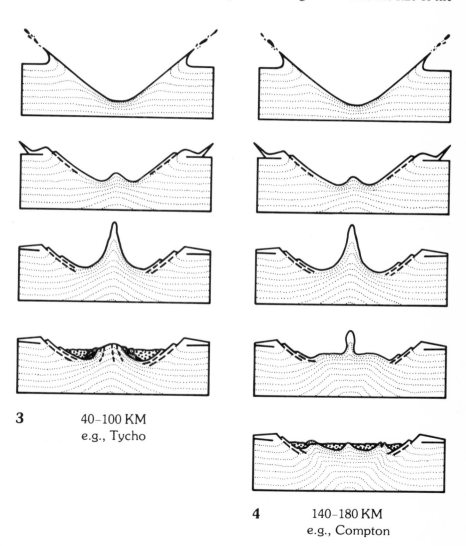

3   40–100 KM
              e.g., Tycho

4   140–180 KM
              e.g., Compton

craters increases. Thus, the small terraces at Schrödinger resemble those of Copernicus, consisting of many short irregular scarps.

Demonstrable megaterraces are not common on the Moon, although a good example exists on the western scarp of the ringed basin Korolev (Fig. 3.19). Multiple rings appear in several terrestrial craters (Ries, West Clearwater, Manicouagan, Gosses Bluff, Popigay), but evidence of fluidization appears to be lacking [62]. Other objections to the fluidization or Tidal Wave or Tsunami hypothesis include comments by Hartmann and Wood [63] that the "lunar rings do not have wave-form profiles, but rather appear to be sharply asymmetric fault scarps."

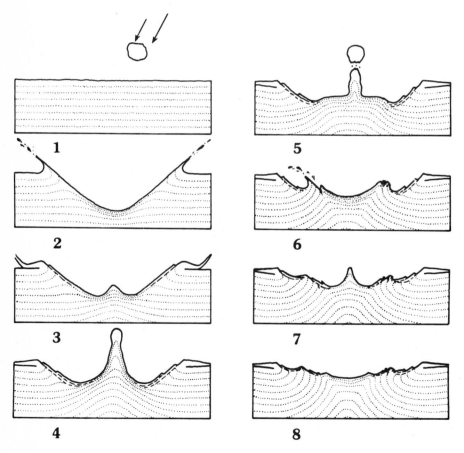

3.21 Cross-sections showing the suggested sequence of events in the formation of the Orientale basin (not to scale). See text for discussion. Symbols as in Fig. 3.20 [47]. (Courtesy J. B. Murray.)

The latest model to join the list of proposed origins for ringed basin structures is the "Oscillating Peak" model [47] (Figs. 3.20 and 3.21). The sequence of events for Orientale is as follows:

(a) The initial impact causes a bowl-shaped crater 850 km in diameter.
(b) Rebound of the center occurs, with some slumping at the crater edge and formation of terraces.
(c) Hevelius Formation is deposited. A central mound is produced. The Cordillera Scarp reaches 920 km in diameter, through terracing.
(d) Gravitational collapse of the central peak begins.
(e) Formation of an inner crater (the Montes Rook ring) occurs due to the collapse of the over-extended central peak.
(f) A second rebound occurs within the Rook crater, forming another central peak, and terracing forms the Rook Mountain ring (620 km).
(g) Collapse of the second central peak forms the Inner Rook ring (480 km) as a complex anticline.
(h) Subsequent flooding with mare basalts at a much later stage causes subsidence and formation of the Inner Basin Scarp (320 km).

Uplift of the mantle beneath the central region, together with the mare basalt fill, account for the mascon, since the mare basalt fill alone is quite inadequate to account for the positive gravity anomaly (Section 7.4.2).

Murray [47] notes that the sequence of events depends on rock behaving in a fluid fashion in very large impacts. This implies that pressures in excess of 500 kbar are involved and are experienced throughout the basin. Evidence from the lunar highland rocks does not seem to favor the development of such extensive shock pressures over the wide area of the basin.

The model draws some support from the terrestrial cratering experiments. The famous example of the Prairie Flat Crater [64, 65] formed a ringed structure in alluvium. The crater is 61 m in diameter and 5 m deep, and was formed by the explosion of a 500 ton TNT sphere. A large central peak crater (Snowball) was formed by the explosion of a hemispherical 500 ton TNT charge. Collapse of the central peak is inferred from displacement of markers, although the formation or collapse of such a peak is obscured from view during the explosion. Since this event occurred in water-saturated alluvium, conclusions may not readily be extrapolated to the Moon.

A somewhat similar model, illustrated in Fig. 3.22, incorporates features of several of the preceding hypotheses [66]. A deep parabolic central cavity forms initially. This rebounds following compression and then collapses due to gravitational forces to form a sequence of rings. A feature of the Dence and Grieve model [66], is that excavation is still occurring at the edge of the cavity, while uplift is taking place in the center (Fig. 3.22, stage b), producing a cavity which resembles a sombrero. Collapse in the rim region enlarges the cavity, reminiscent of the slumping models.

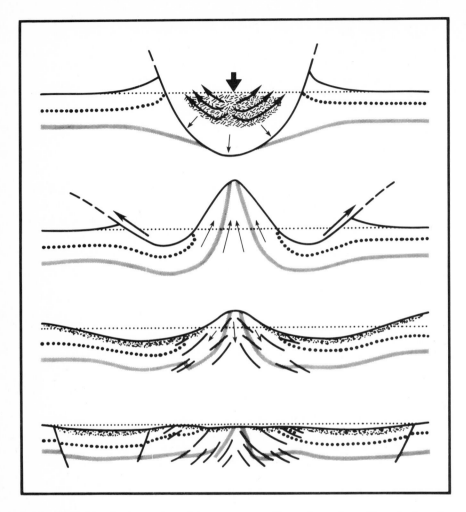

3.22 Model for the formation of ring structures. Top—formation of transient cavity by excavation and compression. Second top—rebound of compressed base of transient cavity above ground surface. Excavation from near surface at periphery is continuing. Second bottom—collapse of uplift peak under its own weight with development of reverse faulting in central area. Major excavation has ceased, although melt and breccias (stipple) are still in motion within cavity. Bottom—final configuration. Rim and uplift have reached equilibrium height controlled by rock strength and gravity. Excess volume in initial uplift, which cannot be accommodated by reverse faulting, appears as a ring. Faulting and slumping have occurred on rim. Movement of material within final crater has ceased [66]. (Courtesy R. A. F. Grieve.)

3.6.1 Very Large Basins

The distribution of lunar basins is shown in Fig. 3.23. Two very large basins occur on the Moon. Procellarum, or Gargantuan basin, has been proposed by Cadogan [57]. It is centered just to the west of the crater Timocharis at 26° N 15° W. Three rings with diameters of 1700, 2400 and 3200 km are identified, thus making this the largest impact event on the Moon. It must be older than 4.2–4.3 aeons. Like the Cheshire Cat, only the ghost of a smile or palimpsest remains of this great event.

The area south of about 45° latitude on the Moon is of unusual interest. Contained within this region are complex basins such as Schrödinger and Antoniadi (Fig. 3.13). The rima or rilles radial to Schrödinger (Rima Planck and Rima Schrödinger) are probably connected with the formation of Schrödinger basin. Antoniadi, with a central peak and a peak ring, is surrounded by an exceptionally large secondary crater field. Twelve ringed basins have been identified in the area, including the South Pole-Aitken or Big Backside basin, which is 2500 km in diameter [67].

The center of this basin is at latitude 56° S, longitude 180°. Its existence was originally postulated by Hartmann and Kuiper [68] from the observation that the Leibnitz Mountains (7–9 km high) were arcuate. Apollo laser ranging indicated the presence of a deep basin [69].

Mare Ingenii (Fig. 6.1) lies totally within this great basin, as does the crater Van de Graff. It is noteworthy that the largest negative gravity anomaly occurs within this basin. The highest concentrations of K, Th and Fe occur at Van de Graff, as well as the lowest lunar far-side concentrations of Ti. The highest magnetic anomalies likewise occur in this vicinity. The diameter of Big Backside basin is 75% of the diameter of the Moon. It can be inferred from the existence of these two giant basins that "collisional fragmentation of moon-sized bodies is difficult" [70].

The Valhalla basin on Callisto and the Caloris basin on Mercury are examples of very large basins at the inner and outer regions of the solar system (see, for example, [71, 72]) showing that cratering on this scale was ubiquitous throughout the solar system.

A tabulation of lunar basins is given in Table 3.1 and in Appendix IV.

3.7 Depth of Excavation of Basins

This is a critical question for our understanding of the composition and evolution of the lunar highland crust. Extrapolation at present from simple bowl-shaped craters to multi-ring basins clearly does not work, since this provides estimates ranging from 20–230 km [59]. Depths greater than

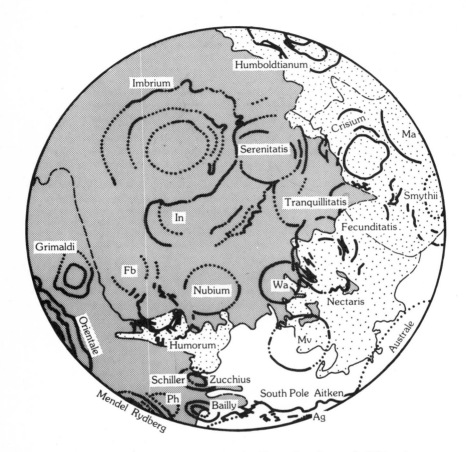

3.23 Distribution of multi-ring basins on the Moon. See Appendix IV for sizes, ages and references. Basin deposits and secondary craters have been identified around 18 basins (shown with names in capitals). (Courtesy D. E. Wilhelms.)

about 60 km are ruled out by the apparent absence of samples from beneath the lunar crust. Such samples should be readily identifiable using present models of the lunar crust, which is chemically distinct from likely mantle materials. If substantial uplift of mantle occurs during large basin formation, as is argued in Chapter 7, then mantle samples might be occasionally excavated by later impacts from shallow depths [49]. Shallow depths of excavation are possibly excluded since the depth estimates of 8–27 km [73] are less than the diameters of the impacting projectiles responsible for the excavation [27]. Estimates of 30–60 km have been made by Grieve [66] with the higher value

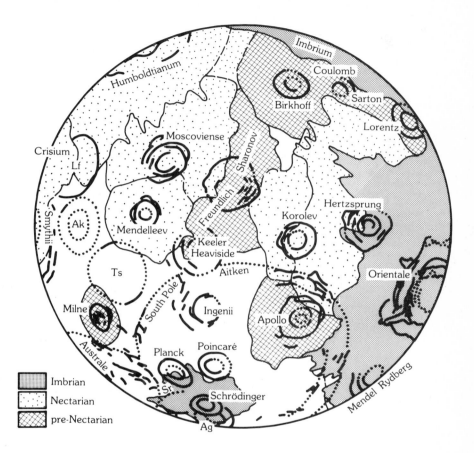

being considered more likely. Evidence of relatively deep excavation comes from the presence of spinel-bearing breccias. One ultramafic spinel cataclasite clast (15445, 177) may indeed come from the uppermost mantle of the Moon [74]. Other samples containing spinel and cordierite from Apollo 15 and 17 are identified as coming from depths of 15–32 km within the highland crust [74]. Some evidence on this question comes from the Ries crater (26 km diameter) (Fig. 3.24). Only 0.1% (wt.) of the ejecta in the continuous ejecta blanket comes from deeper than 600 m [60]. More than 99% of the Bunte breccia is unshocked and emplaced as a cool mass at ambient temperatures. Most of the thermal energy resides in the impact breccias and melts, most of which are retained within the crater structure.

These findings from the Ries crater raise questions about the interpretation of lunar samples with respect to crater ages, basin formation ages,

Table 3.1 The relative age sequence of the younger multi-ring basins, established by superposition relationships.

Source 1	Age Group	Source 2	Age Group	Source 3	Age
Orientale	IV	Orientale	2	Orientale	Imbrian
Imbrium	IV	Imbrium	3	Imbrium	
Crisium	III	Crisium	4	Crisium	
				Humorum	
Humorum	II	Humboltianum	5	Serenitatis	Nectarian[†]
Nectaris	II			Humboltianum	
Grimaldi	II	Grimaldi	6	Nectaris	
Serenitatis	II				
Humboltianum	II	Nectaris	7		Pre-Nectarian*
Smythii	I	Humorum	8		
Fecunditatis	I	Serenitatis	8		
Tranquillitatis	I				
Nubium	I	Smythii	9		
Australe	I				

[†]Additional basins of Nectarian age are: Hertzsprung, Korolev, Mendeleev and Moscoviense. Bailly is probably Nectarian.
*See Appendix for a list of 18 Pre-Nectarian basins and 11 probables.
Sources: 1. Stuart-Alexander, D. E. and Howard, K. A. (1970) *Icarus* 12: 440.
      2. Baldwin, R. B. (1974) *Icarus* 23: 97.
      3. Wilhelms, D. E. (1976) *PLC 7:* 2883; (1980) Multi-Ring Basins Abstracts, 115.

chemistry of basin-forming projectiles, identification of breccias with particular basins and so forth. Impact melt is mostly retained within the crater. The depth of the transient crater at Ries was about 2.5 km. If ejecta blankets are cold, not hot, then the well-documented thermal event at Apollo 14 is not related to the excavation of the Imbrium basin [60, 66, 75]! However, this problem may be resolved if a combination of shocked and unshocked, melted and cold material makes up the lunar ejecta sheets, derived from an already heavily bombarded terrain. Enough occurrences of suevite occur outside the Ries rim (Fig. 3.24) to indicate that melt escapes from the crater. The Ries is an order of magnitude smaller than the smallest lunar ringed basins, so that difficulties of extrapolation arise.

Significant progress in our understanding of excavation and sampling depths has been made since the realization that the transient crater cavity differs from the excavation cavity. Both in turn differ from the final multi-ring basin form [31]. The transient crater may be deep, and the rebound mechanism may bring up deep-seated material in the center, but most of the excavated material comes from shallower depths. The depths of excavation in the highland crust are likely to be in the range 30–60 km [23, 66]. The rebound

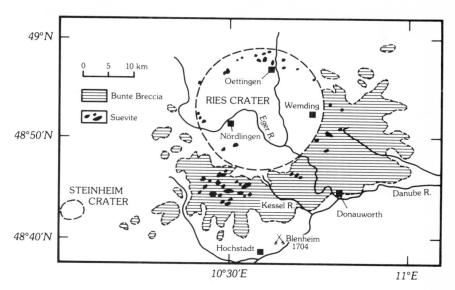

3.24 Sketch map of ejecta deposits at the Ries Crater, Germany. [After Gall, H., et al. (1977) *Geol. Bavarica* 76.]

following the production of the deep transient crater may involve a mantle plug, which does not reach the surface, but provides high density material beneath the basin. (See Section 7.4.2 on mascons.)

3.8 Basin Diameter-Morphology Relationship

Basin morphology apparently varies systematically with basin diameter on the Earth and the Moon. Thus, the sequence from simple bowl-shaped craters, to the appearance of central peaks, central peaks and peak rings (CP basins), to peak ring (PR) basins only, and then to multi-ring basins is well recognized (e.g., [76]). However, two small peak ring basins and two large central peak basins, lying outside the normal sequence, have been identified on Mars [77], so that central peak basins are not restricted to a single diameter interval and peak ring basins may occur at smaller diameters. Two large peak ring basins (Lyot and Herschel) occur at diameters on Mars where only central peak basins are observed on the Moon and Mercury.

The anomalies noted above indicate that morphology may not be simply related to diameter, but is a more complex relationship than previously realized. The new observations on the Saturnian satellites (e.g., craters on Dione and Mimas with large central peaks) must also engender caution. On

Earth, target strength (e.g., crystalline or sedimentary rock) influences the transition from simple to complex crater morphology [78] as well as on Mars [79] and Mercury [80]. In this context, peak rings within Martian and lunar basins are usually composed of short arcs or isolated peaks, but on Mercury, the rings are commonly complete.

The possibility that the "anomalous" Martian basins are caused by cometary rather than asteroidal impact is weakened by the absence of anomalous basins on other planets. The small central peak basins (50–100 km in diameter) appear to be unique to Mars. Most probably they are related to different target characteristics. The subsurface volatile layer on Mars, believed to be responsible for the fluidized craters, central peak craters, and lobate ejecta flows is often considered to be the decisive factor, but Mars is drier than the Earth.

3.9   Impact Melts and Melt Sheets

Impact melts occur commonly in and around fresh lunar craters. Most of them are retained within the crater cavity rather than being ejected. They resemble lava flows, and have been described at 55 lunar craters from 4 to 300 km diameter [81]. The melts mostly occur on crater floors and as small ponds near the crater rims; flow lobes and channels beyond the rim are prominent in

3.25   The crater Tycho (85 km in diameter) showing possible melt sheets on the crater floor (Orbiter V M 125).

craters exceeding about 10 km diameter. Flows are common in craters up to 50 km diameter.

Structures interpreted as sheets of melt-rock appear in Tycho (Fig. 3.25) and Copernicus. At Tycho, the melt sheet appears to have drained down the inner walls of the crater. At Copernicus, some apparent melt sheets have covered slumps, and slumps in other places have overridden melt sheets, indicating that the emplacement of wall slumps and melt sheets is more or less simultaneous.

Caution must be exercised in these conclusions based on photogeological interpretations. Thus, smooth deposits on the floor or Copernicus interpreted as impact melt may be fine-grained debris flows, as appears from data from the Apollo 17 infrared scanning radiometer [82]. This is another reminder of the hazards in photogeological interpretations without close ground control.

With these caveats in mind, the Apennine Bench Formation (Section 2.4.4), of uncertain origin, is here considered to be an impact melt or debris sheet from the Imbrium collision, ponded against the Apennine Mountain Scarp, which is interpreted as the edge of the Imbrium basin excavation. Thus, the Apennine Bench Formation is the equivalent in the Imbrium basin of the knobby facies in the interior of Orientale, as was noted earlier. The chemical features of melt sheets and their relationship to the parental country rock compositions is of much significance and is dealt with principally in Chapter 5 on planetary crusts.

Extrapolation of conditions from one planet to another is fraught with difficulties. On the Earth most melt sheets are derived from impacts into crystalline rocks [25]. Melts produced by impacts into sedimentary rocks are possibly finely dispersed due to expansion of volatiles. Nevertheless, even terrestrial crystalline rock terrains are probably wetter than the Martian subsurface, so that the apparent absence of "lunar-like" melt sheets on that planet cannot be ascribed to a difference in volatile contents between the Moon and Mars. A further effect is that target materials, shocked to pressures well in excess of 500 kbars, may be degassed. Repeated cratering during accretion may therefore enrich volatiles toward planetary surfaces. Cratering in the icy surfaces of the Galilean and Saturnian satellites should produce large amounts of melt and vapor, with widespread dispersal of the melt [25].

3.10 Secondary Impact Craters

These craters have caused much trouble to lunar investigators. Since they are mostly of small size, they have vitiated attempts to secure a sounder statistical base for stratigraphic ages based on crater counting; their use, which predicted very young ages for lunar maria, resulted in some loss of confidence in the simple crater counting method by the lunar science community. When crater populations are restricted to those greater than 2–4 km in diameter, then the method becomes more reliable, at least on lunar mare surfaces. A further complication is that large secondary projectiles result from multi-ring basin formation and may create secondary craters as large as 25 km in diameter.

Various characteristics may be used to distinguish primary craters. Their raised rims are about twice as high above the surrounding terrain as are those of secondaries. The rim-crest circularity is much higher for primaries: 0.81 for a sample of 44 primary craters, but only 0.57 for 29 secondaries [34]. Secondary craters also tend to be shallower, the depth-diameter ratio being 0.13 compared to 0.19 for the primary craters studied above. Herringbone or V-shaped patterns for secondaries are common and are perhaps among the most diagnostic features [83]. The tendency of secondaries to form linear crater chains has caused much confusion with possible volcanic craters. A classic example is Catena Davy (Fig. 3.26) which is a row of about 25 small craters aligned along a distance of about 48 km. These have been identified as

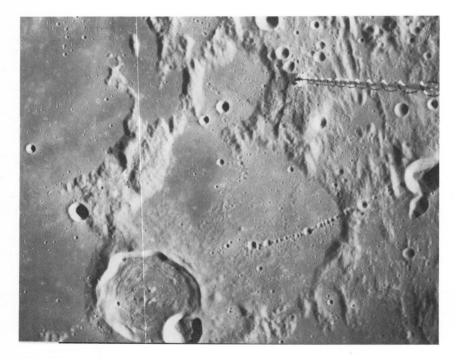

3.26 Catena Davy, a chain of some two dozen craters, formed as a result of secondary impacts, possibly from Orientale. The crater chain is 48 km long. Davy (35 km diameter) is the large complex crater in the southwest corner (AS 16-2198).

3.27 Dark halo craters, some apparently associated with rilles, on the floor of Alphonsus (119 km diameter). The large crater to the south is Arzarchel (97 km diameter). Oceanus Procellarum lies to the right (west) (AS 16-2477).

secondaries [34] and as volcanic vents but are most probably due to the former process [84]. Collapse pits along sinuous rilles add further to the complexity. The number of truly volcanic craters on the Moon appears to be small. Many dark-halo craters were originally thought to be volcanic, but are now known to be impact craters, possibly penetrating through crater ejecta into dark underlying mare basalt (e.g., Copernicus H). The dark-halo craters along the rilles in the large impact crater Alphonsus are more problematic and may be the equivalent of terrestrial cinder cones [34, Fig. 3.27]. Only a visit to them will resolve the question.

There is a wide variation in the population of secondary craters around Martian craters in the size range from 5–50 km diameter [85]. Thus, some craters (e.g., Crater Wx, 34 km diameter, in Chryse Planitia) have extensive crater fields and elongate secondaries, consistent with large ejecta, and ejection angles less than 50°. These features are consistent with impact into a competent target lithology, inferred to be basaltic lava flows [86]. In contrast, the Martian crater Arandas, 45 km in diameter [87] has very few secondaries and the few that exist are nearly circular. These features are interpreted to be a result of meteorite impact into a low strength lithology. Experimental studies indicate high angles of ejection for such lithologies [88], which will produce circular secondary craters with near-circular rims.

Comparisons with lunar craters (e.g., Copernicus, 93 km diameter and Aristarchus, 39 km diameter) and Mercurian craters (e.g., Alencar, 110 km diameter) indicate that the secondary crater population, outside the region of continuous ejecta, is similar for the Moon and Mercury [89].

The large production of secondary craters during basin formation adds further complexity to the scene, and secondary impact by these basin ejecta has had a major modifying effect on the lunar highland landscape [90]. Basin-produced secondary craters may be very numerous even up to diameters of 25 km! This places severe restrictions on simple crater counting statistics. However, such basin secondaries may be useful stratigraphic markers, provided superposition relations among them can be recognized. The abundance of these basin secondaries further reduces the reliability of establishing a relative age sequence by crater counting unless very large areas are sampled.

Secondary craters appear to be common on the Saturnian satellites. Since they have low surface gravity and escape velocities [91] (except Titan), many secondary fragments will escape into orbits around Saturn. Most, however, will be swept up by the parent satellite. Possibly, most of the 10–20 km diameter craters on Mimas, for example, may be secondary craters [92].

When all these factors are taken into account, relative ages of planetary surfaces established by crater counting techniques have yielded remarkable insights into planetary evolution, but it is necessary to be aware of the limitations and problems with the method, as in all other dating techniques.

3.11   Dark-Halo Craters

Dark-halo craters may be either of volcanic or of impact origin. The criteria listed in Table 3.2 may be used to distinguish them [93].

Some small craters (e.g., in Alphonsus) appear to be good examples of possible volcanic craters (Fig. 3.27). They are considered to be due to explosive eruptions caused by the building of localized accumulations of volatiles. In contrast, dark-halo craters such as Copernicus H are undoubtedly of impact origin.

The questions surrounding dark-halo craters are still not properly resolved, and indicate many of the problems with the interpretation of photographs of distant objects. One recalls the difficulties in identification of common objects in close-up views (e.g., safety pins, combs, clothes pegs, etc.).

Table 3.2 Criteria for distinguishing volcanic from impact origins for lunar dark-halo craters (adapted from [93]).

Aspect	Impact Origin	Volcanic Origin
Crater shape	Circular	Often non-circular
Structural association	Random	Often aligned along rilles
Rays	Common	Absent
Ejecta deposit	Uplifted rim, dune facies, secondary craters	No raised rim, smooth ejecta blanket
Depth-diameter ratio	About 1:5 for craters less than 10 km diameter	Usually much less than for fresh impact craters

It should be noted that young craters in basaltic maria are bright, not dark. Craters of the same size adjacent to dark-halo craters frequently do not have dark haloes. There are some dark-halo craters on mare basalts. Accordingly, the conventional explanation that the dark-halo craters in the highlands represent impacts through highland debris into underlying mare basalts may not always be correct. Are the dark-halo craters caused by differences in the projectile compositions? Are they due to the impact of carbonaceous chondrites [94]? Some evidence from spectral studies has confirmed that Copernicus H has indeed excavated mare material [95]. Similar results have been obtained in regions covered by Orientale ejecta (e.g., Crater Schickard). This would indicate that basaltic volcanism was occurring before the Orientale collision, and must revise upwards our estimates of basalt volumes. Nevertheless, special caution is needed in the interpretation of dark-halo craters. Shorty crater at the Apollo 17 site, long expected to be volcanic in origin, was revealed by ground exploration to be of impact origin.

3.12 Smooth-Rimmed Craters and Floor-Fractured Craters

Several craters such as Ritter, Sabine, Lassell, Herodotus and Kopff (the latter in Mare Orientale), 20–45 km in diameter, are comparatively shallow and do not have conspicuous ejecta blankets or radial ridges. The curious case of Kopff is an often-cited example. It has been suggested that these are calderas, but they do not differ in shape from main-sequence craters of impact origin: "They simply do not have the correct shape to be calderas—and exhibit an impact crater morphology" [34]. Modification of the craters by post-impact processes seems a most promising explanation, and flooding by mare basalts seems likely in several cases [96]. This inundation produces the shallow, flat floors and buries the ejecta blankets. Janssen (23 km diameter) is particularly unusual; its floor is 260 m above the surrounding terrain. Wargentin is another example of this rare type of crater. The most likely explanation [34] is that local hydrostatic conditions permitted flooding of the interior of the craters.

Humboldt (Fig. 3.28) and Gassendi (Fig. 3.29) are good examples of craters with bulged and cracked floors. Many other examples are known. A total of 206 such craters with pronounced floor rilles have been listed [96]. In those examples with central peaks, the peak height is usually similar to that in unmodified craters of similar diameter but the crater rim–central peak elevation difference is smaller. This suggests that relative floor uplift has occurred. Since not all craters show this effect, it cannot be due in general to isostatic uplift. Such craters are usually associated with maria (Gassendi is a prime example) so that viscous relaxation of the topography may be a viable

3.28   The floor-fractured crater Humboldt, 207 km in diameter (NASA AS 15-2513).

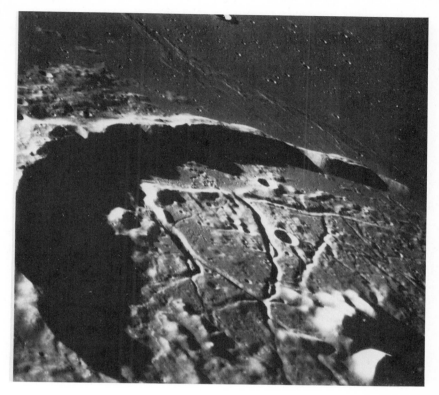

3.29   The floor-fractured crater Gassendi (110 km in diameter) (AS 16-120-19295).

mechanism in some cases [97]. The preferred explanation is that mare flooding and intrusion of sills has caused the uplift [84, 96], but the presence of similar craters on Mars and icy satellites raises further questions.

3.13   Isostatic Compensation of Craters

Isostasy probably does not affect lunar craters less than 30 km diameter [34, p. 53]. Imbrian-age craters up to about 200 km diameter (e.g., Humboldt) have zero Bouger anomalies indicating no uplift of the lunar mantle [98] and the lack of importance of large-scale isostatic processes. For fresh lunar craters, the negative Bouger anomalies (indicating mass deficiencies) can be attributed to a lens of brecciated material with a depth equivalent to about one-third the rim diameter [99]. As noted in Section 3.12, over 200 "floor-fractured" craters have been identified in which post-impact volcanic modification has taken place resulting in apparent rise and uplift of the crater floors [84, 96, 97]. The question of isostasy and multi-ring basins, and the mascon problem, is addressed in Section 7.4.2.

3.14   Linné Crater and Transient Phenomena

Linné was initially recognized as a crater in 1823, but in 1866, Julius Schmidt reported that it had vanished and was replaced by a bright spot surrounding a 500-m-diameter crater. Various other changes were reported and Linné became the center of a controversy. The Apollo 15 mission photographed Linné, which turned out to be a very fresh normal impact crater 2450 m in diameter in western Mare Serenitatis (Fig. 3.4).

The Linné crater controversy is well described by Pike [34, p. 22]: "The detailed geometry and morphology of Linné constitute as compelling an argument for primary impact genesis as occurs anywhere on the Moon. Evidently, optical effects were responsible for the enigmatic changes in Linné, an apt testament to the perils of visual lunar observation near the resolution limit of earth-based telescopes."

3.15   Swirls

Enigmatic light-colored surface markings, named "swirls," have been observed on the far-side lunar highlands [100, 101], on the near side at Reiner Gamma [102, 103] and on Mercury [102]. They appear to have no topographic relief and cross the landscape independent of elevation. On the Moon, they are well developed in three areas: the Reiner Gamma region (5° S, 60° W), near Mare Ingenii (35° S, 180° E) and near Mare Marginis (15° N, 90° E). The origin of these swirls has been extensively debated (e.g., [102]), but they appear to be associated with impact phenomena. Preferred explanations relate them either to the antipodal effects of major basin collisions [100], or to cometary impact effects [102]. The correlation of strong surface magnetization with the Reiner Gamma feature [103] (see Section 7.8.4) lends special interest to the origin of these features.

3.16   Meteorite Flux

The latest production rate for terrestrial Phanerozoic craters with diameters greater than 20 km is about $0.36 \times 10^{-14} km^2/yr.$ [104], in general agreement with previous estimates. Hartmann gave a similar estimate in 1965 [105].

This means that an impact crater with a diameter greater than 20 km will form on the Earth an average of every 30 million years.

Unless it is protected by sedimentary cover, a 20-km-diameter impact structure on the Earth has a lifetime against erosion on stable shield areas of 600 million years. A 10-km-diameter structure will survive about 300 million years, although local conditions may lead to large variations. A total of 13 terrestrial craters have associated meteoritic debris, while a further 78 structures are *probably* of impact origin, as revealed by shock metamorphic effects in the target rocks (Appendix V, Fig. 3.30). The list of these structures is increasing rapidly, and 50 additional structures are possible candidates, for which definitive evidence is currently lacking. Geochemical evidence for high levels of elements such as iridium or nickel in melt-rocks is a useful and definitive index of meteoritic impact [106, 107].

The terrestrial rate may be compared to the lunar cratering rate. The comparison must include the differences in gravitational cross-section, in impact velocities, and in surface gravity. Thus, the Earth will collect 1.24 times as many Apollo objects of the same mass per unit area as the Moon.

The estimated crater production rate for the Moon, using mare craters (with an average age of 3.4×10^9 aeons) is equivalent to a terrestrial rate of $0.37 \times 10^{-14} km^2/yr$. The two estimates are essentially in agreement. However, the mare surfaces themselves indicate that the cratering rate was still decaying at 3.4 aeons and did not become relatively constant until the period between 3.0 and 2.5 aeons. This would mean that the lunar rate between 3.0 and 2.5 aeons was lower than the observed terrestrial rate during the Phanerozoic by perhaps a factor of two. The inference is that the crater production rate has not been constant in the Earth-Moon system during the past 3 billion years, and that the Phanerozoic may have been a time of elevated crater production. This problem may be resolved by better statistics, but may indicate the true state of affairs. It would not seem unreasonable that wide fluctuations occur, and that their frequency would increase with time as more and more objects are swept up by the planets, thereby increasing the randomness of the individual events [108].

The question of whether the craters on the lunar highland surface can be interpreted as representing a saturated surface or a continuing production rate remains unsolved [66]. If the craters represent a production population, then they can be used to establish the relative ages of various surfaces in the highlands (Table 3.3), as well as the average depth of sampling, number of impacts per sample, and the volume of impact melt produced. If the cratering record represents saturation, then it provides only a minimum estimate for ages, and the integrated thermal effects will be severe. If the observed cratering record is a true account of the bombardment history back to 4.5 aeons, then there are severe discrepancies between the amount of impact melt produced and the radiometric age record [109].

Table 3.3 Estimates of ages of geologic provinces on the Moon.[†]

Geologic Province	Crater Density Relative to Average Lunar Mare	Estimated Crater Retention Age (aeons)			Regolith Thickness (m)
		Minimum	Best Estimate	Maximum	
Tycho	0.1	0.09	0.3	0.6	—
Aristarchus	0.2	0.5	0.9	1.3	—
Copernicus	0.3	0.8	1.5	2.2	—
Mare Crisium	0.5	2.1	2.6	3.5 ?	—
Mare Imbrium	0.50	2.2	2.6	3.5	—
Whole Mare Serenitatis	0.65	2.8	3.1	3.7	—
S. Oceanus Procellarum	0.75	3.0	3.3	3.7	3
Mare Fecunditatis	0.93	3.2	3.4	3.8	—
Average of front-side maria	1.00	3.3	3.5	3.8	5
Mare Orientale	1.10	3.3	3.5	3.8	—
Mare Tranquillitatis	1.30	3.5	3.6	3.9	5
Mare Humorum	1.50	3.6	3.7	3.9	—
Mare Nubium	2.5	3.8	3.9	4.0	—
Orientale basin and ejecta	2.5	3.8	3.9	4.0	—
Fra Mauro	3.0	3.9	4.0	4.1	8
Fill in Schrödinger basin	5.1	4.0	4.1	4.2	—
Fill in Mendeleev crater	9	4.1	4.2	4.2	—
Front-side highlands	10	4.1	4.2	4.2	10–15
Heavily cratered highlands and basins	32	4.3	4.4	4.5	} megaregolith
"Pure" highlands	36	4.3	4.4	4.5	

[†] Adapted from *Basaltic Volcanism* (1981) Table 8.6.1.

IMPACT CRATERS OF THE EARTH

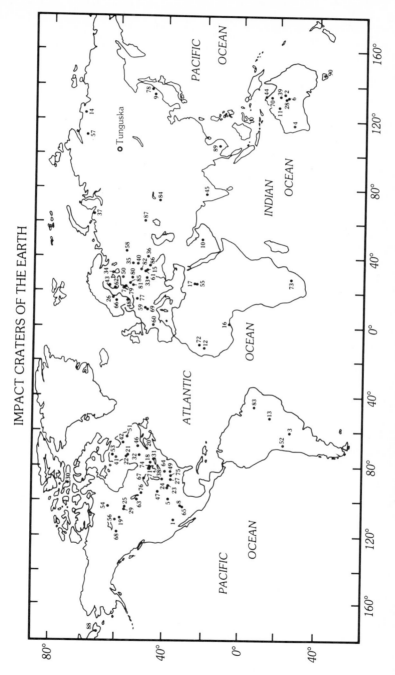

3.30 Distribution of presently known terrestrial impact structures.

There are 84 lunar craters with diameters greater than 161 km [110]. Ten classes may be distinguished, the oldest extending back to the time when saturation bombardment ceased. This date is given by Baldwin as 4.3 aeons (for an age of the Moon of 4.65 aeons!): 4.2 aeons seems to be a better estimate. This division into 10 classes may represent rather fine-tuning, and recognition of five classes may be more realistic [111].

The impact flux between 4.5 and 4.0 aeons is estimated to be 10^3–10^7 times the present flux rate [109]. The effect of this bombardment is to cause enormous rates of megaregolith production [112]. (The megaregolith in the highlands is estimated to be 2.5 km thick [113].) This cratering rate will cause very large amounts of mixing and overturning of crustal surfaces as well as destruction of thin crusts forming over magma oceans [112], and it is possible that we have generally underestimated these effects. "Pristine" lunar highland samples may be in fact very difficult to find. There does not seem to be any evidence in the cratering flux estimates that demands a spike at 3.8 aeons (Fig. 3.29), although this has frequently been suggested from the age and isotopic data (see next section). A preferred explanation is that ages are reset by the cratering events, so that the youngest events dominate. The similarity in surfaces exposed on the various planets and satellites lends credence to this view.

Estimates of the crater density versus age for the Moon, Earth, Mercury, Mars and Venus [108] are given in Fig. 3.31. Estimates of the relative ages of various geological provinces on Mercury and Mars from crater counting statistics are given in Tables 3.4 and 3.5.

3.17 The Cratering Record and the Lunar Cataclysm

Early intense cratering of the Moon was recognized long before the Apollo missions. The absence of landscapes on the Earth which resemble the

Table 3.4 Estimates of ages of geologic provinces on Mercury.[†]

Geologic Province	Crater Density Relative to Average Lunar Mare	Estimated Crater Retention Age (aeons)		
		Mimimum	Best Estimate	Maximum
Caloris Mare	3.7	2.4	3.8	4.0
Caloris whole basin	3.7	2.4	3.8	4.0
S. Polar uplands, intermediate diameter craters	5	3.0	3.9	4.0
S. Polar uplands, large diameter craters	20	4.1	4.4	4.5

[†]Adapted from *Basaltic Volcanism* (1981) Table 8.6.1.

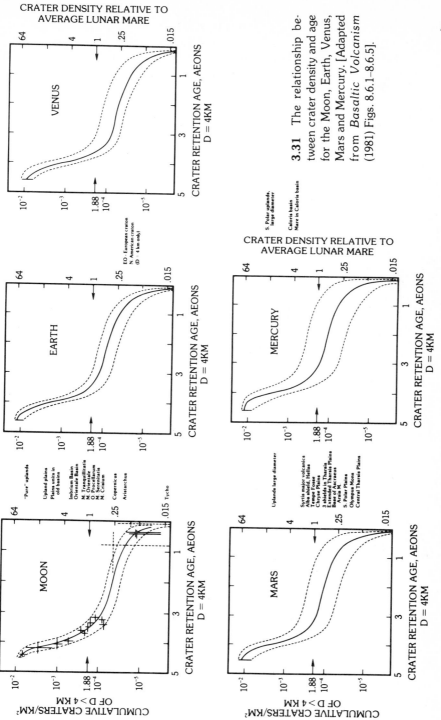

3.31 The relationship between crater density and age for the Moon, Earth, Venus, Mars and Mercury. [Adapted from *Basaltic Volcanism* (1981) Figs. 8.6.1-8.6.5].

Table 3.5 Estimates of ages of geologic provinces on Mars.[†]

Geologic Province	Crater Density Relative to Average Lunar Mare	Estimated Crater Retention Age (aeons)			Map Symbol (Fig. 2.21)
		Minimum	Best Estimate	Maximum	
Central Tharsis volcanic plains	0.1	0.1	0.2	0.4	v
Olympus Mons volcano	0.15	0.1	0.3	0.7	v
Desert at bases of Ascraeus, Arsia Mons volcano	0.35	0.3	0.8	1.7	v
Extended Tharsis volcanic plains	0.49	0.6	1.2	2.7	v
Three shield volcanoes	0.63	0.8	1.7	3.2	v
Elysium volcanics	0.68	0.9	1.9	3.3	pv
Solis Planum volcanic plains	0.90	1.2	2.5	3.5	pm
Heavily cratered plains, small diameter craters	1.4	2.1	3.1	3.7	p
Chryse Planitia volcanic plain	1.4	2.1	3.1	3.7	p
Tyrrhena Patera volcanoes	1.4	2.1	3.1	3.7	pc
Heavily cratered plains, intermediate diameter craters	1.4	2.1	3.1	3.7	pc
Lunae Planum and Tempe Fossae cratered plains	1.5	2.4	3.3	3.8	pc
Syrtis Major Planitia volcanic plains	2.0	2.8	3.6	3.9	pc
Heavily cratered plains, large diameter craters	15	4.0	4.1	4.3	cu

[†]Adapted from *Basaltic Volcanism* (1981) Table 8.6.1.

lunar highlands indicated the likelihood that the intense bombardment preceded the exposed geological record, and may have been responsible for the destruction of a primitive crust. Estimates based on reasonable estimates of the cratering flux indicated that the mare surfaces were old [114]. Fielder [115] assumed a constant cratering flux throughout geological time, which gives ages for the maria of less than 700 million years. Hartmann [116] concluded that the early intense cratering rate was about 200 times that of the post-mare average rate and suggested an age of 3.5 billion years for the age of the mare flooding, essentially the correct age. He noted that "In a field so broad and intrinsically complex as planetary evolution, the application of Occam's razor to selected data is not necessarily the most direct way to correct conclusions." [116, p. 407].

A period of early intense cratering was immediately apparent from the Apollo 11 Preliminary Examination Team age data [117], which provided ages of 3–4 aeons for Mare Tranquillitatis. Since the relative stratigraphic sequence was already established, it was clear that the age of the heavily cratered highlands was about 4 aeons or older.

The relative age sequence of basins, as established by crater-counting and geological criteria, is vitally important for our understanding of the later stages of the great bombardment. Three versions are given in Table 3.1. Imbrium is widely recognized as the second youngest basin, with only Orientale being younger. Mare basalt ages of 3.8 aeons provide a younger limit, for their surfaces bear no trace of Imbrium or Orientale ejecta. The conventional date for Imbrium is 3.82 aeons (Section 5.2) taking the Fra Mauro ages at their face value. Orientale will be only slightly younger. The next question to be addressed is the relative ages of the Crisium, Serenitatis and Nectaris basins. Crisium is generally placed as the next oldest basin after Imbrium and probable secondaries from Crisium have been observed on Nectaris ejecta [90]. Serenitatis is probably dated at 3.86 aeons (Section 5.10). Earlier studies placed Serenitatis among the oldest of the near-side basins (Table 3.1). This apparent old age is interpreted as a result of heavy cratering of Serenitatis ejecta by Imbrium secondaries [90]. Accordingly, Serenitatis becomes one of the youngest of the lunar basins (Table 3.1), and it is no longer necessary to fit in several large basins, including Nectaris between Imbrium and Serenitatis. A currently unresolved question is whether Nectaris ejecta were sampled at the Apollo 16 site, the Descartes Formation being a possible candidate. The total number of Imbrian age basins is thus three, while Nectarian age basins number twelve and Pre-Nectarian basins number twenty-nine.

The cratering evidence does not suggest a spike, but rather a steady flux perhaps declining in the late stages of large objects striking the Moon (only three such craters occur post-Orientale), from Orientale back to the old nearly ruined basin of Al Khwarizmi (Fig. 3.32). The interpretation preferred here is that this flux increases backwards in time and that the earlier basins and

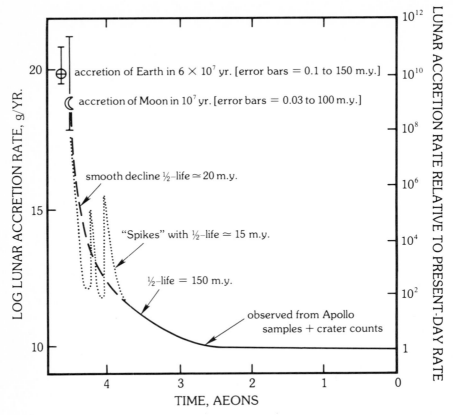

3.32 Schematic reconstruction of accretionary cratering without invoking a "spike" of catastrophic cratering 4 aeons ago. Present rate is observed (lower right) and projected back to 4 aeons ago using known impact rates determined from lunar dating. Additional datum is impact rate required to form Earth and Moon during their estimated formation intervals. Connection in "missing interval" from 4.5 to 4.0 aeons gives plausible planetesimal sweep-up half-lives of the order 20 to 45 m.y. [112]. (Courtesy W. K. Hartmann.)

craters have been obliterated. One predictable effect of a cataclysm should be the presence of many craters in the interval between the formation of Mare Imbrium and Mare Orientale. Baldwin [118] records only four 150-km-diameter craters. To these we may add the smaller Archimedes and Plato. This evidence does not suggest an intense spike in the flux rate, but is consistent with a uniform or declining flux of large impactors [119].

The relative age sequences of the old basins are thus critical for evaluation of the cataclysm concept. This has led to much polarization of opinion, with one author labelling the cataclysm as a "misconception" or "myth" [120]. This topic is addressed again in Section 5.10, where the geochronological evidence from the lunar highland samples is assessed.

3.18   Secondary Cratering and the Cayley Problem

The problem of the relative volumes of primary and secondary ejecta produced by the multi-ring basin collisions is of primary importance for understanding much of lunar geology and geochemistry. This question is another facet of the Cayley problem and has been much debated, particularly in the context of the Apollo 16 samples. The basic question turns on whether the Cayley plains deposits are primary Imbrium ejecta or are reworked from local material by Imbrium secondaries. Relevant evidence comes from terrestrial studies.

At the Ries crater, the ejecta blanket out at two crater radii is 78% locally derived and has 22% primary crater ejecta (Fig. 3.24). The Ries example is particularly instructive, since the ejecta blanket crosses a geological boundary south of the crater, consisting of rock types not exposed at the crater site. Thus, the identification of local versus primary components in the ejecta blanket is simplified [60]. However, the problem is not so simply resolved on the Moon. In the absence of an atmosphere, most transport of material is by ballistic trajectory. In this context, there appears to be little evidence for turbulent flow in the well-exposed Hevelius Formation (Fig. 2.14). On the Moon, ejecta will travel further than on the Earth and with greater velocity. Hence, the ejected blocks possess considerable energy and will produce both secondary craters and extensive secondary ejecta.

The well-known landslide at the Apollo 17 site was probably triggered by Tycho ejecta impacting the top of the South Massif. It is suggested that this process is responsible for much of the valley filling in the highlands. That is, small impacts have transported highland materials into the depressions [121]. Thus, the similarity in chemistry between the surface samples on the Cayley Plains and the adjacent hilly Descartes Formation at Apolio 16 is explained. Certainly there is little difference in chemistry, age, or petrology.

Volumetric calculations tend to favor the secondary cratering hypothesis. Probably not enough ejecta can be derived from the basins to account for the widespread Cayley Formation. However, secondary craters excavate many times their own volume. About twice the original crater mass is deposited by the secondary craters out beyond the thick ejecta blankets [121].

This general explanation tying in the smooth plains to the great ringed basins solves some of the stratigraphic problems. Many lithologically similar units will be of different ages (Fig. 3.23), and much overlapping will occur. Major modifications will occur during the formation of the Orientale basin.

Some of the large post-mare craters may also contribute, as, for example, at the Apollo 17 landslide where the old highland material now overlies the younger mare lavas.

An observation in favor of the local derivation of much of the highland plains-forming units is that the geological mapping shows little relation to the orbital chemistry data [122]. Thus, there is no correlation between either the Al/Si values (as measured by the orbital XRF experiment) or with the orbital gamma-ray thorium values and the mapped distribution of Cayley Formation. For example, in areas showing high gamma-ray values, the amount of Cayley plains-type material may range from 0 to 85% [80]. This lack of correlation between chemistry and geology indicates that the plains-forming units do not originate as one uniform sheet of ejecta, but instead predominantly reflect the local geology, an argument for their local derivation.

This debate is likely to continue. Multi-ring basins are huge. If one formed in the present highland crust, it would excavate and distribute heterogeneous material. The obsession with the Cayley problem has overshadowed other aspects. Much primary basin ejecta also exists and perhaps dominates at the Fra Mauro site (Apollo 14), so providing a date for the Imbrium event. Even the Descartes Formation may be primary Imbrium ejecta piled up against the Kant Plateau, so negating our hopes of sampling primary Nectaris ejecta at the Apollo 16 site. Although the temperatures may be low and shock effects small in much of the ejecta, pods of melt and highly shocked material are probably present. These can provide for the thermal annealing in breccias, age resetting and the magnetic evidence of temperatures in excess of the Curie temperature of iron (750° C), at least locally.

References and Notes

1. Gilvarry, J. J. (1961) *Nature*. 190: 1048; Salisbury, J. W., and Ronca, L. B. (1966) *Nature*. 210: 669; Harrison, E. R. (1960) *Nature*. 188: 1064; Frey, H. (1977) *Icarus*. 32: 235; Green, D. H. (1972) *EPSL*. 15: 263.
2. Dana, T. D. (1846) *AJS*. 2: 235.
3. Spurr, J. E. (1944–49) *Geology Applied to Selenology*, Science Press, Lancaster, Pa.
4. Green, J. (1965) *Ann. N.Y. Acad. Sci*. 123: 385.
5. Fielder, G. (1967) *Lunar Geology*, Dufour, p. 162.
6. Moore, P., and Cattermole, P. J. (1967) *The Craters of the Moon*, Norton, N.Y., p. 10.
7. Gilbert, G. K. (1893) *Bull. Phil. Soc. Wash*. 12: 241.
8. Baldwin, R. B. (1949) *The Face of the Moon*, Univ. Chicago Press; (1963) *The Measure of the Moon*, Univ. Chicago Press.
9. Urey, H. C. (1952) *The Planets*, Yale Univ. Press; (1962) *Phys. Astron. Moon* (ed., A. Kopal), Academic Press N.Y., p. 484.
10. See review by Shoemaker, E. M. (1962) in *Phys. Astron. Moon* (ed., A. Kopal), Academic Press N.Y., p. 283.
11. Whitaker, E. A. (1974) NASA SP 330.

12. El-Baz, F., and Worden, A. M. (1972) NASA SP 289, 25-1.
13. Gault, D. E. (1966) in *Nature of the Lunar Surface*, John Hopkins Press, Baltimore, p. 135; (1964) in *The Lunar Surface Layer*, Academic Press, N.Y., p. 151.
14. See papers in *Shock Metamorphism of Natural Materials* (1968) (eds., French, B. M., and Short, M. N.), Mono Book Corp., Baltimore.
15. Baldwin, R. B. (1965) *A Fundamental Survey of the Moon*, McGraw Hill, N.Y., p. 137.
16. Von Bieberstein, M. (1802) *Untersuchungen über den Ursprung und die Ausbildung der gegenwartigen Anordnung des Weltgebaeudes*, Darmstadt; Gruithuisen, F. von P. (1829) *Analekten-Erd und Himmels-Kunde*, Munchen; see also [4].
17. Muehlberger, W. R., et al. (1980) *Lunar Highlands Crust*, p. 1.
18. Examples include the following works: Green, J. (1971) Copernicus as a lunar caldera, *JGR.* 76: 5719, Erlich, E. N., et al. (1974) General peculiarities of lunar volcanism, *Mod. Geol.* 5: 31; Leonardi, P. (1976) *Volcanoes and Impact Craters on the Moon and Mars*, Elsevier, 432 pp.; Miranova, M. N. (1970) NASA TT 566.
19. Pike, R. J. (1980) USGS Prof. Paper 1046-C.
20. Basic references are: *Impact and Explosion Cratering* (1977) (eds., Roddy, D. J., et al.), Pergamon Press, N.Y.; *Shock Metamorphism in Natural Materials* (1968) (eds., French, B. M., and Short, M. N.), Mono Book Corp., Baltimore.
21. Gault, D. E., et al. (1968) in *Shock Metamorphism of Natural Materials*, p. 87.
22. Dence, M. R., et al. (1977) in *Impact Cratering*, p. 247.
23. Grieve, R. A. F. (1977) in *Impact Cratering*, p. 791; (1980) *Lunar Highlands Crust*, p. 173.
24. Kieffer, S. W. (1977) in *Impact Cratering*, p. 751.
25. Kieffer, S. W., and Simonds, C. H. (1980) *Rev. Geophys. Space Phys.* 18: 143.
26. Oberbeck, V. R. (1977) in *Impact Cratering*, p. 45.
27. O'Keefe, J. D., and Ahrens, T. J. (1978) *PEPI.* 16: 341; (1981) *Rev. Geophys. Space Phys.* 19: 1.
28. Roddy, D. J. (1977) in *Impact Cratering*, p. 185.
29. Stöffler, D., et al. (1980) Multi-ring Basins Abstracts, 89.
30. Melosh, J. H. (1980) *Ann. Rev. Earth Planet Sci.* 8: 65.
31. Croft, S. K. (1980) *PLC 11*: 2347; (1981) *Multi-ring Basins*, p. 207, 227.
32. Fodor, R. V., and Taylor, G. J. (1979) *Impact*, Leisure Books, N.Y.
33. Roddy, D. J. (1978) *PLC 9*: 3891.
34. Pike, R. J. (1980) USGS Prof. Paper 1046-C. This comprehensive work should be consulted for a full and detailed discussion of lunar crater morphology.
35. Cintala, M. J., et al. (1977) *PLC 8*: 3409.
36. Dence, M. R., and Grieve, R. A. F. (1979) LPS X: 292; Dence, M. R., et al. (1977) *Impact Cratering*, p. 2470.
37. Pike, R. J. (1980) *Icarus.* 43:1; (1980) *PLC 11*: 2159.
38. Pohn, H. A., and Offield, T. W. (1970) USGS Prof. Paper 700-C, 153.
39. Settle, M., and Head, J. W. (1979) *JGR.* 84: 3081.
40. Settle, M. (1980) *Icarus.* 42: 1.
41. Gault, D. E., et al. (1975) *JGR.* 80: 2444.
42. Hale, W. S., and Head, J. W. (1979) *PLC 10*: 2623.
43. Milton, D. J., et al. (1972) *Science.* 175: 1199.
44. Wilshire, H. G., et al. (1973) USGS Prof. Paper 599-H, 42 pp.
45. Milton, D. J., and Roddy, D. J. (1972) IGC 24 Sec. 15, 119, Montreal.
46. Hale, W. S. (1980) *Lunar Highlands Crust*, p. 197. The megaregolith, of course, consists principally of basin deposits, with lesser amounts of crater ejecta. Accordingly,

layering is extensively developed, as for example, at Silver Spur at the Apollo 15 site (see Section 5.1.1).

47. Murray, J. B. (1980) *Moon and Planets.* 22: 269.
48. Maxwell, T. A. (1980) Multi-ring Basins Abstracts, 53.
49. Pieters, C., et al. (1981) LPS XII: 833.
50. Moore, H. J., et al. (1974) *PLC 5*: 71; McCauley, J. F. (1977) *PEPI.* 15: 220.
51. Head, J. W. (1974) *Moon.* 11: 327.
52. Baldwin, R. B. (1972) *PEPI.* 6: 327.
53. Hartmann, W. K., and Kuiper, G. (1962) *Comm. Univ. Ariz. Lunar Planet. Lab.* 1: 51.
54. Gault, D. E. (1971) *Lunar Geol. Primer.* 4: 137.
55. Hartmann, W. K., and Wood, C. A. (1971) *Moon.* 3: 3.
56. McCauley, J. F. (1977) *PEPI.* 15: 220.
57. Cadogan, P. H. (1974) *Nature.* 250: 315; Whitaker, E. A. (1981) *Multi-ring Basins,* p. 105.
58. Wilhelms, D. E., et al. (1977) *Impact Cratering,* p. 539.
59. Dence, M. R. (1976) LS V: 165.
60. Hörz, F., and Banholzer, G. S. (1980) *Lunar Highlands Crust,* p. 211.
61. Whitford-Stark, J. L. (1981) *Multi-ring Basins,* p. 113.
62. Hodges, C. A., and Wilhelms, D. E. (1978) *Icarus.* 34: 294.
63. Hartmann, W. K., and Wood, C. A. (1977) *Moon.* 3: 3.
64. Roddy, D. J. (1976) *PLC 7*: 3027.
65. Roddy, D. J. (1977) *Impact Cratering,* p. 185.
66. Grieve, R. A. F. (1980) *Lunar Highlands Crust,* p. 173; Dence, M. R., and Grieve, R. A. F. (1979) LPS X: 292.
67. Wood, C. A., and Gifford, A. W. (1980) Multi-ring Basins Abstracts, 121.
68. Hartmann, W. K., and Kuiper, G. P. (1962) *Comm. Univ. Ariz. Lunar Planet. Lab.* 1: 51.
69. Wollenhaupt, W. R., et al. (1973) NASA SP 330, p. 33.
70. Cadogan, P. H. (1976) *Nature.* 250: 315. See also Hartmann, W. K., and Wood, C. A. (1971) *Moon.* 3: 4; Hodges, C. A., and Wilhelms, D. E. (1978) *Icarus.* 34: 294.
71. Hale, W. S., et al. (1980) Multi-ring Basins Abstracts, 30.
72. Trask, N. J., and Guest, J. E. (1975) *JGR.* 80: 2461.
73. Head, J. W., et al. (1975) *PLC 6*: 2805.
74. Herzberg, C. T., and Baker, M. B. (1980) *Lunar Highlands Crust,* p. 113.
75. Hawke, B. R., and Head, J. W. (1978) LPS IX: 477.
76. Wood, C. A., and Head, J. W. (1976) *PLC 7*: 3629.
77. Wood, C. A. (1980) *PLC 11*: 2221.
78. Grieve, R. A. F., and Robertson, P. B. (1979) *Icarus.* 38: 212.
79. Wood, C. A., et al. (1978) *PLC 9*: 3691.
80. Cintala, M. J., et al. (1977) *PLC 8*: 3409.
81. Hawke, B. R., and Head, J. W. (1977) *Impact Cratering,* p. 815.
82. Mendell, W. W. (1976) *PLC 7*: 2705.
83. Oberbeck, V. R., and Morrison, R. H. (1973) *PLC 4*: 107.
84. Schultz, P. A. (1976) *Moon Morphology,* Univ. Texas Press, Austin.
85. Schultz, P. H., and Singer, J. (1980) *PLC 11*: 2243.
86. Greeley, R., et al. (1977) *JGR.* 82: 8011.
87. Batson, R. M., et al. (1979) *Atlas of Mars,* NASA SP-48, 146 pp.
88. Greeley, R., et al. (1980) *PLC 11*: 2075.
89. Gault, D. E., et al. (1975) *JGR.* 80: 2444.
90. Wilhelms, D. E. (1976) *PLC 7*: 2883.

91. Escape velocities for Mimas are 0.16 km/sec, for Dione, 0.50 km/sec and for Rhea, 0.65 km/sec.
92. Shoemaker, E. M., and Wolfe, R. F. (1981) LPS XIIA: 1.
93. Head, J. W., and Wilson, L. (1979) *PLC 10*: 2861.
94. Although not strictly analogous, it is of interest to recall the Revelstoke event, where black CCl powder was recovered from a frozen lake surface [Folinsbee, R. E., et al. (1967) *GCA*. 31: 1625].
95. Hawke, B. R., and Bell, J. F. (1981) LPS XII: 412.
96. Schultz, P. H. (1976) *Moon*. 15: 241; Pike, R. J. (1971) *Icarus*. 15: 384.
97. Hall, J. L., et al. (1981) *JGR*. 86: 9537. This paper provides the most recent discussion on mechanisms for forming floor fractured craters.
98. Dvorak, J., and Phillips, R. J. (1978) *PLC 9*: 3651.
99. Dvorak, J., and Phillips, R. J. (1977) *GRL*. 4: 380.
100. El-Baz, F. (1972) NASA SP-315, p. 29–93.
101. Evans, R. E., and El-Baz, F. (1973) LS IV: 231.
102. Schultz, P., and Srnka, L. J. (1980) *Nature*. 284: 22.
103. Hood, L. L., et al. (1979) *Science*. 204: 53.
104. Grieve, R. A. F., and Dence, M. R. (1979) *Icarus*. 38: 230.
105. Hartmann, W. K. (1965) *Icarus*. 4: 157.
106. Palme, H., et al. (1978) *GCA*. 42: 313.
107. Grieve, R. A. F., and Robertson, P. B. (1979) *Icarus*. 38: 212.
108. *Basaltic Volcanism* (1981) Section 8.6.
109. Lange, M. A., and Ahrens, T. J. (1979) *PLC 10*: 2707.
110. Baldwin, R. B. (1981) *Multi-ring Basins*, p. 275.
111. Wilhelms, D. E., pers. comm., 1981.
112. Hartmann, W. K. (1980) *Lunar Highlands Crust*, p. 155.
113. Head, J. W. (1976) *PLC 7*: 2913.
114. Kuiper, G. P. (1954) *Proc. Nat. Acad. Sci.* 40: 1104; Urey, H. C. (1952) *The Planets*, Yale; (1960) *Astrophys. J*. 132: 502.
115. Fielder, G. (1963) *Nature*. 198: 1256.
116. Hartmann, W. K. (1965) *Icarus*. 4: 207; (1966) *Icarus*. 5: 406.
117. LSPET (1969) *Science*. 165: 1211.
118. Baldwin, R. B. (1974) *Icarus*. 23: 97.
119. Wetherill, G. W. (1981) LPS XII: 1176; (1981) *Multi-ring Basins*, p. 1.
120. Hartmann, W. K. (1975) *Icarus*. 24: 181.
121. Oberbeck, V. R., et al. (1973) NASA Tech. Memo. TMX 62302. 54 pp.
122. Hörz, F., et al. (1974) LS V: 357.

Chapter 4

PLANETARY SURFACES

4.1 The Absence of Bedrock

A striking and obvious observation is that at full Moon, the lunar surface is bright from limb to limb, with only limited darkening toward the edges. Since this effect is not consistent with the intensity of light reflected from a smooth sphere, pre-Apollo observers concluded that the upper surface was porous on a centimeter scale and had the properties of dust. The thickness of the dust layer was a critical question for landing on the surface. The general view was that a layer a few meters thick of rubble and dust from the meteorite bombardment covered the surface. Alternative views called for kilometer thicknesses of fine dust, filling the maria. The unmanned missions, notably Surveyor, resolved questions about the nature and bearing strength of the surface. However, a somewhat surprising feature of the lunar surface was the completeness of the mantle or blanket of debris. Bedrock exposures are extremely rare, the occurrence in the wall of Hadley Rille (Fig. 6.6) being the only one which was observed closely during the Apollo missions. Fragments of rock excavated during meteorite impact are, of course, common, and provided both samples and evidence of competent rock layers at shallow levels in the mare basins.

Freshly exposed surface material (e.g., bright rays from craters such as Tycho) darken with time due mainly to the production of glass during micro-meteorite impacts. Since some magnetic anomalies correlate with unusually bright regions, the solar wind bombardment (which is strongly deflected by the magnetic anomalies) may also be responsible for darkening the surface [1]. Infrared and radar mapping from the Earth revealed many "anomalies" or "hot spots" on the lunar surface [2]. The regions around young ray craters show high eclipse temperatures and scatter radar signals strongly. These properties correlate with, and are caused by, the block fields surrounding young craters. The anomalies fade into the average lunar surface values with increasing age. A corollary is that fracturing and destruction of fresh rock surfaces by impacts must be a common lunar process [2]. On Mars, the mantling effect of wind-deposited fine material likewise hampers direct observations of bedrock, except perhaps at high elevations. The summit calderas of Olympus Mons (Fig. 2.23b) may be an excellent sampling locality, possibly free of regolith or dust. The megaregolith is dealt with in Chapter 5 on planetary crusts (Section 5.1.2).

4.2   The Extreme Upper Surface

Observations are restricted mainly to the lunar surface. Measurement of conductivity indicates that the top layer is strongly insulating [3]. The exposure of lunar soils to humid terrestrial atmospheres causes large changes in the dielectric constant [4]. Curiously, the electrical properties of the surfaces of Mercury, Mars and Venus appear to be very similar to those of the Moon [3, 5]. This is readily understandable for Mercury which lacks an atmosphere and resembles the Moon in many respects. It also appears that water is not affecting the surface electrical properties of Mars, where the temperatures and pressure at the surface are usually below the triple point for water. Local exceptions may be found on Mars where the daytime temperature rises high enough for liquid water to exist on the surface, and local freeze-thaw conditions may exist. The presence of subsurface ice, permafrost and hydrated minerals may change conductivity at deeper levels [6]. On Venus, in contrast, the temperatures and pressures are far above the critical point for water [3] so that on all the inner solar system bodies (except the Earth), liquid water does not affect the electrical properties of the top surface layer.

The temperature on the lunar surface increases by about $47°$ K in the top 83 cm. The top surface (2–3 cm) is a loosely packed porous layer. Surface temperatures vary considerably. At the Apollo 17 site, the surface reaches a maximum of $384°$ K ($111°$ C) and cools to $102°$ K ($-171°$ C) at the end of the lunar night [2]. The near-surface temperature is $216°$ K ($-57°$ C). At the Apollo 15 site, these temperatures are about $10°$ K lower. The agreement with previous estimates based on terrestrial observations was very close [8, 9].

The temperature range at the Viking sites on Mars was between $150°$ K ($-123°$ C) and $240°$ K ($-33°$ C) [7]; Mercury temperatures range from $93°$ K ($-180°$ C) to $700°$ K ($430°$ C) and surface temperatures on Venus are about $720°$ K ($450°$ C).

The question of the lateral movement of the surface layer or of underlying layers has been extensively discussed (e.g., [10]). Although much material has been redistributed by impact processes (e.g., Figs. 2.1, 2.14), the underlying bedrock is the dominant influence on the composition of the regolith. The truth of this statement is apparent from the chemistry of the soils at the various sites, with minor exceptions [11]. The chemical data from the orbital experiments (Al/Si, Mg/Si, and the gamma-ray Th values) show breaks generally coincident with the mare-highland boundaries [12]. Thus, movement of a surficial layer occurs only on a local scale, and the chemistry and nature of the regolith is dominated by local bedrock components.

Evidence for local stirring of the very top dust layer has been deduced from the light scattering observed at the terminator by the Apollo 17 astronauts [13]. These observations are consistent with effects observed by Surveyor 6 and Lunokhod 2, and are perhaps due to dust levitation caused by temperature changes at the terminator [13, 14]. The effect does not produce any widespread migration of dust, which would blur the sharp mare-highland boundaries observed by the orbital chemical analyses [12]. Chemical and isotopic processes occurring in the upper layer of the regolith are discussed further throughout this chapter.

The importance of the very top layer in cosmic ray, solar flare, and track studies has led to the development of new sampling techniques. One such device consists of a free-floating cloth-covered plate designed to sample the top 100 micron layer. Another, a spring-loaded cloth-covered plate, sampled the top 0.5 mm [15]. Other samples in this category, from the Apollo 16 highland site at Descartes, include top and bottom samples from a boulder and soil samples from shadowed areas. Work on this material is still in progress.

Direct stereo photography with 80 micron resolution at the surface was carried out by the astronauts on the Apollo 11, 12 and 14 missions [16]. The astronauts' comments that the upper surface resembled a beach following a heavy rain shower were interpreted as the effect of micrometeorite bombardment.

4.3 The Regolith

The regolith is the continuous layer, usually several meters in thickness, which covers the entire lunar surface [17]. It is a debris blanket in every sense of the term, ranging from very fine dust to blocks several meters across (Fig. 4.1). Although the finer components (< 1 mm) are often referred to as soil, the distinction from terrestrial soils, formed by the complex interaction of the atmosphere and biosphere on rocks, is complete. The active processes on the Moon at present are cosmic, not planetary. Bombardment of the surface occurs at all scales from impacts which produce craters such as Tycho and Copernicus to erosion by particles producing sub-micron-sized pits, while cosmic ray and solar flare particles produce effects on an atomic scale. The surficial material is saturated with solar wind gases.

4.1   The regolith at Station 8, 3.4 km northeast of South Ray crater (Apollo 16 site). The boulder is the source of gabbroic anorthosite samples 68415 and 68416 (NASA AS-16-108 17697).

4.3.1 Thickness and Mechanical Properties

The thickness of the regolith has been established mainly by observations of those craters that are seen to excavate bedrock, and by direct observations along the edge of Hadley Rille [18]. Here, the base is seen to be irregular. The average thickness of the regolith on the maria is 4–5 m, while the highland regolith is about twice as thick (averaging about 10 m), due both to a longer exposure history and to a more intense meteorite flux.

The regolith, although quite variable locally in thickness, has very similar seismic characteristics at all sites (highlands and maria alike), typically with very low seismic velocities of about 100 m/sec [19] (Fig. 4.2). The value ranges over the different Apollo sites (Apollo 12–17) from 92–114 m/sec, indicating that the processes producing the regolith formed material of uni-

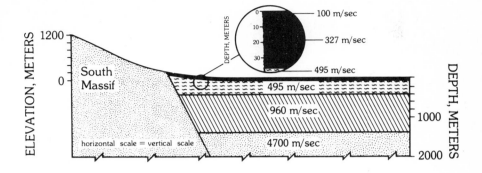

4.2 A cross-section of the regional structure at the Apollo 11 landing site in the Taurus-Littrow valley [19].

form seismic properties moon-wide from different mare and highland source materials. The seismic velocities increase rapidly below the regolith. At the Apollo 17 site the velocity reaches 4.7 km/sec. at a depth of 1.4 km. That is good evidence that the subsurface material is not dust but basaltic flows, fractured near the surface and underlying a few meters of regolith. This conclusion is strengthened by the orbital radar evidence for basin-wide sub-surface reflectors at depths of 0.9 and 1.6 km in Mare Serenitatis, and at 1.4 km in Mare Crisium [20].

There are many local variations in regolith thickness. Perhaps the most extreme example is in the Taurus-Littrow Valley where, near the landing site, thickness ranged from 6.2 to 36.9 m. It can be argued that such changes are due to the local concentration of medium-sized craters (such as Camelot, 700 m in diameter) producing uneven thicknesses due to local throwout, burying smaller craters.

Regolith thicknesses on the Cayley Plains at the Apollo 16 site vary between 6 and 10 m, but may be up to 20 m on the Descartes Mountains [21]. Regolith development on the rims of young craters is very limited. The thickness on the rim of North Ray crater (Apollo 16 site) is only a few centimeters. North Ray crater is 50 million years old [22] (Fig. 4.3).

The bulk density of the soils ranges from 0.9 to 1.1 g/cm^3 at the top surface, but increases with depth to values up to 1.9 g/cm^3. The density increases markedly in the upper 10–20 cm [23, 24]. The porosity of the upper surface is 45%, with higher values around crater rims, as judged from soil mechanics studies of the astronauts' footprints. Thus, the top surface is very loose, due to stirring by micrometeorites, but the lower depths, below about 20 cm, are strongly compacted, presumably by shaking during impacts. The compressibility of the soil is similar to that of ground-up terrestrial basalt.

4.3 North Ray crater, 900 m in diameter, viewed from Station 11 at the Apollo 16 Descartes site. The regolith on the rim of this 50-m.y.-old crater is only a few centimeters thick (NASA AS-16-116 18599).

All of these mechanical properties, vital for manned landings and astronaut surface activities, have been extensively studied [23, 24]. Notwithstanding the differences in composition, particle size and shape, and grain size distribution, from a soil mechanics or engineering point of view the lunar soil does not differ significantly in its behavior from what would be expected on Earth of granular, slightly cohesive terrestrial soil with the same particle size distribution and packing characteristics. The one major difference that might be singled out is the interparticle adhesion of lunar soil, which is demonstrated, for example, by the sharp vertical walls produced by the indentation of the footprints and by the difficulty of handling lunar powders in the laboratory.

4.4   View of the Martian surface from the Viking 1 lander in Chryse Planitia.

The similarity of lunar soil in mechanical properties to those of terrestrial examples is due to the fact that the mechanical properties depend on size, density, and shape of the particles rather than on their chemical composition. The consensus from the soil mechanics studies [24] and from the seismic data [19] is that the mechanical properties of the lunar soils are about the same at all sites. With the exception of two photographs of the Venusian surface, Mars is the only other body in the solar system for which we have some detailed knowledge of the nature and properties of its surface (Fig. 4.4). The clarity of the photographs enables unambiguous interpretations to be made. The surface layer appears to be firmer than that of the lunar regolith [7]. The fine-grained material is strongly adhesive, like the lunar fines. Bulk density is about 1.2 gm/cm^3. Small weakly cohesive clods are common and the soil contains many more large rocks than the lunar regolith. The presence of dunes indicates substantial wind transport and fine material will thus be selectively removed. Remote sampling of such material would not yield the fragments of rocks commonly found in the lunar soils.

4.3.2 Structure

The layered nature of the regolith has now been revealed by many core samples. A total of 26 core samples of varying depths up to 3 m were taken during the missions, along with many scoop samples of soil. The cores provided valuable data after some initial confusion due to the early sampling techniques, which did not return undisturbed cores. The deepest core (294.5 cm) was at Apollo 17. Core tube samples from Apollo 11, 12, and 14 missions produced disturbed samples, but on Apollo 15, 16, and 17 missions, use of thin-walled tubes resulted in almost undisturbed samples of lunar soil with 90–95% core recovery. The greatest insight came from the study of the Apollo 15 deep core, which clearly revealed that the regolith was not a homogeneous pile of rubble. Rather, it is a layered succession of ejecta blankets. An apparent paradox is that the regolith is both well mixed on a small scale and also displays a layered structure. For example, the Apollo 15 deep core tube, 242 cm long, contained 42 major textural units ranging from a few millimeters to 13 cm in thickness. There is no correlation between layers in adjacent core tubes and the individual layers are well mixed.

This paradox is resolved as follows. The regolith is continuously gardened by meteorites and micrometeorites. Each impact inverts the microstratigraphy and produces layers of ejecta, some new and some containing remnants of older layers. The new surface layers are stirred by micrometeorites.

A complex layered regolith is thus built up, but is in a continual state of flux. Particles now at the surface may be buried deeply by future impacts. In this manner, regolith is turned over like a heavily bombarded battlefield. The layering in a core tube has no wider stratigraphic significance. It is local and temporary. The result is that we have the well-stirred and homogeneous regolith, with one portion very like another, uniform moon-wide, as shown by the seismic data, yet layered locally.

These processes have been placed on a quantitative basis [25, 26]. The most important result is that the turnover rate of the regolith decreases rapidly with depth. "While it takes approximately 10^7 yr. to excavate the regolith at least once (with 99% probability) to a depth of 9 mm, it will take 10^9 yr. to excavate to a depth of 7 cm; simultaneously, however, the uppermost millimeters of the regolith has [sic] been turned over approximately 700 times in 10^7 yr. and many thousands of times in 10^9 yr." [26, p. 9].

The upper one millimeter of the regolith is the zone where most reworking and mixing by micrometeorites occurs on these models. There is some indication from measurements of cosmogenic radionuclides that these depth estimates may be too low. Thus, studies employing ^{53}Mn (half-life of 3.7 million years) on the Apollo core samples [27] indicate a mean depth of disturbance, over 10^7 yr. to be about 5–6 g/cm^2 (about 3 cm). The difference between these two estimates depends partly on the mass distribution of the impacting bodies, but also indicates the difficulties in modelling these processes.

The calculations with respect to vertical mixing within the regolith, discussed above, are much more readily treated than are the problems of local and exotic components in the regolith at a particular landing site. Lateral transport is not particularly efficient; otherwise, the regolith would be laterally homogeneous over broad areas. This lack of homogeneity is reflected in the fact that the admixture of mare and highland soils close to geological boundaries is surprisingly small. Mare-highland contacts appear relatively sharp on orbital geochemical data. At the Apollo 15 and 17 sites, steep chemical and petrological gradients in soil compositions occur over distances of a few kilometers. According to early calculations, about five percent of the regolith at a site may be derived from distances greater than 100 km, while 0.5 percent come from distances greater than 1000 km. Fifty percent comes from distances less than 3 km [28–30]. Thus, most of the regolith is of local origin. In accordance with the predictions, components of the Apollo 11 soil sample were derived from highland sources, leading to the identification of the highlands as "anorthositic" [31]. However, Hörz [26, p.11] has drawn attention to the apparent anomaly that most genuine mare soils (e.g., Apollo 11, 12 and Luna 16) contain 10–30% of non-mare material, and that this is constant over wide areas of mare soils, increasing only within 5 km of exposed highlands. A current problem is whether such material is derived from highland material at shallow depths beneath the mare basalts.

4.3.3 Rate of Accumulation

The accumulation rate for the past three aeons of the lunar regolith averages about 1.5 m per aeon [32], 1.5 mm per million years, or about 15 angstrom units per year corresponding to a layer of about 6 oxygen anions. However, such averages are misleading. The accumulation of the regolith occurs as a result of the addition of discrete layers, and is not continuous. The local accretion rate is extremely variable. In comparison with terrestrial erosion and deposition, lunar processes are slow indeed.

The accumulation rate, which is directly related to the meteoritic cratering flux (Section 3.16), steepens appreciably in the period between 3.5 and 4 aeons. The present regolith in the highlands dates from about 3.8 to 4.0 aeons, which marked the close of the intense bombardment, culminating in the impacts that formed the Imbrium and Orientale ringed basins. The accumulation rate for the period 3.5–4.0 aeons is about an order of magnitude greater than in younger epochs. A complicating factor is that the regolith does not grow at a constant rate, even if the meteorite flux is steady [25]. As the regolith increases in thickness, it buffers the bedrock against the smaller impacts. Larger impact events are thus required to generate fresh bedrock debris. With increasing regolith thickness, samples from deeper source areas will be added to the top surface [25].

4.3.4 "Ages" of the Soils

The exceedingly complex mixture represented by the soil contains components of varying ages. This was dramatically emphasized by the apparent paradox from initial investigations that the mare soils had model Rb-Sr ages around 4.6 aeons, although they were lying on and principally derived from

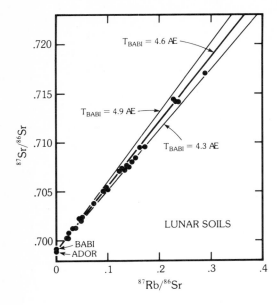

4.5a Rubidium-strontium evolution diagram for lunar soils. The data are without direct age significance since the soils are complex mixtures. The lines T BABI = 4.3 AE, 4.6 AE, and 4.9 AE show the growth of radiogenic strontium as a function of ^{87}Rb from an inital ratio of 0.69898. This is the Basaltic Achondrite Best Initial ratio or BABI. ADOR is the initial ratio for the meteorite Angros dos Reis and is one of the most primitive ratios known. The initial isotopic ratio $^{87}Sr/^{86}Sr$ for the Moon appears to be close to BABI. Note that using this ratio, the lunar soil data cluster about the 4.6 aeon line [33]. (Courtesy G. J. Wasserburg.)

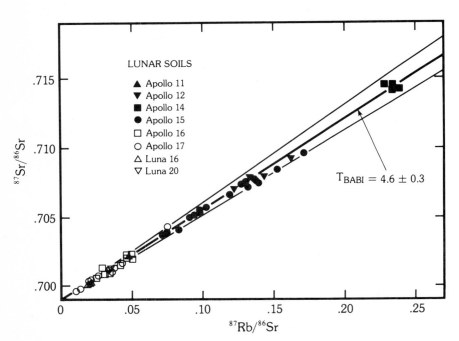

4.5b Rb-Sr evolution diagram for lunar soils. (Courtesy L. E. Nyquist.)

rocks whose crystallization ages, from mineral isochron studies and ^{40}Ar/^{39}Ar data, were 3.6–3.8 aeons [33–35]. This effect is shown in Figs. 4.5a and b where soil age data are shown on a conventional Rb-Sr diagram. There is a strong tendency for the soil data to scatter about a line indicating an age of 4.6 aeons, since the material had an initial ^{87}Sr/^{86}Sr ratio of 0.69898, equivalent to that observed in basaltic achondrites [36]. These model ages from the soils indicate that the principal fractionation of Rb from Sr occurred at about 4.6 aeons and that the amount of fractionation during partial melting to produce the mare basalts was smaller.

Experience from later missions showed that some soils had model ages greater than 4.6 aeons (Fig. 4.6a and b). There is so much evidence indicating formation of the Moon and the rest of the solar system at about 4.6 aeons ago

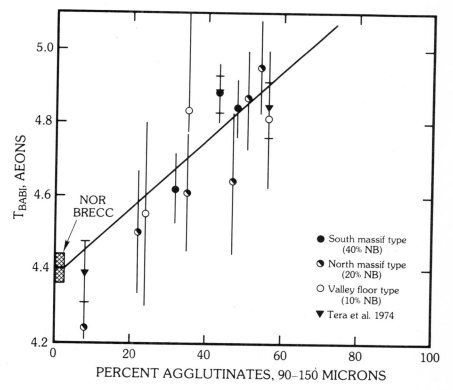

4.6a Model ages for soils as a function of agglutinate content of the soils. As the agglutinate (fused glassy particles) content increases, so does the apparent age of the soils. T BABI is the age (in aeons) based on an initial ^{87}Sr/^{86}Sr of 0.69898. NOR BRECC is noritic breccia. The amount of fused glassy particles in the soils is a direct index of the amount of micrometeorite bombardment and hence of soil maturity. (Courtesy L. E. Nyquist.)

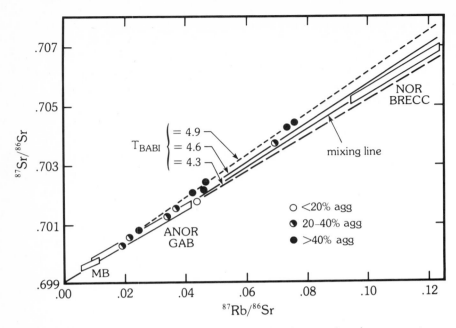

4.6b Rb-Sr evolution diagram for Apollo 17 soils, showing that the apparent age increases with increase in agglutinate (agg) content. MB = mare basalt. ANOR GAB = anorthositic gabbro. (Courtesy L. E. Nyquist.)

(Section 9.2), that apparent ages older than this value indicate that alternative explanations for the data are required. It should be emphasized that there are "no known rock types which can be added to the soils to produce high model ages greater than 4.6 aeons" [37]. This fact strongly argues against explanations that the exotic component is merely a finely divided KREEP component disseminated over the Moon.

Since the soils have arisen by meteorite impact processes repeated many times, it is perhaps not surprising that the delicate interrelationships between radiogenic ^{87}Rb and its daughter product ^{87}Sr might be disturbed occasionally. Thus, small losses of volatile Rb relative to Sr during melting and glass formation at elevated temperatures provide a way to produce old apparent ages [37] as is shown by the correlation between age and agglutinate formation.

These problems with age dating of the lunar soils provide a salutary lesson, not appreciated before the Apollo missions. This dearly bought experience enables us to obtain meaningful ages from clasts separated from soils as was amply demonstrated by the analysis of fragments from Luna 16, 20 and 24 samples. This lesson will be a valuable one in future sampling missions.

4.3.5 Soil Breccias and Shock Metamorphism of Lunar Soils

Soil breccias, consisting of weakly indurated mixtures of rock fragments (Fig. 4.7), glass and fine soil, achieved an undue early prominence by comprising about half of the initial Apollo 11 sample return. Experimental work has shown that single meteorite impacts into soil can produce these breccias [38]. Using a 20 mm powder gun to simulate shock pressures between 150 and 730 kbars, the lunar soil was compacted and partially remelted to form agglutinates and soil breccias. *In situ* melts of agglutinates and melts along grain edges occur at low pressures. The highest pressure shock melts are closest in composition to the bulk soil composition. Propagation of shock waves from a single impact in soils causes collapse and shear of grains, collapse of pore spaces, and compaction which indurates the soil at low pressures (150–180 kbars) with less than 5% melting. These resemble soil breccias. As the pressure increases, the amount of intergranular melt and shock melting increases. Above 650 kbars, 30–75% of the sample melts, producing a pumiceous glass. Thus, the formation of soil breccias does not require sintering in hot base surge deposits [39] or repetitive impacts [40].

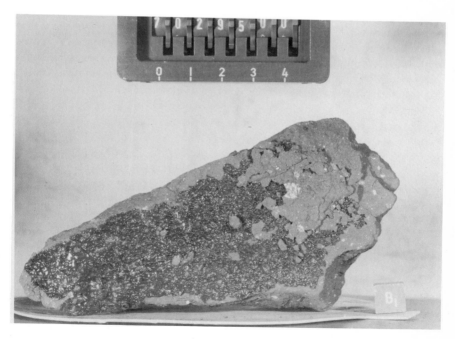

4.7 Soil breccia 70295 showing glass coating (NASA S-73-17192).

4.4   Glasses

Many varieties of glasses are found in the lunar soils. The existence of the glasses is mostly due to melting during meteorite impact, but some important exceptions occur. The emerald green glass from Apollo 15 provides us with a primitive basaltic composition. The orange glasses from Taurus-Littrow (Apollo 17) contribute to our knowledge of the dark mantle and of the mechanism of mare basalt eruption. These and other glasses of possible volcanic origin are discussed in Chapter 6 on basaltic volcanism (Section 6.2.3). The emphasis here is on the common and abundant impact-derived glasses.

Because the more spherical varieties of the glasses are abundant and generally homogeneous, they are suited to chemical analysis by electron microprobe techniques. This enables estimates to be made of the frequency of various chemical compositions, which may then be related to the parental rocks. This approach is important for the heavily cratered and brecciated highlands, where terrestrial petrographic experience and techniques have been less readily adapted to lunar studies than was the case with the mare basalts. However, in these studies, care must be taken to avoid bias and the identification of mixtures as primary rock types. There are many non-homogeneous glasses, while the agglutinates (Section 4.4.1) are particularly heterogeneous.

The relative abundance of impact-produced and volcanic glasses is difficult to determine. Even the absence of siderophile element meteoritic signatures does not guarantee a volcanic origin, on account of the small volume of the impacting body relative to the amount of glass produced.

4.4.1 Agglutinates

These are glass-bonded aggregates consisting of glassy, rock and mineral fragments (Figs. 4.8, 4.9). They form during micrometeorite impact into soil [41], and may be in part the remnants of small glassy craters produced during the impacts. Their abundance in a soil is an index of exposure to the micrometeorite bombardment and hence of soil maturity. Soils that have low agglutinate contents contain evidence of a shorter exposure to cosmic radiation than do soils with a high agglutinate content [42]. The formation of agglutinates offsets the grinding up of the soil particles by the micrometeorite bombardment. Eventually an equilibrium grain size in the soils of about 60 microns is reached by these two competing processes. The agglutinate content thus increases with the maturity of the soil.

In general, the average composition of the glassy portion of agglutinates is that of the soil in which they were formed, providing the soil is mature. Accordingly, such compositions may be used in complex regoliths to deter-

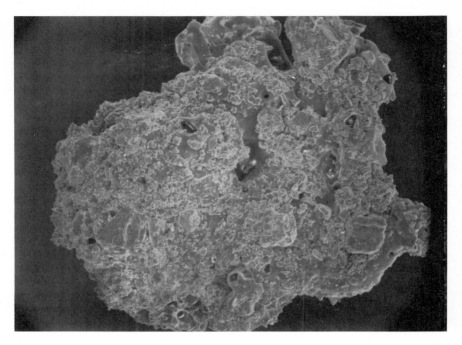

4.8 SEM photo of an Apollo 12 agglutinate, consisting of lithic, mineral and glassy debris bonded together by inhomogeneous glass. Width of photo is 1 mm. (NASA S.70 34983. Courtesy D. S. McKay.)

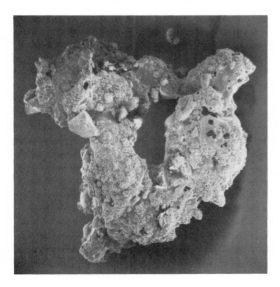

4.9 SEM photo of a 1-mm Apollo 12 agglutinate particle. (NASA S71-24575. Courtesy D. S. McKay.)

mine fossil soil compositions. For example, at the Apollo 17 site, it is possible to calculate the contribution to the regolith from the landslide at the South Massif. Pre-Camelot crater soils and pre-central cluster soils can also be identified by this approach [43]. These considerations place some restrictions on mixing calculations for soils. It is clear that such chemical calculations should always be accompanied by petrographic examination of the soils, particularly in areas such as the Apollo 17 regolith where agglutinates formed from mature soil compositions may be mixed with freshly excavated material [43].

For some time, it was thought that major elemental fractionations occurred during the formation of agglutinates [44]. Detailed analysis of agglutinate fractions separated by magnetic separation techniques showed increases in the concentration of the ferromagnesian elements (Fe, Mg, Ti) and decreases in the abundances of Na, Ca, Al and Eu, typical constituents of plagioclase. These effects were attributed to selective melting of pyroxene during agglutinate formation, relative to the more refractory plagioclase. However, like other impact-produced partial melting scenarios on the Moon, this effect is not real. The magnetic separation processes used to separate the agglutinate fractions were biased not only toward magnetic agglutinates, but also select magnetic non-agglutinates (e.g., pyroxenes, ilmenites and olivines with included Fe-Ni grains). The non-magnetic residue was selectively enriched in non-magnetic agglutinates, including plagioclase compositions. Accordingly, the proposed chemical fractionations were an artifact [45]. It has been proposed that agglutinates form by preferential fusion of the finest ($<$10 micron) fraction of the soils [46] but this model requires much testing before it can be substantiated.

4.4.2 Impact Glasses

Many different forms have been described. The spheres are typically about 100 microns in diameter, but range widely in size. Ellipsoidal shapes, dumbbells, teardrops, and rods, are common (Figs. 4.10, 4.11). These are the typical rotational shapes assumed by splashed liquids. Some of the spheroidal forms are flattened, indicating that they were plastic when they landed. There is a great abundance of angular fragments. Many of these are broken pieces of the more regular forms. Others occur as large irregular masses ($>$10 g) coating rocks or as linings in pits clearly produced by impact of small particles. The outer surfaces of the spherules commonly have small craters.

A wide range in color is shown by the glasses from colorless through pale yellow, green, brown, orange to red, and black. The colors show a clear relation to refractive index and to composition. Table 4.1 shows the interrelationship of color, refractive index and density. (See also Section 6.2.3 on the emerald green and orange glasses.) The color of the glasses is clearly reflected

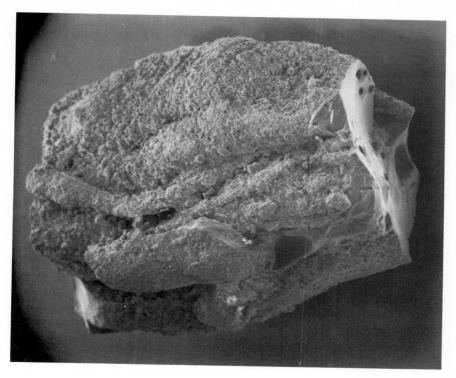

4.10 Ropy glass fragment, 1 mm long, of KREEP glass (12033). (NASA S.70 17107. Courtesy D. S. McKay.)

in the chemistry. The lighter colored glasses are similar in composition to the feldspathic or anorthositic fragments, whereas the more numerous red, yellow and dark brown glasses resemble the bulk analyses of the rocks and the fine material.

Tiny spherules of iron are present in many of the glasses. They are normally less than 30 microns in diameter and are especially frequent in the heterogeneous glasses. Some of these spherules contain about 10% Ni as well as troilite and schreibersite, which are common minerals of iron meteorites.

Others, probably the majority, are derived by auto-reduction of Fe^{2+} (see Section 4.5.1). These contain nickel-free iron and troilite derived from the lunar rocks, forming immiscible droplets in the glass melt. The chief evidence for an external meteoritic origin for the nickel-rich iron spheres is their nickel content, since nickel is very depleted in the parental rocks. These nickel-iron spheres resemble those found in terrestrial glasses that have been formed at meteorite craters by the fusion of country rock by the impacting iron-nickel meteorite. The presence of these globules indicates that most of the glasses

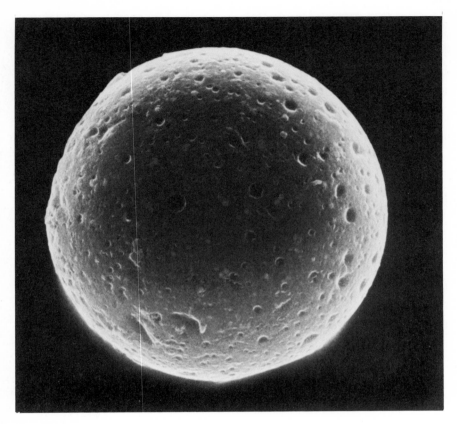

4.11 SEM photo of Luna 24 glass sphere, showing some superficial resemblance to Tethys. The pits are due to removal of included iron spherules, and not to impact. (NASA S.78-36991. Courtesy D. S. McKay.)

Table 4.1 Relationships between color and other properties of lunar glasses.[†]

Color	Refractive Index	TiO_2 (%)	FeO (%)	Density
Colorless, transparent	1.50–1.60	0–0.1	0–1.6	< 2.7
Light yellow, green to light green, transparent	1.59–1.65	0.1–2	4–10	2.7–2.8
Dark green, transparent	1.65	0–1	7–16	2.8–2.9
Yellow-brown, transparent	1.65	1–2.5	8–14	2.7–2.8
Light to dark brown and red-brown, transparent	1.65–1.75	3–8	9–16	2.8–3.0
Dark brown to opaque	1.75	7–12	15–25	3.0–3.25

[†]From Frondel, C. (1970) *PLC 1:* 450.

bear a genetic relationship to meteorite impact. Most of the surface features, such as small craters, pits, splashes of glass (Figs. 4.12a and b) and iron beads are the result of meteorite impact. There are no major differences in chemical composition between the angular glass fragments and the spheroidal forms.

The range in composition found among the glasses can be due to several causes:

(a) Melting of whole rocks and bulk samples of fine material, giving the range in composition found for the rocks.

(b) Melting of individual mineral phases by impact of small particles.

(c) Selective vaporization. The overall evidence seems to indicate that this process was not effective in altering the composition to any marked degree. Studies from terrestrial glasses support this contention. Thus, at the Lonar Crater, India, the glasses are identical in composition to the parental basalts [47] supporting similar conclusions for impacts into terrestrial sedimentary rocks [48]. Evidence of rather thorough mixing to produce homogeneous glasses from diverse target rocks occurs at large terrestrial impact sites, such as Manicouagan [49]. This question becomes important in studying the highland samples, where most of the evidence for primary rocks has to be inferred.

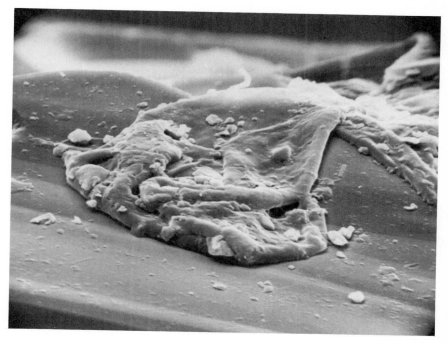

4.12a Glass splash on a lunar olivine crystal. The width of the "lava flow" is 100 microns. (NASA S.71 24637. Courtesy D. S. McKay.)

4.12b   Glass coating Apollo 16 lunar breccia 64435 (NASA S.72-39674).

(d)   Selective melting of specific minerals; e.g., under shock conditions, feldspar melts at lower pressures than does pyroxene or olivine.

Although there are some similarities in form between the spherules and chondrules observed in chondritic meteorites, detailed textural studies of chondrules [50, 51] do not provide many direct analogies, and do not directly support an origin for chondrules in a lunar-type regolith. The chondritic meteorites are, of course, different in composition from the lunar breccias, and there is ample evidence from the Apollo studies that the chondritic meteorites are not derived from the Moon.

A final observation concerning the resemblance between some glass forms and microfossils deserves wide circulation: "The abundant spheroids and ovoids are similar in shape to some algal and bacterial unicells, and the smaller ones are comparable in size. Indeed, if such particles were coated with carbon, they would make impressive pseudomicrofossils . . . .This is not to propose that there are or were solid glass Protozoa on the Moon, but to add one more warning about a too-ready interpretation of exotic objects as of vital origin on the basis of gross morphology alone . . . . This warning deserves emphasis. Elsewhere on the lunar or Martian surface may be lifelike artifacts that will be harder to discriminate from the real thing" [52].

4.4.3   Tektites

One of the achievements of the Apollo missions was to solve the long-standing scientific question of a terrestrial versus lunar origin for tektites [53]. Before Apollo 11, it was the scientific consensus that tektites were melted and splashed material formed during large cometary or meteorite impact events. Whether the impact took place on the Earth or the Moon was the topic of an intense scientific debate. These impact events were large-scale since the tektites were sprayed over immense distances. In the Australian strewn field, tektites occur across the entire southern half of the continent. The numerous occurrences in Southeast Asia are related to the same event, and the occurrence of microtektites in deep-sea cores (within the zone marking the 700,000-year reversal of the Earth's magnetic field: Bruhnes-Matuyama reversal) has extended the strewn field to over 4% of the Earth's surface.

Four separate tektite-strewn fields are known: bediasites (North America, 34 million years); moldavites (Czechoslovakia, 14 million years); Ivory Coast (1.3 million years); and Southeast Asian and Australian fields (0.7 million years). No normal geological processes (e.g., volcanic action) are capable of explaining this distribution. Neither do these small objects come from deep space, for they record no evidence of any exposure to cosmic radiation. These facts have restricted their origin to the Earth or the Moon.

Of immediate interest to the tektite problem was the glass present in the lunar soil. Although generally small, these forms resemble the primary tektite forms. The lunar glass, terrestrial volcanic glass, and protected tektite surfaces all show similar features, such as bubble pits, domal gas blisters, abraded and ablated surfaces, and spatter. These features are evidence for similar formational processes, but the consensus among workers on the lunar glasses is that they do not resemble tektites in refractive index, chemistry or infrared spectral characteristics.

The chemistry of tektites reflects that of the parent material (Table 4.2), and losses during formation appear to be restricted to elements and compounds more volatile than cesium. Terrestrial impact glasses provide small-scale analogues of tektite-forming events [48] and indicate that only the most volatile components are lost during fusion. The present composition of tektites can accordingly be used to infer the composition of the parent material.

Tektite chemistry is totally different from that observed in lunar mare basalts. These possess Cr contents that are two orders of magnitude higher than tektites, distinctive REE patterns with large Eu depletions, high Fe and low SiO_2 contents, low K/U ratios, and many other diagnostic features, which are not observed in the chemistry of tektites.

Typical highland compositions with high-Al and Ca and low K concentrations are distinct from those of tektites. Lunar samples with characteristic KREEP-type chemistry have contents of U and Th similar to tektites. How-

Table 4.2 Comparison of lunar and tektite major element compositions.†

	Average Lunar Highlands (1)	Lunar high-SiO$_2$ glasses		Bediasite (4)	Moldavite (5)	Indochinite (6)	Javanite (7)	Australite (8)	Bottle-green micro-tektite (9)
		(2)	(3)						
SiO$_2$	44.9	75.8	77.0	73.5	80.7	72.7	73.2	73.8	50.5
TiO$_2$	0.60	0.53	0.5	0.87	0.32	0.78	0.64	0.67	0.86
Al$_2$O$_3$	24.6	11.4	12.0	15.9	9.6	13.6	11.2	11.4	16.7
FeO	6.6	2.5	0.5	5.0	1.93	4.60	5.70	4.31	4.53
MgO	7.0	0.25	0.10	1.38	1.59	2.14	3.75	1.92	20.7
CaO	15.8	1.8	0.5	0.06	2.13	1.98	2.10	3.99	4.81
Na$_2$O	0.45	0.35	1.0	1.30	0.37	1.05	0.98	1.23	0.05
K$_2$O	0.11	6.4	8.0	1.73	3.60	2.62	2.09	2.28	0.04
Σ	100.0	99.0	99.6	99.7	100.2	99.5	99.7	99.6	98.2
K$_2$O/MgO	0.016	25.6	80	1.25	2.3	1.22	0.56	1.19	0.002

† Data in wt.%

1. From Table 5.6.
2. Apollo 11, Roedder, E., and Weiblen, P. W. (1971) *PLC 2*: 522.
3. Apollo 14, Glass, B. P. (1977) NASA Report NGR 08-001-029.
4–5. Baker, G. (1959) Mem. Nat. Mus. Vic. 23.
6–8. Taylor, S. R., and McLennan, S. M. (1979) *GCA*. 43: 1551.
9. Frey, F.-A. (1977) *EPSL*. 35: 45.

ever, they retain the lunar characteristics of high Cr and Eu depletion.

Three minor lunar occurrences analogous in composition to K-rich terrestrial granites are (a) late-state glassy K-rich mesostasis found in lunar basalts; (b) immiscible globules of granitic composition; and (c) minor amounts of granitic glasses in lunar soil. All these compositions are typified by high potassium values, typically 6–8%. The Mg contents are very low. None of these compositions resembles those of tektites, which in any event are not closely related to terrestrial granites. In a study of 500 glass particles from lunar soils, Glass [55] found only one spherule with > 60% SiO_2 ($62\%SiO_2$, $4\%K_2O$), illustrating the rarity of such compositions on the Moon. Evidence for a lunar origin of tektites has been adduced from studies of lunar liquid immiscibility by Roedder and Weiblen [54], who concluded that "any such surface mass (of granitic composition derived from liquid immiscibility) would also remove one of the obstacles to a possible lunar origin for tektites." However, the data do not encourage this speculation. A comparison of K/Mg and K/Na ratios for australites, with the microscopic lunar granitic material, shows that the potash granites have ratios a order of magnitude higher than tektites, while the high silica glasses have ratios 30 to 50 times those of tektites.

The water content of tektites, although low (~80 ppm) is nevertheless at least four orders of magnitude higher than the ppb levels found in a few lunar samples. The consensus is that even these levels are not indigenous to the Moon, but represent terrestrial, meteoritic or cometary contamination (see Section 5.5).

The REE patterns observed in tektites and in bottle-green microtektites are identical to those of terrestrial sedimentary rocks [48]. This fact constitutes strong evidence for a parental terrestrial sedimentary rock, and mineralogical studies indicate that the parent material of tektites was "a well-sorted, silt-size sedimentary material" [56].

The major element chemistry of the bottle-green microtektites remains enigmatic, not resembling that of any common terrestrial or extraterrestrial sample. There is some resemblance between the major element chemistry of bottle-green microtektites and the silicate portion of the mesosiderite meteorite, Estherville, but this interesting similarity does not extend to the rare-earth elements. The Estherville silicate REE pattern is nearly flat at about 3 times chondritic levels, with a small negative Eu anomaly, thus differing from bottle-green microtektite patterns by an order of magnitude.

The microtektite compositions are sufficiently variable to lead Mason [57] to query whether they are related to tektites. Their composition is characterized by high concentrations of refractory elements Mg, Al, Ca and low concentrations of alkali elements in comparison to tektites. They also possess sedimentary-type REE patterns [58]. These characteristics indicate that the bottle-green microtektites may represent a refractory residue following severe heating of the source material. Some of the compositions could represent material condensed from a vapor phase. During these extreme conditions, few element signatures characteristic of the source rock will survive. However, the REE are notably refractory and could be expected to preserve the evidence of their parental material. Such complex processes may account for their unusual compositions.

One of the characteristic features of the lunar rocks is their great age. The lava floods occurred between 3.8 and 3.2 aeons ago. No younger rocks are known at present, although some very minor activity may have occurred. The uplands were formed between 4.4 and 4.0 aeons ago. All these ages are much greater than those observed for the parent material of tektites. As noted earlier, the tektite-forming events all occurred in the Tertiary or Pleistocene on Earth, as dated by K-Ar or fission track methods. The melting episodes have not seriously disturbed the Rb-Sr or Sm-Nd isotopic systems, enabling ages of the parent material to be assessed. The Ivory Coast tektite data fall on a 2.0×10^9 year isochron defined by the Bosumtwi crater rocks, indicating derivation from that material. The Southeast Asian and Australian tektite Rb-Sr data indicate a Mesozoic age for the parent material. There thus appears to be an order of magnitude difference between the lunar ages and those indicated for the parent material of the Southeast Asian tektites. Recent studies of Sm-Nd isotopic systematics confirm these suggestions. The data are consistent with a terrestrial sedimentary precursor derived from continental crust material for the Australasian tektites [59].

The large young ray craters indicate a frequency of large-impact events on the Moon at about one per hundred million years, about an order of magnitude less frequent than the tektite events. If Tycho were the source of tektites, it should be 0.7 million years old. The current estimate for the age of Tycho is about 100 million years (Section 4.10). Crater counting statistics suggest even older ages.

The $\delta^{18}O$ values for tektites range between +9 and +11.5 per mil. The lunar $\delta^{18}O$ values are low, ranging from +4 for ilmenite to +7 for cristobalite. There is a consensus that the oxygen isotope data are unfavorable to the lunar tektite hypothesis and Taylor and Epstein [60] conclude that "suitable parent materials are rare or non-existent on the lunar surface."

The absence of exposure to cosmic-rays precludes origins from outside the Earth-Moon system, either from elsewhere in the solar system, or outside. The evidence supports the origin of tektites by cometary (or meteorite) impact on terrestrial sedimentary rocks. The chemical, isotopic, and age evidence from the lunar samples enable us to reject the lunar impact hypothesis. A fitting epitaph has been provided by Schnetzler [61]: "The lunar origin of tektites, a controversial and stimulating theory on the scientific scene for almost 75 years, died on July 20, 1969. The cause of death has been diagnosed as a massive overdose of lunar data."

O'Keefe [62] has revived the theory that tektites come from lunar volcanoes, proposing to launch them from a lunar volcano "which should be powered by hydrogen." This useful fuel, which provided some of the thrust needed for the Apollo lunar missions, is probably scarce on the Moon, except in the lunar soil where it is trapped from the solar wind (e.g., [63]). The composition of lunar rocks is characterized by a depletion in volatile elements and an enrichment in refractory elements. The presence of hydrogen in the amounts needed for volcanic activity in the Moon would be a scientific fact of great significance and would reverse most of the present conclusions about the origin and evolution of the inner solar system (see Chapter 9). Kozyrev [64] reported H_2 emission from the central peak of Aristarchus, which he ascribed to volcanic activity. The central peaks of large impact craters on the Moon are not volcanoes, but result from the impact process, and characterize craters with diameters greater than 10–20 km (e.g., [65]). Kozyrev's observation joins a long list of transient lunar phenomena, observed using earth-based telescopes, but not from the Apollo missions, and which should be treated with caution (cf. Crater Linné controversy, Section 3.14).

Several difficulties attend this volcanic hypothesis. Most of the objections to the lunar-impact theory—based on chemical, age and isotope data—still apply. Our sampling of lunar volcanic rocks does not encourage the view that lavas of tektite composition might be erupted. Such eruptions would need to be among the recent lunar events (<50 million years ago), to account for the tektite ages. If they are powerful enough to accelerate tektite glass beyond the escape velocity (2.4 km/sec) of the lunar gravitational field, then tektite glass or material of tektite composition should occur in the uppermost portions of the lunar regolith. Although many diverse compositions occur among the returned lunar samples, tektite compositions are not among them.

Spherules of meteoritic origin occur in tektites [66]. This and the presence of shocked mineral inclusions indicative of shock pressures far in excess of 100 kbars [57] seems particularly difficult to ascribe to volcanic pressures but highly supportive of an impact origin. The apparent absence of young volcanic vents, or of glassy ejecta of tektite composition on the lunar surface, comprises insuperable problems for the lunar volcanic model. To this may be added the lack of reasonable petrogenetic models capable of generating large volumes of material of tektite composition on the Moon. We may safely conclude that tektites do not come from the Moon.

All the evidence summarized above points unequivocally to an origin for tektites by meteoritic, cometary or asteroidal impact on terrestrial sedimentary rocks.

4.5   Chemistry and Petrology of the Regolith

The average chemical composition of the fine-grain-size (< 1 mm) regolith at the various sites is given in Table 4.3. There are considerable variations

Table 4.3a Major element composition of soils from the Apollo landing sites.[†]

Apollo:	11	12		14	15		16		17
	10084	12001	12033	14163	15221	15271	64501	67461	70009
SiO_2	41.3	46.0	46.9	47.3	46.0	46.0	45.3	45.0	40.4
TiO_2	7.5	2.8	2.3	1.6	1.1	1.5	0.37	0.29	8.3
Al_2O_3	13.7	12.5	14.2	17.8	18.0	16.4	27.7	29.2	12.1
FeO	15.8	17.2	15.4	10.5	11.3	12.8	4.2	4.2	17.1
MgO	8.0	10.4	9.2	9.6	10.7	10.8	4.9	3.9	10.7
CaO	12.5	10.9	11.1	11.4	12.3	11.7	17.2	17.6	10.8
Na_2O	0.41	0.48	0.67	0.70	0.43	0.49	0.44	0.43	0.39
K_2O	0.14	0.26	0.41	0.55	0.16	0.22	0.10	0.06	0.09
MnO	0.21	0.22	0.20	0.14	0.15	0.16	0.06	0.06	0.22
Cr_2O_3	0.29	0.41	0.39	0.20	0.33	0.35	0.09	0.08	0.41
Σ	99.8	101.0	100.8	99.8	100.5	100.4	100.3	100.8	100.5

[†]Values given in wt.%.

Sources: Laul, J. C., and Papike, J. J. (1980) *PLC 11:* 1307.
Laul, J. C., et al. (1978) *PLC 9:* 2065.

in chemistry, which reflect the nature of the underlying bedrock. Thus the Apollo 11 soils have high-Ti contents and other characteristics of the Apollo 11 mare basalts. In contrast, the Apollo 16 site, deep within the highlands, is dominated by the high-Al and Ca, and low-Fe and Mg contents of the highland rocks.

The chemical compositions of the soils have been interpreted generally in terms of mixing models. Basic to the problem is the identification of the "end-members" involved in the mixing models. This problem is complex in detail at all sites.

During the formation of the regolith, a wide sampling of the lunar surface occurs. The astronauts' sampling of rocks was necessarily restricted. The aluminous mare basalts, for example, are almost certainly underrepresented in the collections. Thus, it is somewhat surprising that we get such a good match between the regolith and the whole-rock compositions at the mare sites. The more pulverized highlands present less of that particular problem, but put the problem back into a megaregolith stage. Average chemical compositions for the lunar regolith reflect the bedrock differences to be discussed in later chapters and indicate the truth of the comment that there is no large-scale lateral transfer of surface material between highlands and maria.

Increasingly detailed studies have been undertaken on regolith samples, and in particular, much effort has been expended on core studies [67, 68]. Samples restudied include many well-known soils (10084, 12001, 12033, 14163, 15221, 15271, 64501 and 67461).

The petrographic examination clearly indicates the effects of local derivation of much of the material. In terms of pyroxene compositions, 10084

Table 4.3b Trace elements in soils from the Apollo landing sites.[†]

Apollo:	11	12		14	15		16		17
	10084	12001	12033	14163	15221	15271	64501	67461	70009
Rb	3.2	23	14	14.6		5.7	2.0		
Ba	170	430	600	800	240	300	130	60	120
Pb	1.4		4.0	10		2.8			
Sr	160	140	160	170	120	130	170	170	210
La	15.8	35.6	50	67	20.5	25.8	10.8	4.7	7.9
Ce	43	85	133	170	54	70	28	12	28
Nd	37	57	85	100	36	45	19	7.2	23
Sm	11.4	17.3	22.8	29.1	9.7	12	4.8	2.0	8.1
Eu	1.60	1.85	2.45	2.45	1.30	1.50	1.05	1.00	1.76
Tb	2.9	3.7	4.9	5.9	2.0	2.6	1.0	0.45	1.9
Dy	17	22	30	36	12	—	6.0	2.8	11.4
Ho	4.1	5.0	7.2	8.6	2.9	3.9	1.4	—	2.9
Tm	1.6	1.8	2.6	3.2	1.1	1.4	0.55	0.25	—
Yb	10.0	13.0	17	21	6.9	8.5	3.4	1.6	7.1
Lu	1.39	1.85	2.45	3.00	0.97	1.20	0.49	0.22	1.1
Eu/Eu*	0.37	0.30	0.30	0.22	0.38	0.35	0.60	1.38	0.59
Y	99	—	160	190	86	—	—	—	—
Th	2.1	5.40	8.50	13.3	3.0	4.6	1.85	0.83	0.95
U	0.54	1.7	2.4	3.5	—	1.2	0.4	—	0.23
Zr	320	—	760	850	—	390	—	—	—
Hf	9.0	11.8	16.6	23	6.7	8.6	3.3	1.6	6.6
Nb	118	—	44	46	—	25	—	—	—
V	70	110	100	45	80	80	20	20	100
Sc	60	40	36	22	21	24	8.0	7.8	57
Ni	200	310	210	330	273	220	380	—	—
Co	28	43	34	33	41	41	20	9	32
Cu	10	7.2	8	8	—	9	—	—	—
Zn	23	—	14	34	—	21	—	—	44
Li	10	18	24	27	—	—	—	—	—
Ga	5.1	4.2	3.1	8.3	—	4.4	—	—	6.3
Au, ppb	2.4	2.6	—	5.4	—	4	14	—	3
Ir, ppb	6.9	11	—	14	—	9	12	—	—

[†]Values given in ppm except where noted.

resembles the Apollo 11 basalts, 12001 is similar to the Apollo 12 olivine and ilmenite basalts. The pyroxenes from soil 14163 correspond to those of Apollo 14 low-K Fra Mauro basalts. Soils 15221 and 15271 have pyroxenes similar to intermediate-K Fra Mauro basalts. Those from 64501 and 67461 are similar to anorthositic rocks. The high-K soil from Apollo 12 (12033) has a pyroxene chemistry similar to Apollo 14 KREEP basalts, confirming its exotic origin. Intrasite mixing of soils is minor. This accords with models of regolith production as a sucession of ejected sheets of debris from impacts, with stirring (gardening) by micrometeorites of the upper layers only. The two Apollo 16 soils do not show any real differences. The finer fraction of the soils does not contain any preferential enrichment of *exotic* material, except in the Apollo 17 soils, where highland material is enriched in the finer fractions [67, 68]. Average glass compositions are given in Table 4.4.

There is a 20–30% highland component in the Apollo 11 soil, most of which is observed in petrographic studies. The nearest highlands are over 50 km away, and lateral transport is not very efficient. It is suggested that this material is mainly derived by excavation by meteorite craters from beneath the mare basalts. Accordingly, these must be thin [69, 70]. However, we must recall that all the mare basalt landing sites (except Apollo 12) are near the edges of basins, where the basalt flows are thin anyway. Apollo 12, in Oceanus Procellarum, is in an irregular maria, which by definition is thin. Nevertheless, the presence of a thin layer of "light grey fines," 12033, is clearly exotic to the Apollo 12 site, and is usually identified as coming from Copernicus.

Other explanations, such as the occurrence of an earlier stratum of aluminous mare basalts, have been postulated. According to Labotka [67], "There is no evidence, nor any need for the widespread aluminous mare basalts" (e.g., [71]).

The chemistry of the regolith at different sites has been surveyed most recently by Laul and Papike [72]. They find that a major chemical discontinuity appears in the < 10 micron fraction. This comprises 10–15% of the bulk soil and is apparently more feldspathic and enriched in highland components, relative to the coarse fractions of the soils. In addition, the KREEP component (Section 5.4.3) is distributed in the soils on a moon-wide basis. Figure 4.13 shows the REE patterns in lunar soils. The bulk composition of highland soils was modelled by Korotev and co-workers [73], using end-member components of KREEP, FAN (ferroan anorthosite), MAF (mafic components) and HON (highlands olivine norite) creating a Disney-like assemblage of characters. Although a 5–10% mare component in the fine fractions of the Apollo 16 soils has been claimed [72], this is probably an artifact of the mixing calculations. There is less than 0.1% mare basalt in the fractions greater than one millimeter in size. The presence of a large fraction of mare basalt at the Descartes site far removed from the maria does not accord with other evidence. The soil composition at the Apollo 16 site is of interest on account of possible differences between the Cayley and the Descartes Formations [74]. However, the soils do not show any distinction in chemical composition [75]. ". . . There are no significant differences between Cayley and Descartes material at the A-16 site . . . " [75, p. 1354]. A general review of the petrology of lunar soils is given by Heiken [76].

4.5.1 Metallic Iron in the Lunar Regolith

The regolith is enriched in metallic Fe relative to its ultimate source material. The origin of this metallic iron is of interest because of its effect on

Table 4.4a Average composition of glass types in Apollo 15 fines.[†]

	Mare basalt					Highland basalt (%)	Fra Mauro basalt			"Granite" 1 (%)	"Granite" 2 (%)
	Green glass (%)	Mare 1 (%)	Mare 2 (%)	Mare 3 (%)	Mare 4 (%)		Low K (%)	Moderate K (%)	High K (%)		
SiO_2	45.43	45.70	44.55	43.95	37.64	44.35	46.56	49.58	53.35	73.13	62.54
TiO_2	0.42	1.60	3.79	2.79	12.04	0.43	1.25	1.43	2.08	0.50	1.18
Al_2O_3	7.72	13.29	11.77	8.96	8.46	27.96	18.83	17.60	15.57	12.37	15.73
Cr_2O_3	0.43	0.33	0.26	0.46	0.48	0.08	0.20	0.17	0.12	0.35	0.03
FeO	19.61	15.83	18.83	21.10	19.93	5.05	9.67	9.52	10.25	3.49	6.67
MgO	17.49	11.72	8.84	12.30	10.49	6.86	11.04	8.94	5.77	0.13	2.51
CaO	8.34	10.41	10.46	9.02	8.81	15.64	11.60	10.79	9.57	1.27	6.86
Na_2O	0.12	0.30	0.34	0.27	0.54	0.19	0.37	0.74	1.01	0.64	0.98
K_2O	0.01	0.10	0.13	0.05	0.13	0.01	0.12	0.47	1.11	5.97	3.20
Total	99.57	99.28	98.97	98.90	98.52	100.92	99.66	99.24	98.83	97.85	99.70
No. of analyses	187	67	21	26	6	36	82	90	29	2	1
Percentage of total analyses	34.2	12.2	3.8	4.8	1.1	6.6	15.0	16.5	5.3	0.4	0.2

[†]From Reid, A. M., et al. (1972) *Meteoritics.* 7: 406.

Table 4.4b Glass compositions in soil from the Apollo 16 site, Descartes.

	Anorthosites		Highland basalt		LKFM	MKFM	High Mg	Mare		Green Glass
	(1)	(2)	(3)	(4)	(5)	(6)	(7)	(8)	(9)	(10)
SiO_2	44.9	48.8	42.0	45.4	46.7	50.8	45.4	46.2	42.0	44.1
TiO_2	0.04	0.08	0.35	0.43	0.75	2.54	0.19	1.76	7.92	0.37
Al_2O_3	35.8	33.4	30.9	27.6	21.5	15.6	15.4	14.3	10.0	7.8
Cr_2O_3	—	—	0.04	0.07	0.11	0.15	0.25	0.29	0.42	0.33
FeO	0.18	0.22	3.46	4.62	6.89	11.1	8.46	14.9	17.0	21.1
MgO	0.12	0.05	6.24	6.13	10.4	7.13	21.1	10.3	10.4	16.7
CaO	19.0	16.3	17.3	15.6	12.9	10.2	9.55	11.0	10.1	8.4
Na_2O	0.55	1.82	0.12	0.42	0.31	0.85	0.17	0.36	0.39	0.13
K_2O	0.03	0.11	0.01	0.05	0.08	0.65	0.01	0.10	0.10	0.03
Total	100.62	100.78	100.42	100.32	99.64	99.02	100.53	99.21	98.33	98.96
No. of Analyses	103	7	30	111	29	13	2	9	2	3
Percentage of total analyses	33	2.2	9.7	36	9.3	4.2	0.6	2.9	0.6	1.0

Source: Ridley, W. I., et al. (1973) *PLC 4*: 309.

Table 4.4c Glass compositions from soils from the Apollo 17 site, Taurus-Littrow Valley.

	Mare						Non-Mare			
	Orange glass		High-Ti		VLT		Anor-thosite	Highland basalt	MKFM	LKMF
	(1)	(2)	(3)	(4)	(5)	(6)	(1)	(2)	(3)	(4)
SiO_2	39.2	39.8	41.7	37.8	46.7	46.6	44.5	45.4	51.0	45.9
TiO_2	8.9	8.9	8.6	11.4	0.72	0.85	0.01	0.39	1.56	1.57
Al_2O_3	5.9	7.9	10.8	8.9	10.4	12.5	35.5	25.4	18.9	19.3
Cr_2O_3	0.67	0.58	0.30	0.39	0.60	0.51	—	0.12	0.13	0.21
FeO	22.4	22.0	17.4	21.8	18.4	18.0	0.20	5.6	8.3	9.2
MnO	0.28	0.29	0.22	0.23	0.24	0.27	—	0.04	0.09	0.10
MgO	14.6	11.3	9.2	9.0	12.7	10.5	0.12	8.0	7.0	10.9
CaO	7.2	8.3	10.9	9.0	9.7	10.5	18.7	14.5	11.2	11.9
Na_2O	0.36	0.41	0.31	0.67	0.18	0.14	0.51	0.24	0.68	0.24
K_2O	0.09	0.09	0.07	0.15	0.03	0.04	0.05	0.06	0.53	0.11
Total	99.6	99.57	99.5	99.34	99.67	99.91	99.59	99.75	99.39	99.43
No. of Analyses	94	11	10	8	42	14	5	64	10	30
Percentage of total Analyses	33	3.8	3.5	2.8	15	4.9	1.7	22	3.5	10

Source: Warner, R. D., et al. (1979) *PLC 10*: 1437.

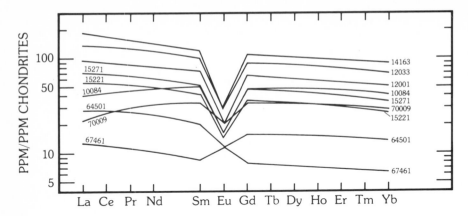

4.13 Chondrite normalized rare-earth element (REE) patterns in soils from the Apollo 11, 12, 14, 15, 16 and 17 landing sites. Note the wide diversity in patterns. Data from Table 4.3b.

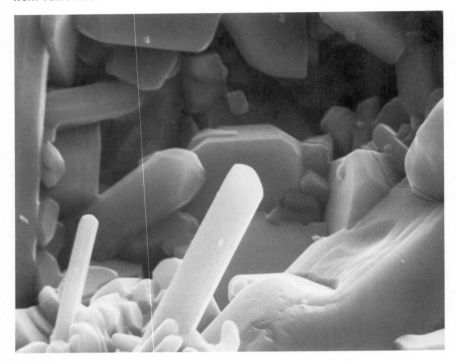

4.14 SEM photograph of a vug lined with crystals in an Apollo 14 lunar breccia. The large pyroxene crystal in the foreground is 8 microns long. Euhedral crystals of apatite, pyroxene and plagioclase form an open network. (NASA S.73-34448. Magnification 3000X. Courtesy D. S. McKay.)

lunar magnetic properties (see Section 7.8). There is general agreement that this "excess" metal is produced by lunar surface processes [77–79].

There are three sources of metallic iron particles in the regolith [80]. These are: (a) Metal particles produced from reduction of ferrous iron induced by exposure. These are mainly associated with agglutinates, and range in size between 40–330 Å. (These particles are derived by reduction of Fe^{2+} in silicate and oxide phases). (b) Metal particles from micrometeorites involved in the formation of agglutinates. These are larger, being mainly >330 Å in diameter, and comprise generally less than about 30% of the metallic fraction. (c) Metallic particles derived from the source rocks for the soils. These are also in general > 330 Å in diameter.

4.5.2 Volatile Element Transport on the Lunar Surface

There is a considerable body of evidence which indicates volatile element mobility, such as the occurrence of vapor phase crystals in cavities and vugs in breccias (Figs. 4.14, 4.15). Such movement of chemical elements on the Moon

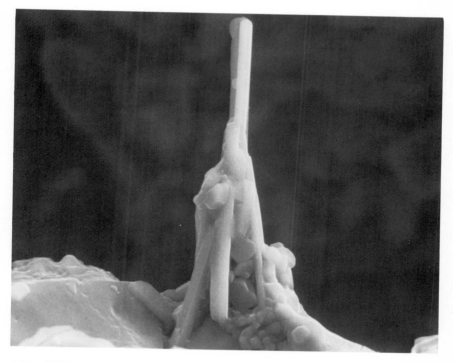

4.15 SEM photo of an isolated pyroxene crystal in a vug in an Apollo 14 breccia. (NASA S.73-34441. Magnification 6000X. Courtesy D. S. McKay.)

may occur by the following processes:

(1) fire-fountains;

(2) solar-wind sputtering;

(3) redistribution of volatiles during meteorite impact.

Volatile elements such as Cd, Zn, In and Ga increase by factors of 10–20 from coarse (500 micron) to fine grain (< 5 micron) sizes. This surface corre-lated effect indicates selective movement of volatiles during regolith evolu-tion. Trace siderophile elements peak in the 80–300 micron grain size (typical of agglutinate fractions) so that agglutinate formation must involve incorpo-ration of the siderophiles from the meteorite bombardment [81]. Indeed, very moderate temperatures can lead to volatilization and movement of elements on the lunar surface [82]. Hg is probably volatile during lunar daytime conditions when the temperature reaches 130° C, with recondensation at night when the temperature drops to –170° C. Therefore, Hg might be expected to be ubiquitous on regolith grains. Temperatures as low as 250° C cause mobili-zation of other elements under laboratory conditions [82]. The following elements are enriched on the surfaces of regolith grains: Au, Br, Cd, Ga, Ge, Hg, In, Pb, Sb, Te and Zn. Stepwise heating experiments reveal the following sequence of volatility: Hg≫Cd>Zn>Se>In>Br≫Ag.

These elements are enriched both on the surfaces of mineral grains and especially on agglutinates. These results must engender caution about the interpretation of volatiles on the green and orange glass spheres (Section

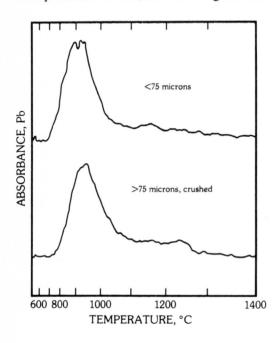

4.16a Thermal release profiles of Pb in rock sample 66095 showing that most of the Pb in both < 75 μm and >75 μm frac-tions is released below 1000°C before the sample melts and therefore is surface Pb.

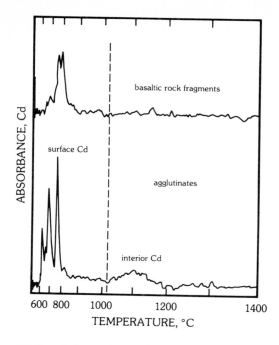

4.16b   Cd release profiles of basaltic rock fragments and agglutinates from 150–420 μm size fraction of submature mare fines 75081. Note very little or no interior Cd is present in basaltic rock fragments, whereas some interior Cd is present in agglutinates.

6.2.3) as being uniquely indicative of a volcanic origin [83] since micrometeorite impact will readily volatilize these elements. In this context, volatiles were deposited on the orange glass spheres from Apollo 17, after the spheres had been broken. This is interpreted to have occurred during the formation of

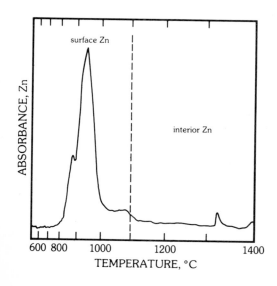

4.16c   Thermal release profile of Zn in a pristine anorthosite 65325 showing that most of the Zn is on surfaces.

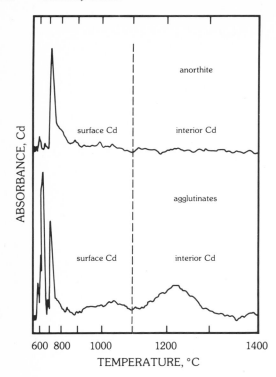

4.16d Thermal release profiles of Cd in anorthite grains and agglutinates from 150–420 μm size fraction of the most mature fines sample 65701. Cd in the anorthite fraction is almost all on grain surfaces. On the other hand, a fair amount of interior Cd as well as surface Cd is present in the agglutinates [85]. (Courtesy R. M. Housley.)

Shorty crater. This event must also have redistributed the volatile elements [84, 85]. Large meteorite impacts will provide enough energy to promote widespread volatile transport across the face of the Moon, although this is restricted to the very surface. However, since most of our returned samples come from this environment, caution should be exercised about the fundamental petrological significance of volatile coatings on glass spheres. The sample investigated by Wegmüller [82] (75080, from the rim of the Camelot crater) contained very few orange or black spherules. Some Cd was volatile at 250°C, Zn and Se at 450°C and Br and In at 650°C. The nature of the volatile compounds is not yet established (Fig. 4.16).

The effects on isotopic systematics, particularly for lead, but also for Rb distribution, need also to be borne in mind. Loss of volatile Rb relative to involatile Sr causes old apparent ages, as are observed on some agglutinate samples [37]. Na and K as well as Rb are mobile [86]. K/Rb ratios are altered at temperatures above 1000–1200°C. Some spherules have surface enrichment of Na and K [87]. Sulfur is even more volatile with 12–30% being lost at temperatures as low as 750°C [86].

Such movement of volatile elements on the lunar surface raises questions about the existence of traps in permanently shadowed regions. Water might be trapped as ice in craters in polar regions [88]. The source of the water could be from solar wind hydrogen as a source for reduction of FeO, carbonaceous chondrites or cometary impacts, since no unequivocal identification of indigenous water has been made. Other suggestions include the possibility of trapped frozen gases, for example, radon-222 derived from radioactive decay of uranium [89].

4.5.3 The Meteoritic Component

Addition of meteoritic material to the Moon is derived from three sources: the ancient heavy bombardment, young crater forming events, and micrometeorites in addition to possible cometary material. In this section, the nature and composition of the meteorite contribution in the lunar regolith is explored, while the question of the nature of the basin-forming projectiles is discussed in Section 5.6. The investigation of this component was aided by two geochemical factors [90]. The first was the strong depletion of siderophile elements (e.g., Ni, Ir, Au, Re) in the lunar rocks. Such depletions are to be expected on the surfaces of differentiated planets; however, the extreme depletion of the Moon in volatile (and chalcophile) elements (Ag, Bi, Br, Cd, Ge, Pb, Sb, Se, Te, Tl, Zn) enabled the use of these elements, which are relatively abundant in C1 meteorites, as indexes of the meteoritic component. Typical data are shown in Fig. 4.17 for mature soils, with high surface exposure ages. Such soils from all sites give the same result. The meteorite component appears to be similar to C1 chondrites, and the concentration level is 1.5–2% by weight. There is some variability among the volatile constituents, probably due to their mobility in the soils, hence producing a sampling problem. It is of interest that this chemical signature virtually eliminates all the other meteorite classes (irons, stony irons, ordinary chondrites, achondrites) from consideration [90, 91]. Mean influx rates have been calculated as 2.4×10^{-9} g/cm^2/yr. [90] and 2.9×10^{-9} g/cm^2/yr. [91]. Discrepancies between these estimates and those calculated from the microcrater population (0.2×10^{-9} g/cm^2/yr.) [92], which are an order of magnitude less, are discussed in the next section.

4.6   Microcraters and Micrometeorites

The lunar surface is a particularly fine recorder of micrometeorite impacts. This arises from the ubiquitous presence in the regolith of glasses, whose smooth surfaces provide an ideal recording surface for micrometeorite impacts. Typically, these will contain about 5000 microcraters (with diameters of a few microns) per square centimeter. The number of craters decreases with increasing diameter. These microcraters or "zap pits" which occur on rock and

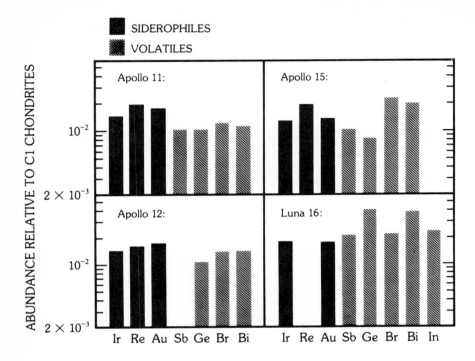

SIDEROPHILES
VOLATILES

4.17 All mare soils are enriched in "meteoritic" elements, relative to crystalline rocks. Net meteoritic component is obtained by subtracting an indigenous lunar contribution estimated from crystalline rocks. Abundance pattern is flat, with siderophiles and volatiles almost equally abundant. The meteoritic component has a primitive C1 composition [90]. (Courtesy E. Anders.)

mineral surfaces, as well as on the glass spheres, are spectacular (Fig. 3.1). They range in size from less than one micron to more than one centimeter.

Information on the nature of the projectiles provides data on micrometeorite flux and on the nature and composition of "interplanetary" or "cosmic" dust. The microcraters larger than about 3 microns typically consist of a glass-lined pit, and a "spall" zone concentric to the pit (Fig. 4.18). Spallation in this zone will sometimes leave the glass-lined pit standing on a pedestal of "halo" material. The hypervelocity nature of the impacts indicates that "primary" cosmic dust particles [93] are involved rather than "secondary" particles resulting from lunar impacts [94]. Most of these microcraters are compatible only with impact velocities of the dust grains greater than 3.5 km/sec, the actual average velocity being about 20 km/sec.

What conclusions can be drawn about the nature of the particles from the crater morphology? The densities lie within the range 1–7g/cm^3, peaking in

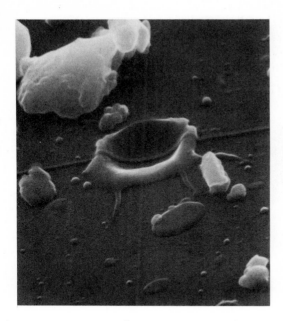

4.18 One-micron sized zap pit on a lunar glass spherule. (NASA S.73-18445. Courtesy D. S. McKay.)

the range 1–2 g/cm³. Thus, there is a lack of iron particles, and the composition is similar to that of Type I carbonaceous chondrites. These data are consistent with the properties of "Brownlee" particles collected by high-flying aircraft [95].

Therefore, the "model" micrometeorite is non-porous, equant in shape, with a density of 1–2 g/cm³ with C1 element abundances, including finely dispersed carbon. This observation has implications for theories of cosmic dust origin and early planetary condensation and accretion. The abundance of "frothy" rims on impact pits is consistent with impacts of hydrated phyllosilicates, common constituents of carbonaceous chondrites. The higher density objects (5 g/cm³) are probably magnetite grains. The most frequent mass range for the particles is from 10^{-7} to 10^{-4} g.

Craters from 1 mm to 100 microns in diameter correspond to a mass ranging from 10^{-3} to 10^{-6} g. Those craters with diameters from 100 to 0.1 microns are caused by particles in the mass range 10^{-6} to 10^{-15} g. Those in the size range 10^{-6} to 10^{-2} g contribute the most energy and are responsible for most of the damage to the rock surfaces. The production rate is about five craters per square centimeter per million years (with diameters > 0.05 cm). Thus, surfaces are effectively saturated at about one million years, and this limits the microcrater technique for exposure-age estimation. The impacts cause erosion, ionization, vaporization, and lateral small-scale transport but little vertical mixing. The erosion rate is of the order of 1 mm per million years [96].

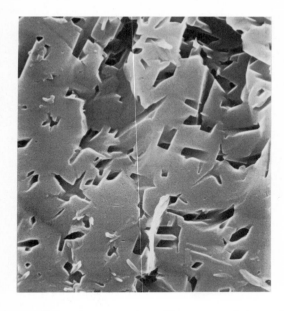

4.19 Etched pits, about one micron across, caused by solar flare iron group nuclei, in an Apollo 14 feldspar. The rectangular outlines of the etched pits are presumably due to crystallographic factors. (NASA S.72-55177. Courtesy D. S. McKay.)

The microcratering effects extend down from those caused by micrometeorites to sputtering by solar wind ions (Fig. 4.19). Table 4.5 indicates the density of microcraters spanning this range. The erosion rate due to the sputtering is about 10^{-8} cm (1Å) per year in the near surface (< 0.1 cm) regions, or about 0.1 mm per million years. This is about one-tenth of the mass erosion rate caused by micrometeorites.

Table 4.5 Microdensities due to micrometeorite impact and solar wind sputtering.

	Crater Diameter	Crater Density (per cm^2)
Micrometeorites	$\geqq 0.3$ mm	20
	$\geqq 0.1$ mm	70
Solar wind sputtering	$\geqq 1$ micron	1000–3000
	$\geqq 0.25$ micron	10,000

The impact velocities of meteoritic dust particles vary from 2.4 km/sec to 74 km/sec. Meteorites larger than 50 g should produce a detectable signal on the lunar seismometers [97]; however, the seismometers have not detected increased flux rates on the Moon at times when meteorite showers have been observed striking the Earth. Since most showers of micrometeorites consist of particles ranging from 1 to 10^{-6}g (F. Hörz, pers. comm, 1981), the failure to detect increased flux rates on the Moon is most likely due to the fact that only the larger events are recorded (about one signal per day due to meteorite impact [97]).

4.7 Irradiation History of the Lunar Surface

The lunar surface is not excelled as a detector and recorder of solar and cosmic radiation over geological time scales. It is much superior to meteorites, since it has been in its present orbit for a long time, whereas the orbits and dynamical evolution of meteorites are uncertain, extending probably inside the orbit of Venus and out to the asteroid belt. The orbits of two chondritic meteorites, Pribram and Lost City, photographically recorded during entry, indicate an elliptical orbit extending to the asteroid belt. These data help to dispose of the myth that meteorites come from beyond the solar system.

Three types of radiation interact with the lunar surface. In order of increasing energy, these are: (a) solar wind, (b) solar flares, and (c) galactic cosmic rays. The composition of this radiation is similar, being mainly composed of protons, with about 10% helium nuclei and about 1% of nuclei heavier than helium. There are, however, dramatic differences in energy, by many orders of magnitude, and, accordingly, in the effects on the lunar surface. In contrast, the number of particles is inversely proportional to their energy (Table 4.6).

4.7.1 Solar Wind

The solar wind has an average velocity of 400 km/sec with an energy of about 1 keV/nucleon (Table 4.6). The average density at one Å is about 10 ions/cm^3. This plasma, which continually flows outward from the sun, probably represents the composition of the solar corona. Ions are directly implanted into the surfaces of lunar materials, to depths of about 500 Å. The solar wind also causes some surface damage, resulting in some fine-scale rounding and production of an amorphous coating about 400 Å thick on the surfaces of grains [98]. The abundances of the rare gases in the solar wind, relative to hydrogen, approximate the average solar photosphere and coronal abundances [99, 100]. The composition of the solar wind, and its significance for the history of the sun, is discussed in Section 4.11.

4.7.2 Solar Flares

These have higher energies than those of the solar wind, typically in the range of 10 keV to 100 MeV/nucleon (Table 4.6). They occur as short bursts of radiation and there is some correlation with the eleven-year sunspot cycles.

Table 4.6 Nuclear particle effects in lunar samples and meteorites.[†]

Radiation Source	Proton Flux $(P/cm^2/sec)$	Energy	Typical Penetration Distance	Major Observable Effects
Solar wind	3×10^8	1 keV/nuc	300 Å	Direct implantation (e.g., surface correlated rare gases) Re-implantation of lunar atmospheric species (e.g., ^{40}Ar excess in lunar soils) Radiation damage (e.g., amorphous layers on lunar dust grains)
Solar flares	10^2	< 1 MeV/nuc to ≥100 MeV/nuc many more low energy than high energy particles	millimeter to centimeter	Radionuclide production (e.g., ^{26}Al, ^{53}Mn) Track production (principally tracks produced by slowing down VH nuclei) Electronic defects (e.g., thermoluminescence)
Galactic cosmic rays	1	≥100 MeV/nuc typically~3 GeV/nuc	centimeter to meter	Radionuclide production Stable isotope production (e.g., ^{21}Ne, ^{15}N) Nuclear effects due to buildup of nuclear cascades with depth (e.g., N-capture in Gd) Tracks (spallation recoils in addition to slowing down heavy nuclei)

[†]Adapted from Walker, R. M. (1980) *Ancient Sun*, p. 11.

Solar flares produce three identifiable effects: (1) directly implanted ions, (2) cosmogenic nuclides and (3) nuclear tracks.

The Apollo missions were well timed. Large solar flares occurred in November, 1968, and April, 1969, before the Apollo 11 mission in July, 1969. Large flares were absent before the Apollo 15 mission. A small solar flare, which enhanced the particle flux by a factor of 10^3, occurred during the Apollo 16 mission. In August, 1972, before the December, 1972, Apollo 17 mission, the most intense solar flares observed during the past fifteen years occurred. Tracks produced by solar flares dominate the upper one millimeter of lunar materials that have been exposed directly to sunlight. These track densities are high (10^7–10^9/cm^2).

4.7.3   Galactic Cosmic Rays

These very high energy particles (10^2–10^4 MeV/nucleon) are derived from outside the solar system (Table 4.6). These particles are not implanted in the lunar surface materials since they undergo nuclear interactions due to their high energies. Cosmic ray tracks are produced by the VH ions (Z = 18–28) and VVH ions (Z > 28), while the lighter nuclei (protons and He nuclei) produce both stable and radioactive isotopes.

There are two main classes of phenomena which result from the interaction of cosmic rays with the lunar surface: (a) solid-state damage as a result of penetration of ionizing particles, producing etchable tracks, and (b) production, through nuclear interactions, of new isotopic species. These effects have been pursued by different investigators using etching and microscopic techniques to study the tracks produced, and mass spectrometric and radiochemical procedures to investigate the cosmogenic isotopes.

4.8   Physical Effects of Radiation

The solar wind causes only minor damage, as noted earlier, producing amorphous surface layers a few hundred angstrom units thick on grain surfaces. Helium nuclei are thought to be mainly responsible [101–103]. The time needed to produce this amorphous layer is about 100 years. Solar flares, in contrast, produce high track densities, sometimes exceeding 10^{11} tracks/cm^2. These are observed even in the deepest core samples, indicating that this material was once at the surface. Such tracks are also observed in gas-rich meteorites [104], although both track densities and the number of grains irradiated by solar flares are less, consistent with much lower rates of surface overturn (gardening) on asteroidal surfaces compared with the lunar regolith.

The tracks produced by galactic cosmic rays are mostly caused by the heavier ions, dominated by the VH ions (Z = 18–28) of which iron is the most abundant species [105]. The track densities fall off steeply with depth, but tracks due to cosmic rays may reach depths of 20 cm or so, in contrast to the predominance of tracks from low energy solar flare ions in the upper one millimeter. Track lengths differ in differing minerals, in the order feldspar > pyroxene > olivine. The tracks are produced only toward the end of the penetration.

Workers have used either the total etchable track length or the track etch rate to identify the nature of the particles. The track length is, in principle, simply related to atomic number, but it has proven difficult to apply in practice [105,106]. Measurement of the track etch rate has been more useful and indicates that there appears to be no sharp threshold ionization value. Tracks from light ions ($Z<18$) etch more slowly than those from heavier ions, and they are more readily erased by thermal annealing [105, 106].

The interpretation of nuclear track data is model dependent "due to erosion of grains by impact and sputtering, or to fading of the tracks over long intervals of time" [107]. This erosion is of the order of one angstrom per year, by sputtering, and one millimeter per million years by micrometeorite impact. Thus, only those surfaces that escape the sputtering and erosion processes yield meaningful data about exposure ages. The intensity of the irradiation is shown by these few uneroded surfaces. A millimeter-sized crystal at the bottom of a vug from an Apollo 15 rock [108], estimated to have been uncovered and exposed about 2×10^4 years ago, has a track density of 5×10^{10} tracks/cm^2 on its surface, and some soil grains that have escaped sputtering have surface track densities of 10^{11}/cm^2. Unshielded rock and grain surfaces have track densities typically about 10^8 tracks/cm^2. Because of experimental difficulties, the error in track counts from different investigators is on the order of $\pm20\%$ [105].

The cosmic-ray-produced tracks need to be distinguished from fission tracks due to ^{238}U and to ^{244}Pu. High uranium concentrations can contribute more than 10^8 fission tracks/cm^2 at grain boundaries or in inclusions. Their non-uniform distribution, differing etch pit characteristics, and the low fission track density in common minerals (where uranium is low in abundance) enable them to be distinguished from the cosmic ray tracks. Tracks due to the spontaneous fission of ^{238}U are commonly observed, but those from ^{244}Pu (half-life of 82×10^6 yr.) are difficult to find in lunar rocks. Most of the mare basalts are too young to contain measurable amounts of decay products. A slightly more favorable environment exists in uranium-rich mineral phases in the older lunar rocks. Tracks due to ^{244}Pu fission occur in a whitlockite crystal from the Fra Mauro breccias (14321) [109]. This sample was heavily shielded from cosmic rays until about 25 million years ago.

The possible existence of superheavy elements has been much debated. The principal evidence adduced was the presence of long fission tracks. These >20 micron tracks, which were identified as being due to extinct superheavy elements, are probably "fresh" tracks due to cosmic ray nuclei. There is no definitive evidence for the existence of the superheavy elements around Z = 114 [107].

Despite all the complications, the track record produces very useful results. A principal conclusion in this section is that the galactic cosmic ray flux has been effectively constant over the past 50 million years. It becomes difficult to extrapolate this conclusion much beyond this age from the track data because of the increasingly important effects of erosion [110].

4.9 Chemical and Isotopic Effects

The reaction of the lunar surface materials with the solar and galactic radiation causes many changes in chemical abundances. Hydrogen in lunar soils is nearly devoid of deuterium, which is destroyed in nuclear reactions in the sun. Accordingly, the hydrogen is of solar, not lunar (or terrestrial) origin [111].

4.9.1 Carbon, Nitrogen and Sulfur

The overall average carbon content of lunar rocks is 30 ppm, but the soils average 115 ppm [112, 113]. Of this, about 5–10 ppm may be accounted for from meteoritic or cometary sources and the remainder is derived from direct implantation by the solar wind. The finest grain sizes contain the highest concentrations, and this "surface correlated" effect is usually considered to be evidence of an extra-lunar component. However, this concept should not be applied too widely, since the surfaces of many grains are coated with glass splashes. Unlike the rare gases, carbon is reactive and this has made study of the isotope ratios difficult on account of fractionation effects.

Nitrogen, like carbon, is mainly of extra-lunar origin. The average concentration in igneous rocks is less than one ppm [114, 115]. Higher values [116, 117] are due to atmospheric contamination. The soils range from 50 to 100 ppm and average 82 ppm. The low content of nitrogen in lunar rocks is consistent with the overall depletion of the Moon in volatile elements. Nitrogen appears to be retained more quantitatively than the noble gases, H or C [118] and is thus a better measure of the integrated solar wind flux. A significant change in the isotopic composition has occurred, with $^{15}N/^{14}N$ decreasing at a rate of 15% per billion years. The only viable hypothesis appears to be that of a secular change in the solar wind [119, 120].

Sulfur is relatively abundant in lunar rocks, averaging about 1000 ppm, where it is present as troilite (FeS). An extra-lunar component has been identified in the soils, which have about the same abundance levels [116, 121]. The isotopic fractionation of sulfur in lunar soils indicates relative enrichment in the heavy isotopes [122] similar to that of the heavy isotopes of oxygen,

Table 4.7 Comparisons of rare gas contents (cm³ STP/g) and elemental ratios of regolith fines, the solar wind and the terrestrial atmosphere.

Site	^{4}He 10^{-2}	^{20}Ne 10^{-4}	^{36}Ar 10^{-4}	^{84}Kr 10^{-8}	^{132}Xe 10^{-8}	$(^4$He$/^{20}$Ne$)$	$(^{20}$Ne$/^{36}$Ar$)$	$(^{36}$Ar$/^{84}$Kr$)$ $\times10^3$	$(^{84}$Kr$/^{132}$Xe$)$
Apollo 11	11–25	20–31	3.3–4.1	16–38	2.1–10	55–104	5.2–9.2	1.0–2.2	2.1–8.6
Apollo 12	4–38	7–61	1.2–3.1	4–20	1.1–2.6	55–72	4.9–6.6	1.2–3.1	3.5–8.5
Luna 16	18	34	5.4	22	8.5	53	6.3	2.5	2.6
Apollo 14	5–9	9–16	2.4–4.4	9–24	1.4–4.6	52–64	3.0–4.2	1.6–2.8	3.6–12
Apollo 15	4–10	7–22	0.9–4.1	4.4–24	0.6–3.3	38–58	4.4–7.7	1.5–2.9	4.8–8.1
Apollo 16	0.6–5.1	2.4–13	1.3–6.0	4.5–34	1–6.5	26–50	1.6–3.2	1.5–3.2	2.3–7.2
Luna 20	3.81	10.1	2.88	10.9	2.25	32–39	2.4–3.5	2.6–3.0	4.8–6.6
Apollo 17 ("mare")	12–29	14–45	1.6–6.2	3.7–16	1.3–2.4	53–117	4.9–9.3	2.2–4.3	2.8–6.5
Apollo 17 ("highland")	5.9–16	12–28	2.5–4.9	10–18	1.8–2.9	40–64	4.4–5.8	2.4–2.8	5.4–7.2
Solar wind composition experiment	—	—	—	—	—	570 ± 70	28 ± 9	—	—
Terrestrial atmosphere	—	—	—	—	—	0.3	0.5	—	—

Sources: Walton, J. R., et al. (1973) *PLC 4*: 2086.
Heymann, D. (1977) *PCE* 10: 47.
Bogard, D. D., and Nyquist, L. E. (1972) *PLC 3*: 1804.
Geiss, J., et al. (1972) NASA SP 315, 14-1.

silicon and potassium. Sputtering by micrometeorite bombardment is usually considered as the dominating process.

4.9.2 The Rare Gases

The concentration of rare gases in the lunar fines is typically in the range 0.1–1.0 cm^3 STP/g, corresponding to about 10^{19}–10^{20} atoms/cm^3. These are very large amounts and were one of the surprises encountered during the Apollo 11 Preliminary Examination Team study. The detailed composition of the gases is given in Tables 4.7 and 4.8 [123].

Table 4.8 Isotopic composition of trapped rare gases in the lunar regolith compared with solar wind and terrestrial atmospheric values.

	^{4}He/^{3}He	^{4}He/^{20}Ne	^{20}Ne/^{22}Ne	^{22}Ne/^{21}Ne	^{20}Ne/^{36}Ar
Bulk lunar fines	2300 – 2800	96 ± 18	12.4 – 12.8	31 ± 1.2	
Solar wind	2350 ± 120	570 ± 70	13.7 ± 0.03	30 ± 4	28 ± 9
Terrestrial atmosphere	7 × 10^5	0.3	9.8 ± 0.08	34.5 ± 1.0	0.5

Sources: Bogard, D. D., and Nyquist, L. E. (1972) *PLC 3*: 1804.
Geiss, J., et al. (1972) NASA SP 315, 14-1.
Heymann, D. (1977) *PCE* 10: 49.
Eberhardt, P., et al. (1972) *PLC 3:* 1821.

The location of such large volumes of gas is intriguing. The gas occurs in bubbles, typically 50–100 Å in diameter in soil grains, under high (~5000 atm) pressures [108]. The penetration of the solar wind as shown by surface etching is on the order of 1000 Å, while the solar flare ions penetrate to millimeter depths. The rare gas contents increase with decreasing grain size.

On the Moon, there is no evidence of rare gases that might be residual from a primitive atmosphere. This is consistent with the general depletion of volatile elements, and, based on the low abundances of elements such as Pb and Tl, it may safely be concluded that the primordial rare gas content of the Moon was effectively zero and that any primitive isotopic signature is swamped by the later effects.

Thus, the large quantities of rare gases found, particularly in the lunar soils, must be of secondary origin. The rare-gas inventory records a complex variety of origins. Although these relate to differing processes (e.g., solar wind trapping, cosmic ray interaction, fission products, and radioactive decay), it is convenient and customary to treat the rare-gas studies as a group, which reflects the historical development of the studies.

There are several distinct sources for the rare gases observed in lunar surface samples. These include the following:

(a) trapped solar wind ions;

(b) implanted solar flare ions;

(c) isotopes produced by solar proton spallation and by high energy cosmic ray interactions (e.g., ^{3}He, ^{21}Ne, ^{38}Ar);

(d) isotopes from radioactive decay (e.g., ^{40}Ar from ^{40}K);

(e) fission-produced isotopes (e.g., ^{131}Xe from ^{244}Pu fission);

(f) neutron capture produced isotopes (e.g., ^{131}Xe from ^{130}Ba).

Trapped He and Ne are richer in soils from the maria compared to those of the highlands. This effect appears to be due to the superior trapping efficiency of iron- and titanium-bearing minerals (e.g., ilmenite compared with plagioclase [123]). He and Ne are generally depleted relative to the heavier gases (Ar, Kr and Xe) compared to solar abundances.

4.9.3 The Argon-40 Anomaly

There are about two orders of magnitude excess of ^{40}Ar in the lunar soils compared with the amount expected from solar wind trapping and ^{40}K decay. Early suggestions were that ^{40}Ar atoms, resulting from the ^{40}K decay, escape from the lunar surface and are ionized by solar radiation [124, 125]. After ionization, the particles are then subject to rapid acceleration by the magnetic fields associated with the solar wind and strike the lunar surface with energies of 100–1000 keV within a few seconds. At low energies, they are neutralized but not trapped, and hence recycled. Ions with impact energies greater than 1 keV are trapped in the fines, thus building up an excess of ^{40}Ar.

The apparent absence of excess ^{40}Ar in near-surface regions of grains, and the apparently low amounts of ^{40}Ar released at low temperatures, led to alternative suggestions that potassium, volatilized during impact, coated the grain surface and produced the excess ^{40}Ar [126, 127].

It has also been argued that most of the excess argon was implanted 3–4 aeons ago and that calculations based on the present-day flux, solar wind, and magnetic fields are not relevant [128]. However, it is probable that the initial suggestions with regard to the mechanism for trapping argon are correct [124, 125].

The meteoritic gas-rich breccias greatly resemble lunar soil breccias [104, 130]. The exposure of meteorites to cosmic ray bombardment produces cosmogenic (= spallogenic) noble gases, especially ^{3}He, ^{21}Ne and ^{38}Ar, which may be used to calculate exposure ages, assuming a production rate of 2.39×10^{-8} ^{3}He STP/g/million years (see Section 4.10). The significant differences in rare gas contents among the atmospheres of the Earth, Venus and Mars are discussed in Section 4.13 [130].

4.9.4 Cosmogenic Radionuclides

A wide variety of nuclides are produced by the solar flare and galactic cosmic ray interaction with the lunar surface [106, 107, 131], ranging from ^{52}Mn with a half-life of 5.6 days to the geologically more useful ^{53}Mn with a half-life of 3.7 million years (Table 4.9). At the surface, the solar flare protons account for most production, but below about one centimeter, secondary particles from the galactic cosmic rays dominate. There is accordingly a strong depth dependence. There were no observable differences at the Apollo 16 site between samples collected from the Cayley and the Descartes Formations. This result is consistent with the absence of chemical variation observed at the site. Work on some of the larger rocks shows that the contribution from galactic cosmic rays can be separated at depth from the solar cosmic radiation [132, 133].

Production of ^{236}U (half-life of 2.34×10^7 yr.) and ^{237}Np (half-life of 2.14×10^6 yr.) by solar proton reactions with ^{238}U have been reported [134]. The difference in the half-lives of ^{236}U and ^{237}Np provides a monitor of solar cosmic ray activity by comparison of the ratios. Although no direct evidence of plutonium was found, the fission track evidence in uranium-rich minerals such as whitlockite points to its presence [109].

4.9.5 Thermal Neutron Flux

The low-energy thermal neutrons produced by the cosmic ray bombardment result in anomalous isotopic abundances for those nuclides with large thermal neutron capture cross sections. Those most commonly measured are ^{158}Gd (from ^{157}Gd), ^{156}Gd (from ^{155}Gd), and ^{150}Sm (from ^{149}Sm). The neutron flux with depth was measured by a neutron probe, using the ^{10}B capture rate and ^{235}U induced fission rate, on the Apollo 17 mission [135, 136]. The neutron fluxes measured on the returned samples are low compared with those expected for a well-mixed regolith; the data are consistent with non-uniform mixing, in that the material toward the base has undergone more irradiation. Since the depth of penetration of neutrons is of the order of a meter or so, this is a reasonable conclusion, consistent with regolith formation models which add fresh layers of material to the top.

4.10 Exposure Ages and Erosion Rates

The record from nuclear tracks and the rare gas data can be employed to determine exposure ages of rocks and soils, providing valuable evidence for

Table 4.9a Radionuclides commonly detected in lunar samples and meteorites.

Radionuclide[†]	Half-life (yr.)	Targets
^{3}H	12.33	O, Mg, Si
^{10}Be	1.6×10^6	O, Mg, Si
^{14}C	5730	O, Mg, Si
^{22}Na	2.60	Mg, Al, Si
^{26}Al	7.3×10^5	Al, Si
^{36}Cl	3.0×10^5	Ca, Fe
^{37}Ar	0.095	Ca, Fe
^{39}Ar	269	K, Ca, Fe
^{46}Sc	0.23	Ti, Fe
^{48}V	0.044	Ti, Fe
^{53}Mn	3.7×10^6	Fe
^{54}Mn	0.86	Fe
^{55}Fe	2.7	Fe
^{56}Co	0.215	Fe
^{59}Ni	8×10^4	Fe, Ni
^{60}Co	5.27	Co, Ni
^{81}Kr	2.1×10^5	Sr, Y, Zr

[†] ^{10}Be, ^{36}Cl, ^{39}Ar, ^{46}Sc, and ^{60}Co are produced mainly by high energy galactic cosmic rays: ^{56}Co principally by solar cosmic radiation the remainder are produced by both type of cosmic rays.

Sources: Reedy, R. C. (1980) *Ancient Sun,* p. 370.
    Lal, D. (1972) *Space Sci. Rev.* 14: 25.

Table 4.9b Some long-lived radionuclides, potentially useful in cosmic ray studies.

Radionuclide	Half-life (yr.)
^{42}Ar	33
^{44}Ti	47
^{63}Ni	100
^{32}Si	~300
^{91}Nb	~800
^{93}Mo	$\sim 3.5 \times 10^3$
^{94}Nb	2.0×10^4
^{41}Ca	1.3×10^5
^{233}U	1.6×10^5
^{60}Fe	$\sim 3 \times 10^5$
^{237}Np	2.1×10^6
^{129}I	1.6×10^7
^{236}U	2.3×10^7
^{92}Nb	3.3×10^7
^{146}Sm	1.0×10^8
^{40}K	1.28×10^9

Sources: Reedy, R. C. (1980) *Ancient Sun,* p. 370.
    Lal, D. (1972) *Space Sci. Rev.* 14: 25.

lunar chronology. A very full discussion of the problems of interpretation of the irradiation record in this context is given by Burnett and Woolum [137].

4.10.1   Nuclear Track Data

Much information on exposure ages and regolith history has been obtained from the track studies. Track production at shallow depths (< 0.5 cm) is dominated by solar flares, while at greater depths, galactic cosmic ray tracks predominate [105, 106]. The "sun tan" exposure time is defined as the period during which a rock was at the surface of the Moon (usually less than 3 million years), while the "sub-decimeter" exposure time is the period during which a rock was within 10 cm of the surface (typically 1–100 million years).

High track densities, indicating surface exposure, are found throughout the core samples. The early interpretations were of a high rate of regolith turnover, but it quickly became clear that the regolith was stratified, and models involving deposition of thin layers, stirring of the surface, and burial by subsequent layers were adopted and so "throwout" models became preferred to "gardening" models [138]. This became clear from the study of the Apollo 15 deep core which contained forty-two strata, in which high track densities were distributed throughout the 242-cm core length and were independent of depth.

Many micron-sized particles with very high track densities have an amorphous surface layer from the solar bombardment. Each layer in the core contains grains throughout that were once at the "very surface," indicating good mixing of each layer before burial. Stirring takes place to a depth of a few centimeters. Deeper stirring is rare. Instead, overturning of layers by cratering occurs. Surface layers seem to survive for periods of from 1 to 50 million years before burial, as indicated by cratering studies (Section 4.3.2).

Exposure ages of rocks lying on the surface have been measured (Table 4.10). Most (80%) have a complex exposure history. Erosion by the micrometeorite bombardment and solar wind sputtering exposes fresh surfaces at rates of about one millimeter per million years.

Information about the exposure of old highland breccia fragments may be obtained from the track data, but this is offset by annealing due to heating during breccia formation. Because of this, the use of mare basalt rocks or individual grains in the regolith is preferred in cosmic ray studies. The tracks are largely inherited from the original parent materials which must have had exposure at the very top surface. Grains with track densities of $\geq 10^8/cm^2$ occur within the breccia matrix. The track densities correlate with metamorphic grade. Low track counts due to thermal annealing have occurred [139]. Tracks are absent in the well-recrystallized breccias (e.g., 15418, 61016, 66055, 67015) [140]. Tracks vanish first from the glasses during annealing, then

Table 4.10 Surface residence times for rocks apparently exposed in one orientation.

Rock	Mass (g)	Surface Residence Time (m.y.)
12018	787	1.7
12038	746	1.3
12022	1864	~10
62295	251	2.7
74275	1493	2.8
67915	*	50[†]
68815	*	2[†]
76315	*	21[†]

* Boulder chip.
[†] Surface residence time = total cosmic ray exposure time.
From Burnett, D. S., and Woolum, D. S. (1977) *PCE* 10: 87.

generally from olivine, pyroxenes, and feldspar in that order with increasing temperature.

There are a number of similarities between the lunar breccias and the gas-rich meteorites (e.g., Kapoeta, Fayetteville), which also contain grains exposed before incorporation in the meteorite with track densities of $\geq 10^{10}/cm^2$. Preserved solar flare tracks are seen even in the most recrystallized chondrites (e.g., LL6 grade [140, 141]), suggesting that even the most metamorphosed chondritic meteorites correspond only to the less metamorphosed lunar breccias [142]. The data seem consistent with the formation of the gas-rich meteorites in a regolith-type environment less extreme than that of the lunar surface, possibly on asteroid surfaces [140]. In this context, observations of the asteroid Toro suggest that it has a lunar-type regolith [143].

4.10.2 Rare Gas Data

The exposure history of the regolith and of surface rocks was one of the first applications of the rare gas studies. Of the various age methods, the most reliable are ^{81}Kr-^{83}Kr, ^{21}Ne, and ^{38}Ar ages [144]. The ^3He ages are subject to loss of He by diffusion. A full review of the data is provided by Burnett and Woolum [137] with appropriate cautionary tales.

The impact events dated by the ^{81}Kr technique are given in Table 4.11. The dating, of course, depends on the proper sampling of the ejecta blanket. Much confusion arose over the age of South Ray (Apollo 16), one of the freshest craters on the lunar surface. The ages of about 2 million years quoted in Table 4.11 are from rocks, and are in accord with the geological criteria [146]. Many of the soils, interpreted as South Ray ejecta, are far removed from the crater rim. South Ray is 640 m in diameter. The astronaut traverses came no closer than about 3 km, and fine ejecta from South Ray at about five

Table 4.11 Ages of lunar craters from ^{81}Kr studies.

	^{81}Kr Age (m.y.)	Reference*
Apollo 14: Cone Crater	25	1
Apollo 16: North Ray Crater	50	1
South Ray Crater	2.0	1
Apollo 17: Camelot Crater	90	2
Central cluster†	109	2
Shorty Crater	19	3

†If the central cluster of craters at Taurus-Littrow and the bright mantle landslide are due to secondaries from Tycho, this age provides a date for Tycho.

*References:
1. Burnett, D. S., and Woolum, D. S. (1977) *PCE* 10: 76.
2. Drozd, R. J., et al. (1977) *PLC 8:* 3027.
3. Eugster, O., et al. (1977) *PLC 8:* 3059.

crater diameters distant may not have been sampled [146]. Based on this interpretation, the astronauts sampled only scattered blocks from South Ray. A possible explanation is that most of the light-colored ejecta at this distance from the South Ray crater rim is derived from old thick regolith at the target site. Accordingly, it preserves an "older" age than that of the South Ray event (F. Hörz, pers. comm., 1981).

An equally interesting set of dates was obtained at the Apollo 17 site, where a site-wide event apparently occurred at about 100 million years ago [147]. This is equated with secondaries from Tycho, which accordingly provides one date for that event. The age of Camelot crater, thought to be older from photogeological evidence, is not distinguishable. A reevaluation of the photography has reconciled the data [148].

The exposure ages of regolith soils from the various missions typically show averages of about 400 million years with much spread in the data. Individual soil grains have been measured with ages up to 1.7 aeons [145]. In general, the exposure ages of rocks are complex, with ages ranging from 1 to 700 million years. A list of surface residence times of rocks which were apparently only exposed in one orientation is given in Table 4.10. The polymict brecciated chondrites, which may have originated in a young asteroidal regolith, have exposure ages less than about 50 million years [130].

4.11 Solar and Cosmic Ray History

The long-term stability of the sun is one of the fundamental scientific questions. Many locations on the lunar surface have not changed on a macroscale for 3 to 4 billion years, in great contrast to the terrestrial surface.

Accordingly, the lunar surface provides us with areas that have been exposed to solar radiation, although the record is complex in detail and complicated by little-understood problems, both within the sun, and on the lunar surface. Nevertheless, a useful amount of information is available and a picture is beginning to emerge of the history of the sun. A comprehensive account of the present state of knowledge is given in *The Ancient Sun: Fossil Record in the Earth, Moon and Meteorites* [149]. Although the detailed history of the sun lies outside the scope of this book, a synopsis of the present understanding, particularly as gathered from the lunar data, is useful. Superimposed on the short-term solar variability, perhaps best known through the 11- and 22-year cycles, are longer scale variations of which the best established is the Maunder Minimum period of reduced solar activity, when no sunspots, for example, were observed from about 1645 to 1715 A.D. [150].

The direct terrestrial record extends back several thousand years, using ^{14}C in trees, varves and ^{10}Be in polar ice [151–153]. However, the solar record in the lunar samples provides the possibility, in principle, of extending the record back to more than 4 billion years ago, within a few hundred million years of the formation of the sun. Comparative studies of microcraters and of solar flare tracks indicate that there is no convincing evidence for a change in the dust flux or the solar particle flux for the past 10^4–10^6 years. The long term Fe/H ratio is the same as the modern ratio [154] although some contrary opinions exist as to whether there was an episode of higher solar flare activity 2×10^4 years ago [155].

The solar flare and galactic cosmic ray fluxes in lunar samples and meteorites provide some very useful constraints on the history of the ancient sun [102]. Solar flare activity was present at least 4.2 aeons ago from the meteoritic record [156, 157]. The energy spectrum of the solar flare particles has not changed appreciably in the last 4 aeons. Thus, the track density profiles in ancient lunar and meteoritic samples is close to that recorded on a glass filter exposed at the Surveyor 3 site for three years (1967–70) and recovered during the Apollo 12 mission [158] (Fig 1.3). The enrichment of heavy particles in solar flares at low energies, relative to solar photosphere compositions, appears also in ancient flare records. There does not seem to be any good evidence for a higher incidence of solar activity during the early history of the Moon, as recorded in the lunar samples.

Evidence for the constancy of the galactic cosmic ray flux, composition of cosmic rays and of the energy spectrum may likewise be deduced from the track record, for the past 50 million years. The rare gas record [98] for the solar wind is similar for both present and ancient times, and the isotopic composition of Ar, Kr and Xe appear to have been invariant over the past 3–4 aeons [159]. This is in striking contrast to the record for nitrogen isotopic variations. The ^{15}N/^{14}N ratio has changed at a rate of 150 per mil per aeon for the past 2.5 aeons at least. This effect must reflect changes in the solar photosphere or corona, since there is probably no indigenous nitrogen in the Moon (see Section 4.9.1). At present there is no theoretical explanation for this remarkable observation [118, 119].

Apart from this interesting fact, the overall impression is of relatively uniform solar conditions at least for 4 aeons. Unfortunately, there is not space here to discuss the interesting discrepancy between the geological evidence for the presence of liquid water on the surface of the Earth at 3.8 aeons [160] and theoretical models for solar evolution, which predict a solar luminosity of only about 70% of present day values [161]. Possibly, an early atmospheric greenhouse effect operated on the Earth if the astrophysical calculations are correct.

4.12   The Lunar Atmosphere

During the Apollo 17 mission, mass spectrometric measurements were made of the tenuous lunar atmosphere [162–165], which is, in fact, a collisionless gas. The primary components detected result from either the solar wind or from radioactive decay, and degassing of the spacecraft. Hydrogen, helium, neon and argon have been detected. The abundances are compatible with a solar wind source, except for ^{40}Ar, derived from radioactive decay of ^{40}K. During the lunar night, the temperature falls below $100°K$, and the components from the degassing of the spacecraft mainly condense, thus allowing measurement of the indigenous components. Argon shows a predawn enhancement, consistent with the release of this condensible gas at the sunrise terminator. No components due to any volcanic activity have been detected and there is no evidence of any contribution to the present atmosphere from transient events, unless they are connected with degassing of ^{40}Ar.

A proposal of the existence of a lunar atmosphere 5×10^7 times the present atmosphere in the past 10^8 yr., deduced from the micrometeorite record [166], does not seem to be in accord with most of the other evidence [165].

4.13   Rare Gases and Planetary Atmospheres

The atmospheric compositions of Mars [167] and Venus [168, 169] have provided several surprises and important constraints on models of planetary evolution [131]. The rare or noble gases are separated into those produced at least partly by radiogenic decay (e.g., ^4He, ^{40}Ar, ^{129}Xe) and the primordial rare gases present in the solar nebula. Commonly measured isotopic species include ^3He, ^4He, ^{20}Ne, ^{36}Ar, ^{38}Ar, ^{84}Kr and ^{132}Xe. It is conventional to divide

Table 4.12 Rare gas contents of the atmospheres of Venus, Earth and Mars.

	^{36}Ar (Atmosphere) g/g	$^{40}Ar/^{36}Ar$	$^{20}Ne/^{22}Ne$	$^{20}Ne/^{36}Ar$
Venus	2.4×10^{-9}	1.0	14	0.3
Earth	3.5×10^{-11}	296	9.8	0.52
Mars	2.0×10^{-13}	3000	—	0.38
Solar Wind	—	< 1	13	28

the primordial rare gases into "solar" and " planetary." The "solar" rare gas composition is that derived from the sun via the solar wind. These abundances are well represented in the lunar soils (see Section 4.9.2). The "planetary" rare gas component is that observed in meteorites and is accordingly thought to represent the relative isotopic and elemental abundances in the solar nebula before accretion of the planets.

The rare gas compositions of the atmospheres of Venus and Mars given in Table 4.12 provide interesting new data. The principal results are that Venus has about 70 times the abundance of Ne and Ar observed in the terrestrial atmosphere, while Mars has much lower abundances than the Earth. These results are the reverse of what might have been expected from simple condensation models for planetary formation. The isotopic composition of neon in Venus, as shown by the $^{20}Ne/^{22}Ne$ ratios, is higher than that of the Earth, and is similar to the values observed in the solar wind. This indicates a differing source for neon in the Earth and Venus, once thought to be very similar planets. The abundance of ^{36}Ar in the Martian atmosphere is so low (about 180 times less than that of the Earth) that it is probably not due to a low rate of degassing. Evidence of this is given by the abundance of ^{40}Ar, derived from ^{40}K, in the Martian atmosphere. Mars probably has a low abundance of potassium (Table 8.5) and calculations based on $^{40}Ar/^{36}Ar$ ratios indicate that the bulk ^{36}Ar content of Mars is about an order of magnitude less than that of the Earth.

These data have caused us to revise our ideas on planetary evolution. There is not space here to discuss the many theories advanced to account for these data (see refs. [130] and [170–173]) but some tentative conclusions may be drawn. The high abundances of neon and argon in Venus may represent an early solar wind contribution to small planetesimals which accrete to form Venus. The low abundances of the planetary rare gases in Mars are likewise a primitive feature, due to the accretion of Mars from planetesimals depleted in neon, argon, krypton, and xenon. The new rare gas data lend support to the concept of planetary accretion of a heterogeneous collection of planetesimals (see Section 9.13).

4.14   Organic Geochemistry and Exobiology

A complete account of the organic geochemistry investigations on lunar samples was given previously [174]. Little further work has been done on this subject and the interested reader is referred to the previous summary.

The account of the search for life in the lunar samples is given in the same reference [174] and need not be repeated here. It should be noted that the absence of life on the Moon was in accordance with the prediction of C. Huyghens in 1757 that "the Moon has no air or atmosphere surrounding it as we have, [and I] cannot imagine how any plants or animals whose whole nourishment comes from fluid bodies, can thrive in a dry, waterless, parched soil" [175]. The search for life on Mars has been dealt with exhaustively in other references [176, 177] and will not be treated here. The effective absence of organic compounds at the parts per billion level in the Viking organic mass spectrometric experiment seems decisive [178]. This contrasts with the development of life on Earth [179].

References and Notes

1. Hood, L. L., and Schubert, G. (1980) *Science*. 208: 49.
2. Thompson, T. W., et al. (1974) *Moon*. 10: 87; (1981) *Icarus*. 46: 201; See also Pettengill, G. H. (1978) *Ann. Rev. Astron. Astrophys*. 16: 265 for an extended review of radar observations of planets and satellites. See also Schultz, P. H., and Mendell, W. (1978) *PLC 9*: 2857 for a discussion of orbital infrared observations.
3. Keihm, S. J., et al. (1973) *EPSL*. 19: 337; (1973) *PLC 4*: 2503; Strangway, D. W., and Olhoeft, G. R. (1977) *Phil. Trans. Roy. Soc*. A285: 441.
4. Olhoeft, G. R., et al. (1975) *PLC 6*: 3333.
5. See Tang, C. H., et al. (1977) *JGR*. 82: 4305 for surface electrical properties of Mars.
6. See Hess, S. L., et al. (1977) *JGR*. 82: 4559 for a description of Martian meteorology.
7. See Moore, H. J., et al. (1977) *JGR*. 82: 4497 for a description of Martian surface properties.
8. Saari, J. M. (1964) *Icarus*. 3: 161.
9. Mendell, W. W., and Low, F. J. (1970) *JGR*. 75: 3319.
10. Gold, T. (1971) *PLC 2*: 2675.
11. The extreme case is the presence of the layer of "light gray fines" (12033) at Apollo 12, a KREEP-rich layer of exotic origin.
12. The XRF experiment samples only the top few microns of the surface. The gamma-ray experiment looks somewhat deeper (10–20 cm). Adler, I., et al. (1973) *PLC 4*: 2783; Metzger, A. E., et al. (1973) *Science*. 179: 800.
13. McCoy, J. E., and Criswell, D. R. (1974) *PLC 5*: 2991.
14. Berg, O. E. (1978) *EPSL*. 39: 377.
15. Hörz, F. (1973) NASA SP 315, 7–24.
16. Gold, T. (1970) *Icarus*. 12: 360; (1971) *Apollo 14 PSR*, p. 239.
17. A major source of information on the lunar regolith is given in *The Moon* (1975), Vol. 13, p. 1–359, which contains the proceedings of a conference on the lunar regolith held

at the Lunar Science Institute, Houston, Texas, November 1974. See also an excellent review by Langevin, Y., and Arnold, J. R., (1977) *Ann. Rev. Earth Planet. Sci.* 5: 499. See also Regolith Conference Abstracts, LPI. Nov. 1981.

18. NASA SP 289 (1972) 5–23.
19. Cooper, M. R., et al. (1974) *Rev. Geophys. Space Phys.* 12: 291.
20. Peeples, W. J., et al. (1978) *JGR.* 83: 3459.
21. Freeman, F. J. (1981) USGS Prof. Paper 1048, Chap. F.
22. Eberhardt, P. (1973) *Moon.* 8: 104.
23. Houston, W. N., et al. (1974) *PLC 5*: 2361; Carrier, W. D., et al. (1973) *PLC 3*: 3213; Houston, W. N., et al., ibid., 3255; Mitchell, J. K., et al., ibid., 3235; Mitchell, J. K., et al. (1973) *PLC 4*: 2437.
24. Cherkasov, I. I., and Shvarev, V. V. (1975) *Lunar Soil Science* (Trans. N. Kaner), Keter Publishing House, Jerusalem, 170 pp.
25. Quaide, W., and Oberbeck, V. (1975) *Moon.* 13: 27.
26. Hörz, F. (1977) PCE 10: 3.
27. Nishiizumi, K., et al. (1979) *EPSL.* 44: 409.
28. Schmitt, R. A., and Laul, J. C. (1973) *Moon.* 8: 190.
29. Schonfeld, E., and Meyer, C. (1972) *PLC 3*: 1415.
30. Shoemaker, E. M., et al. (1971) *PLC 1*: 2399.
31. Wood, J. A., et al. (1970) *PLC 1*: 965.
32. Bhandari, N., et al. (1972) *PLC 3*: 2811.
33. Papanastassiou, D. A., and Wasserburg, G. J. (1970) *EPSL.* 8: 1, 269; (1971) *EPSL.* 11: 37, 12: 36; (1972) *EPSL.* 13: 368; 17: 52.
34. Cliff, R. A., et al. (1972) *JGR.* 77: 2007; Mark, R. K., et al. (1973) *PLC 4*: 1785; Murthy, V. R., et al. (1972) *PLC 3*: 1503.
35. Wetherill, G. W. (1971) *Science.* 173: 389.
36. The $^{87}Sr/^{86}Sr$ ratio of 0.69898 is the Basaltic Achondrite Best Initial (or BABI) ratio [33].
37. Nyquist, L. E., et al. (1973) *PLC 4*: 1839.
38. Schaal, R. B., and Hörz, F. (1980) *PLC 11*: 1679.
39. McKay, D. A., et al. (1970) *PLC 1*: 673.
40. Chao, E. C. T., et al. (1971) *JGR.* 75: 7445.
41. McKay, D. S., et al. (1972) *PLC 3*: 988.
42. McKay, D. S., et al. (1971) *PLC 2*: 755.
43. Taylor, G. J., et al. (1978) *PLC 9*: 1959.
44. Adams, J. B., and McCord, T. B. (1973) *PLC 4*: 163; Rhodes, J. M., et al. (1975) *PLC 6*: 2291.
45. Via, W. N., and Taylor, L. A. (1976) *PLC 7*: 393.
46. Papike, J. J. (1981) LPS XII: 805.
47. Stroube, W. B., et al. (1978) *Meteoritics.* 13: 201.
48. Taylor, S. R., and McLennan, S. M. (1979) *GCA.* 43: 1551.
49. Simonds, C. H., et al. (1978) *JGR.* 83: 2773.
50. Gooding, J. L., et al. (1980) *EPSL.* 50: 171.
51. Gooding, J. L., and Keil, K. (1981) *Meteoritics.* 16: 17.
52. Cloud, P. (1970) *PLC 1*: 1794.
53. Taylor, S. R. (1973) *Earth Sci. Rev.* 9: 101.
54. Roedder, E., and Weiblen, P. W. (1970) *PLC 1*: 801.
55. Glass, B. P. (1976) *PLC 7*: 679.
56. Glass, B. P., and Barlow, R. A. (1979) *Meteoritics.* 14: 55.
57. Mason, B. H. (1979) *Smith. Contrib. Earth Sci.* 22: 14.
58. Glass, B. P. (1972) *JGR.* 77: 7057; Frey, F. A. (1977) *EPSL.* 35: 43.

59. Shaw, H. F., and Wasserburg, G. J. (1981) LPS XII: 967.
60. Taylor, H. P., and Epstein, S. (1970) *PLC 1*: 613.
61. Schnetzler, C. C. (1970) *Meteoritics.* 5: 221.
62. O'Keefe, J. A. (1970) *Science.* 168: 1209.
63. Epstein, S., and Taylor, H. P. (1973) *PLC 4*: 1559.
64. Kozyrev, N. (1963) *Nature.* 198: 979.
65. Pike, R. J. (1980) USGS Prof. Paper 1046 C.
66. Chao, E. C. T., et al. (1962) *Science.* 135: 97; (1964) *GCA.* 28: 971.
67. Labotka, T. C., et al. (1980) *PLC 11*: 1285.
68. Laul, J. C., and Papike, J. J. (1980) *PLC 11*: 1307.
69. Rhodes, J. M. (1977) *Phil. Trans. Roy. Soc.* A285: 293.
70. Hörz, F. (1978) *PLC 9*: 3311.
71. Hubbard, N. J. (1979) *PLC 10*: 1753.
72. Laul, L. C., and Papike, J. J. (1980) *PLC 11*: 1307. It should be noted that both Ni and Co data reported in this study are probably contaminated through the use of rhodium plated nickel sieves.
73. Korotev, R. L., et al. (1980) *PLC 11*: 395.
74. Muehlberger, W. R., et al. (1980) *Lunar Highlands Crust*, p. 1.
75. Kempa, M. J., et al. (1980) *PLC 11*: 1341.
76. Heiken, G. (1975) *Rev. Geophys. Space Phys.* 13: 567.
77. Nagata, T., et al. (1974) *PLC 5*: 2815.
78. Cisowski, C. S., et al. (1974) *PLC 5*: 2841.
79. Chou, C. L., and Pearce, G. W. (1976) *PLC 7*: 779.
80. Morris, R. V. (1980) *PLC 11*: 1697.
81. Boynton, W. V., et al. (1976) *EPSL.* 29: 21.
82. Wëgmuller, F., et al. (1980) *PLC 11*: 1763.
83. Butler, P. (1978) *PLC 9*: 1459.
84. Cirlin, E. H., et al. (1978) *PLC 9*: 2049.
85. Cirlin, E. H., and Housley, R. M. (1979) *PLC 10*: 341; (1981) *PLC 12*: 529.
86. Gibson, E. K. (1977) PCE 10: 57.
87. Kurat, G., and Keil, K. (1972) *EPSL.* 14: 7.
88. Arnold, J. R. (1979) *JGR.* 84: 5659.
89. Fremlin, J. H. (1979) *Nature.* 278: 598.
90. Morgan, J. W., et al. (1977) NASA SP 370, 659.
91. Wasson, J. T., et al. (1975) *Moon.* 13: 121.
92. Schneider, E. (1973) *PLC 4*: 3277.
93. Hörz, F., et al. (1971) *JGR.* 76: 5770; Hartung, J. B., et al. (1973) *PLC 4*: 3213; Morrison, D. A., et al., ibid., 3235; Neukum, G., et al., ibid., 3255; Schneider, E., et al., ibid., 3277; Morrison, D. A., et al. (1972) *PLC 3*: 2767; Neukum, G., et al., ibid., 2793; Gault, D. E., et al., ibid., 2713; Hartung, J., et al., ibid., 2735; Hörz, F. (1975) *Planet. Space Sci.* 23: 151.
94. Frondel, C., et al. (1970) *PLC 1*: 445; Fredriksson, K., et al., ibid., 419.
95. Brownlee, D. E., et al. (1977) *PLC 8*: 149.
96. Hörz, F., et al. (1975) *PLC 5*: 2397.
97. Duennebier, F. K. (1976) *Science.* 192: 1000.
98. Borg, J. (1980) *Ancient Sun*, p. 431.
99. Pepin, R. O. (1980) *Ancient Sun*, p. 411.
100. Marti, K. (1980) *Ancient Sun*, p. 423.
101. Walker, R. M. (1975) *Ann. Rev. Earth Planet. Sci.* 3: 99.
102. Crozaz, G. (1977) PCE 10: 197.
103. Crozaz, G. (1980) *Ancient Sun*, p. 331.

104. Macdougall, J. D., et al. (1974) *Science*. 183: 73.
105. Price, R. B., Fleisher, R. L., and Walker, R. M. (1975) *Nuclear Tracks*, Univ. Calif. Press.
106. Lal, D. (1972) *Space Sci. Rev*. 14: 45.
107. Herrman, G. (1979) *Nature*. 280: 543.
108. Phakey, P. P., et al. (1972) *PLC 3*: 2905.
109. Crozaz, G., et al. (1972) *PLC 3*: 1623.
110. Crozaz, G. (1980) *Ancient Sun*, p. 311.
111. Epstein, S., and Taylor, H. P. (1970) *PLC 1*: 1085.
112. Eglinton, G., et al. (1974) *Topics in Current Chemistry*. 44: 88.
113. Pillinger, C. T. (1979) PCE 11: 61.
114. Becker, R. H., and Clayton, R. N. (1975) *PLC 6*: 2131.
115. Becker, R. H., et al. (1977) *PLC 8*: 3685.
116. Gibson, E. K. (1977) PCE 10: 57.
117. Muller, O. (1979) PCE 11: 47.
118. Clayton, R. N., and Thiemens, N. H. (1980) *Ancient Sun*, p. 463.
119. Kerridge, J. F. (1980) *Ancient Sun*, p. 475.
120. Goel, P. S., et al. (1975) *GCA*. 39: 1347.
121. Kerridge, J. F., et al. (1975) *PLC 6*: 2151.
122. Thode, H. G., and Rees, C. E. (1979) *PLC 10*: 1629.
123. Heymann, D. (1977) PCE 10: 45.
124. Heymann, D., and Yaniv, A. (1970) *PLC 1*: 1261.
125. Eberhardt, P., et al. (1970) *PLC 1*: 1037.
126. Baur, H., et al. (1972) *PLC 3*: 1947.
127. Signer, P., et al. (1977) *Phil. Trans. Roy. Soc.* A285: 385.
128. Yaniv, A., and Heymann, D. (1972) *PLC 3*: 1967.
129. Schultz, L. (1979) PCE 11: 39.
130. McElroy, M. B., and Prather, M. J. (1981) *Nature*. 293: 535. This article contains a review of the noble gases in the terrestrial planets.
131. Reedy, R. C. (1980) *Ancient Sun*, p. 370.
132. Wahlen, M., et al. (1972) *PLC 3*: 1719.
133. Reedy, R. C., and Arnold, J. R. (1972) *JGR*. 77: 537.
134. Fields, P. R., et al. (1973) *PLC 4*: 2123.
135. Burnett, D. S., and Woolum, D. S. (1974) *PLC 5*: 2061.
136. Russ, G. P. (1973) *EPSL*. 16: 275.
137. Burnett, D. S., and Woolum, D. S. (1977) PCE 10: 63.
138. Arrhenius, G., et al. (1971) *PLC 2*: 2583.
139. Hart, H. R., et al. (1972) *PLC 3*: 2831; Hutcheon, I. D., et al., ibid., 2845, 2863; Dran, J. C., et al. (1972) *PLC 3*: 2883; Yuhas, D. E., et al. (1972) *PLC 3*: 2941.
140. Macdougall, J. D., et al. (1973) *PLC 4*: 2319.
141. Van Schmus, W. R., and Wood, J. A. (1967) *GCA*. 31: 747.
142. Crozaz, G., et al. (1974) *PLC 5*: 2475.
143. Dunlap, J. L., et al. (1973) *Astron. J*. 78: 491.
144. Marti, K. (1967) *Phys. Rev. Lett*. 18: 264; Lugmair, G. W., and Marti, K. (1972) *PLC 3*: 1891; Marti, K., et al. (1973) *PLC 4*: 2037.
145. Kirsten, T., et al. (1972) *PLC 3*: 1865.
146. McKay, D. S., and Heiken, G. H. (1973) *PLC 4*: 41.
147. Drozd, R. J., et al. (1977) *PLC 8*: 3027.
148. Lucchitta, B. K. (1979) *Icarus*. 37: 46.
149. *The Ancient Sun: Fossil Record in the Earth, Moon and Meteorites* (1980) (eds.,Pepin, R. O., et al.), Pergamon, 581 pp.

150. Those readers with a historical bent will reflect on the curious fact that the Maunder Minimum of solar activity 1645–1715, coincided with the life of the Sun King (Louis XIV, 1638–1715), a correlation no doubt of use to Francophiles and Francophobes alike.

151. Stuiver, M., and Grootes, P. M. (1980) *Ancient Sun*, p. 165.

152. Fairhill, A. W., and Yorg, I. C. (1980) *Ancient Sun*, p. 175.

153. Raisbeck, G. M., and Yiou, F. (1980) *Ancient Sun*, p. 185.

154. Zinner, E. (1980) *Ancient Sun*, p. 201.

155. Zook, H. A. (1980) *Ancient Sun*, p. 245.

156. Macdougall, J. D., and Kothari, B. K. (1976) *EPSL*. 33: 36.

157. Goswami, J. N. (1980) *Ancient Sun*, p. 347.

158. Crozaz, G., and Walker, R. M. (1971) *Science*. 171: 1237.

159. Marti, K. (1980) *Ancient Sun*, p. 423.

160. Windley, B. F. (1977) *The Evolving Continents*, Wiley.

161. Newkirk, G. (1980) *Ancient Sun*, p. 293.

162. NASA SP-330.

163. Hodges, R. R., et al. (1973) *PLC 4*: 2854; (1974) *Icarus*. 21: 415.

164. Hoffman, J. H., et al. (1973) *PLC 4*: 2865.

165. Hodges, R. R., et al. (1974) *Icarus*. 21: 415 gives a comprehensive review of the lunar atmosphere.

166. Chernyak, Y. B. (1978) *Nature*. 273: 497.

167. McElroy, M. B., et al. (1977) *JGR*. 82: 4379.

168. Hoffman, J. H., et al. (1980) *JGR*. 85: 7882.

169. Donahue, T. M., et al. (1981) *GRL*. In press.

170. Anders, E., and Owen, T. (1977) *Science*. 198: 453.

171. Pollack, J. B., and Black, D. C. (1979) *Science*. 205: 56; Pollack, J. B., and Yung, Y. L. (1980) *Ann Rev. Earth Planet. Sci.* 8: 425.

172. Wetherill, G. W. (1981) *Icarus*. In press.

173. Hostetler, C. J. (1981) *PLC 12*: 1387.

174. Taylor, S. R. (1975) *Lunar Science: A Post-Apollo View*, Pergamon, N.Y., p. 110–114.

175. Huyghens, C. (1757) *Cosmotheoros*.

176. The biological experiments are described in (1977) *JGR*. 82: 4659–4677.

177. An extended and readable account is given by H. S. F. Cooper (1980) in *The Search for Life on Mars*, Dial Books.

178. Biemann, K., et al. (1977) *JGR*. 82: 4641.

179. Schopf, J. W., ed. (1982) *Origin and Evolution of the Earth's Earliest Biosphere*, Princeton Univ. Press.

Chapter 5

PLANETARY CRUSTS

The terrestrial planets have developed crusts that differ in chemical composition from their interiors, and from their bulk composition. This was remarked upon long ago for the Earth. This process shows remarkable diversity among those planets and satellites of which we have detailed knowledge. When the lunar samples were examined, the crustal composition of the Moon was found to be so highly differentiated that models involving heterogeneous accretion were invoked, plastering on a layer of refractory material as the last episode of forming the Moon. Such models were in direct contrast to earlier views that the Moon might be a primitive object. The surface of Mars, Mercury and Venus likewise turned out to be different from reasonable estimates of their bulk compositions and accordingly the study of planetary crusts received impetus. Crustal development is an expression of planetary differentiation, which in turn is driven by mass, volatile content, radioactive heat sources, initial accretion energy and many other factors. Crusts may develop early or grow slowly through time. In the examples with which we are familiar—Earth, Moon, Mars, Mercury, Venus, Galilean and Saturnian satellites (in decreasing order of knowledge)—their crustal compositions depart, sometimes to an extreme degree, from our concepts of solar nebula compositions.

Models involving the late accretion of differentiated material have fallen into disfavor. It has now become clear that such surficial crusts on planets may arise in two ways, either as a consequence of early melting and differentiation, or by derivation from the planetary mantles by partial melting long after accretion, this time being measured in billions of years. The highland

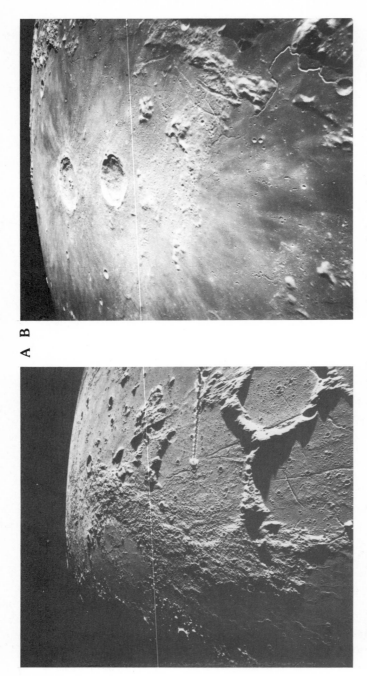

5.1 The four Apollo landing sites where the lunar highland crust was directly sampled. **a.** Apollo 14, on the ejecta blanket from Imbrium. The spacecraft boom points to the center of Fra Mauro crater, 95 km diameter. The landing site was northwest of the crater rim (NASA AS-16-1420). **b.** Apollo 15 site, showing Hadley Rille, 1 km wide, adjacent to the Apennines Mountains. The prominent craters are Autolycus (diameter, 39 km) and Aristillus (55 km diameter)(NASA AS 15–1537).

C D

c. The Apollo 16 site showing the smooth plains of the Cayley formation to the left of the landing site and the hilly Descartes formation to the right (east). The bright white dot is South Ray Crater just south of the landing site. The diameter of Dollond B, the large crater floored with Cayley formation, in the northwest quadrant, is 37 km in diameter (NASA A–16 metric 539, 540). **d.** The Apollo 17 landing site in the Taurus-Littrow Valley, bottom left. The landslide from the South Massif is clearly visible. Littrow is the large degraded crater (31 km diameter) north-northeast of the landing site. Mare Serenitatis occupies the left side of the picture. The large crater, north center embayed and flooded with basalt, is Le Monnier, site of Lunokhod 2 mission. The large crater beyond Le Monnier is Posidonius, 95 km in diameter. Note the dark basalts near the Apollo 17 site, and the wrinkle ridges and rilles in the mare.

crust of the Moon represents a well-studied example of the first type, while the continental crust of the Earth is a familiar example of the second process.

The continental crust of the Earth, as distinct from the oceanic crust, is uniquely useful in providing a platform on which the later stages of evolution could occur and from which we can study the development of the planets [1]. Our understanding of the composition and evolution of the Earth's crust is slowly growing. It is an interesting commentary that we understand the formation of the lunar crust to a rather better degree. Accordingly, in this chapter, a large section is devoted to the highland crust of the Moon, both because it differs in nearly all respects from the continental crust of the Earth and because it provides us with a key to understanding the crusts on other planets.

The section on the continental crust of the Earth is brief, since adequate extended accounts are available (e.g., [2]). It would be a disservice to the reader to attempt a summary of these, and only some highlights are noted with an emphasis on chemical composition and crustal evolution which are especially relevant here.

5.1   The Lunar Highland Crust

The highland crust forms the oldest accessible area on the Moon and is saturated with ringed basins and large craters (Chapter 3). All of the original crustal structure has been obliterated. The complexity of the highland samples constitutes a severe test for the scientific method of inductive reasoning, proceeding from the details to construct a general theory. Workers on these samples, which have endured countless meteorite impacts, must guard against the charms of the deductive approach, deriving the details from a general theory in the manner of medieval scholars, before the strength of the inductive approach was realized. "Reality cannot enter . . . as an afterthought. Either it is shown due respect at the start, or it forces its way into the picture later, taking revenge on those who tried to ignore it" [3]. It is possible that the surface which we sampled dates from about 4.2–4.3 aeons (Fig. 5.1), the earliest age at which the growth of a crust solidifying from the magma ocean could outstrip destruction by the intense bombardment. The debate over the question of the ultimate origin of the highland crust, late-accreted refractory addition or derived by differentiation from within the planet, was resolved principally by geochemical data, in favor of the latter alternative.

5.1.1   Thickness and Density

Haines and Metzger [4] have used the orbital geochemical values, the lunar sample and the seismic data, to derive values for highland crustal thickness and density. Crustal densities are calculated using the new orbital values for Fe and Mg [5]. Knowing the average elevation, and assuming isostatic equilibrium, the densities provide estimates of crustal thickness for thirty-five highland regions.

The average weighted density for the lunar highland crust obtained in this manner is 2.933 ± 0.007 g/cm^3 [4]. The near-side highlands have an average density of 2.95 g/cm^3. The far-side highlands are less dense, averaging 2.92 g/cm^3. The weighted average thickness is 73.4 ± 1.1 km, with the near-side crust averaging 64 km (with a lower Th content) while the far-side highlands average 86 km in thickness. The near-side highlands are thus 22 km thinner than the average far-side highlands. Recent revisions of the seismic data may indicate a thickness of 50 km for the near-side crust (Y. Nakamura, pers. comm., 1981). Accordingly, the average lunar crust may be closer to 60 km in thickness. This comprises 10% of lunar volume and is the conservative value adopted in most geochemical balance calculations (see Sections 5.9 and 8.4). The geochemical data indicate that lunar isostasy is controlled mainly by crustal thickness rather than by density (Fig. 5.2). The iron content is of prime importance due to its abundance (6.5–7.0% FeO) and to its high atomic weight. Haines and Metzger [4] make a number of clearly stated assumptions: (a) The crust is uniform with depth, which is one interpretation of the seismic data, and (b) the composition of the regolith is the same as that of the underlying crust. This implies that meteoritic contamination, lateral transport, and movement of very fine soil particles do not affect this assumption. Metzger noted that any correlation between elevation and chemical composition is evidence that the composition of the regolith, which was sampled by the orbital geochemical experiment, is linked to that of the crust " . . . the correspondence between chemistry, elevation, and crustal thickness provides direct proof that much of the highland surface is representative of the underlying crust" [5]. A striking observation is that all large unfilled highland basins have Fe contents similar to that of the surrounding highlands [6], which confirms crustal uniformity to the depths to which the basins have sampled. It is further assumed that (c) the upper 20 km is fractured and has a density 0.93 times that of solid rock at depth, lunar gravity being too low to compress rock at that depth [7]. P-wave velocities are less than 6 km/sec. down to 25 km, but increase to 6.8 km/sec. over about the next kilometer and then slowly increase to 7 km/sec. at 60 km depth. The reduced seismic velocities in the upper 25 km are ascribed to fracturing and brecciation.

The abrupt velocity increase at 25 km is here interpreted to indicate an absence of microcracks below that depth. The very slow increase in velocity from 25 to 60 km (only about 0.2 km/sec) is evidence of a uniform material; this velocity (~7 km/sec) is close to the intrinsic velocity of lunar samples. A final assumption that (d) the highlands are isostatically compensated is suggested by elevation and gravity data [7–9], although possibly the crustal mass

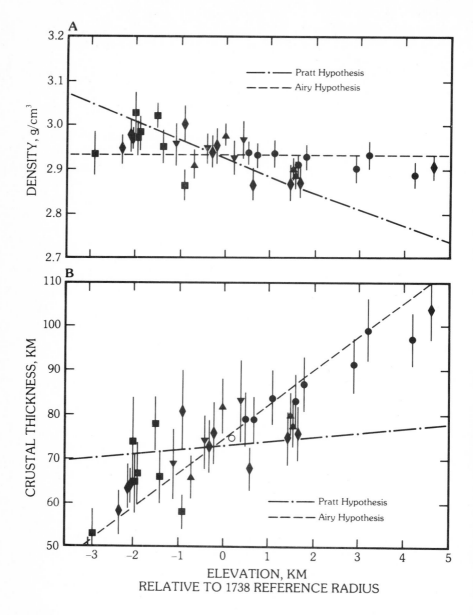

5.2a Plot of crustal density versus elevation for 33 highland regions. Trend lines are shown for the Pratt and Airy theories. Data fit the latter [4]. **b** Plot of crustal thickness versus elevation of 33 highland regions, showing that the data fit the Airy hypothesis better then the Pratt hypothesis [4].

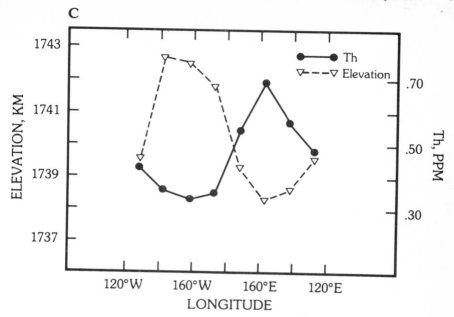

5.2c The lunar thorium concentration from orbital gamma-ray data shows an inverse correlation with elevation [5].

is not fully compensated [10]. If the crust is not compensated fully, then the crustal thickness increases for uncompensated mass deficiencies, or thins for uncompensated mass excesses [10].

Figure 5.2 shows the density and thickness versus elevation relationships for the highlands. Elevation is related to thickness, more than to density, supporting the Airy hypothesis of isostatic adjustment [11]. The alternative Pratt hypothesis [12] states that variations in elevation are due to differences in density of adjacent blocks. The geophysical data for elevation, gravity, or seismic velocity do not distinguish between these two hypotheses, nor do the data require that either hypothesis account for the lunar case. The orbital geochemical data provide a test for the hypotheses and suggest that the difference in crustal thickness is responsible for most of the differences in elevation, with only a minor contribution from the density variations. The variation in highland crustal thickness between near and far-sides is possibly a relic of early convection processes in a cooling and crystallizing magma ocean (see Section 5.11). It is considered generally to be the main contributor to the center of figure-center of mass offset [13] (Section 7.3).

The problems of studying the lunar highland crust and other early planetary surfaces center on the effects of the great bombardment. How has

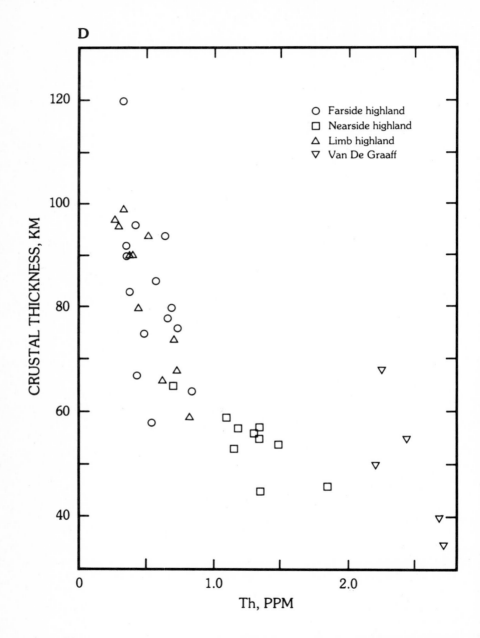

5.2d Thorium concentrations versus crustal thickness for far-side, limb, and near-side highland regions including the regions in and around Van de Graaff [5]. (Courtesy A. L. Metzger.)

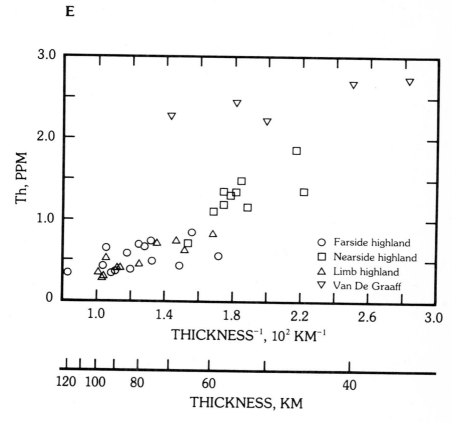

E

5.2e Data in Fig. 5.2d replotted against the *inverse* of the crustal thickness. If equal amounts of Th were incorporated into each crustal block and were similarly distributed with depth through each block, the relationship would show a straight line with an intercept of zero Th at infinite thickness [5]. The data approximate this theoretical relationship.

this affected the chemistry, mineralogy, petrology and age of the samples? Is any primitive crust left and, if so, can we identify samples from it? A basic question is how deeply is the crust affected by the bombardment, and from what depth do we have samples. This question was discussed in Section 3.17 from which it appears that some samples at least came from depths of from 12–32 km, and that the great basin collisions excavated to depths of 30–60 km, with the latter figure being more likely. Another effect of the formation of multi-ring basins is that mantle plugs may form beneath the centers of the basins. Such a hypothesis is required to account for the mascons (Section

7.4.2). Accordingly, the base of the highland crust may be highly irregular due to this effect, and mantle material may be excavated by subsequent impacts.

5.1.2 The Megaregolith

The present 10-m-thick regolith in the highlands overlies a zone of brecciation and fracturing due to the intense early bombardment of the highland crust. All surfaces older than about 3.8 aeons are saturated with craters 50–100 km in diameter. Most estimates for the thickness of rubble produced during these events fall within the range of 1–3 km. Some of the older mare surfaces also have been affected, and many of the early, now buried, lava flows are probably much broken up by the declining stages of the bombardment. Such fracturing would account for much of the scattering of seismic signals observed in the upper 25 km and concentrated in the upper 2 km.

A detailed model study of the evolution of the megaregolith over the earth-facing side of the Moon [14] predicts an average thickness of about 2 km, although thicknesses are less than 1 km over about 50% of this area. This study has an important corollary in that it provides evidence that the deep megaregolith will affect crater geometry by providing a two-layer structure in the highlands.

A distinct but related question deals with brecciation of the highland crust beneath the multi-ring basins. This possibly extends to depths of 20–25 km, accounting for the seismic velocity data. Additional evidence for a deep megaregolith in the highlands comes from a study of blocky craters. Fresh highland and mare craters exceeding 12 km in diameter have similar infrared and radar signatures [15], but mare craters less than 12 km in diameter exhibit anomalies which are correlated with the presence of fresh rubble ejecta, consistent with the excavation of coherent basaltic rock layers at shallow depths. In contrast, such infrared and radar signatures were less conspicuous on fresh highland craters, consistent with ejecta blankets of previously pulverized material. From the observation that highland craters greater than 12-km diameter have blocky rims, a value of 2-km thickness for the megaregolith results, which is consistent with other estimates.

This ancient megaregolith might be expected to contain some of the oldest accessible fragments of highland crust. The feldspathic fragmental or "light matrix" breccias, excavated by North Ray crater at the Descartes site might represent such material. Indeed, there is support for this view from geochronology. Very old ^{40}Ar–^{39}Ar ages (4.25 aeons) are reported for breccias [16]. The overlapping ejecta blankets probably account for the layering observed in the highlands, such as that exposed at Silver Spur and observed in other areas [17, 18].

5.2 Breccias

The possible complexities in materials subjected to repeated cratering episodes were fully revealed in the samples returned by the Apollo missions. Over 60% of the samples returned from the highlands of the Moon are breccias, most of the remainder being classified as impact melt rocks. Major efforts have been expended by petrologists and geochemists to decipher the record preserved in these breccias. The complexity of the topic, where several generations of breccias may occur within a single hand specimen, frequently accompanied by glass and remelted crystalline fragments, might daunt the most accomplished petrologist. Further problems arise as to the geological location of the material. Which of the Apollo 14 samples from Fra Mauro represent Imbrium basin ejecta and which locally derived material? Is the Apollo 15 Apennine Front composed of Serenitatis ejecta? Are the Taurus-Littrow samples from Apollo 17, on the rim of the Serenitatis basin, ejecta from that event [19] or are they derived from a complex multiple-impact sequence of events? The Apollo 16 Descartes site is located within several old barely recognizable craters, but the surface material is almost certainly younger, related to either the Nectaris or Imbrium collisions. The Cayley Formation could represent Imbrium-derived material, while the Descartes Mountains might be primarily derived from the Nectaris basin [20] or be Imbrium material piled against the Kant Plateau, or be dominated by older material ploughed up by Imbrium, Serenitatis, Crisium, Nectaris (or even Orientale) secondary projectiles.

Were the ejecta blankets hot or cold? The Apollo 14 samples were first interpreted as being derived from a hot ejecta blanket. A current view is that the percentage of Imbrium derived debris at the Apollo 14 landing site is 10–20%[20] but such interpretations are heavily based on terrestrial analogies. Nevertheless, melts from the basin impacts appear to be widespread, and occur, for example, at a distance of 630 km from the Cordillera Scarp of the Orientale basin (Section 3.9). Thus, although most of the ejecta blanket is probably at low temperature, the melted material may be more widely distributed than believed earlier, perhaps as pods or lenses within the cooler mass. As in so many scientific debates where the data are at the limit of resolution, there is some truth on either side. These questions are addressed throughout this chapter in an attempt to reach a consensus.

An important advance has been the creation of a classification system of lunar breccias [21], given in Table 5.1. Soil breccias, which form part of the classification, are discussed in Section 4.3.5. The incredible diversity and complexity of lunar highland breccias is well reflected in the terminology. A list of previous classifications, too long to reproduce, is given by Stöffler et al. ([21], Table 5, p. 62). This useful table correlates the previous terminology with that adopted here [21, 22] (Table 5.1). The production of breccias in

Table 5.1a Highland breccia classification.[†]

Breccia Group	Breccia Class	Main Textural Characteristics	Typical Examples
Monomict	Cataclastic rock	intergranular in-situ brecciation of a single lithology	60225, 65015, 67955, 72415, 78527
	Metamorphic (recrystallized) Cataclastic rock	intergranular in-situ brecciation of a single lithology and partial recrystallization	67955
Dimict	Dimict breccia	intrusive-like, veined texture of very fine-grained crystallized melt breccia within coarse-grained plutonic or metamorphic rock types	61015, 62255, 64475
Polymict	Regolith breccia or soil breccia	clastic regolith constituents including glass spherules with brown vesiculated matrix glass	10018, 14313, 15205
	Fragmental breccia	rock clasts in a porous clastic matrix of fine-grained rock debris (mineral clasts) with or without melt particles	14063, 14082, 67115, 67455, 67475
	(Crystalline) melt breccia or impact melt breccia	rock and mineral clasts in an igneous-textured matrix (granular, ophitic, subophitic, porphyritic, poikilitic, dendritic, fibrous, sheaf-like etc.); may be clast-rich or clast-poor	14310, 14312, 15455, 62235, 62295, 68415, 78235
	(Impact) glass or glassy melt breccia	rock and mineral clasts in a coherent glassy or partially devitrified matrix, with or without clasts	60095, 61016, 72215, 77135
	Granulitic breccia	rock and mineral clasts in a granoblastic to poikiloblastic matrix	15418, 67955, 73215, 77017, 78155, 79215

[†] Adapted from Stöffler, D., et al. (1980) *Lunar Highlands Crust*, p. 54, 57 (Tables 1 and 2).

Table 5.1b Shock metamorphic characteristics of lunar highland breccias.[†]

Rock Type	Degree of Shock	Estimated Peak Pressure (kbars)	Textural Characteristics
Non-porous rocks (igneous and metamorphic rocks, monomict breccias, dimict breccias, crystalline melt breccias, impact glass, granulitic breccias)	unshocked	50	primary rock texture and intragranular texture of mineral grains unchanged
	weakly shocked	5–300	intragranular fracturing and mosaic texture in all mineral constituents; planar deformation structures in feldspar pyroxene and olivine; primary rock texture unchanged
	strongly shocked	300–450	mosaic texture and planar deformation structures in mafic minerals; plagioclase changed to maskelynite, diaplectic or thetomorphic glass; primary rock texture unchanged
Porous rocks (regolith breccias, fragmental breccias)	unshocked (friable)	0–30	highly porous aggregation of breccia constituents
	shocked (coherent)	>30 mostly >100	moderate, small or lacking porosity in the matrix; mosaicism and fracturing of minerals; intergranular melt

[†]Adapted from Stöffler, D., et al. (1980) *Lunar Highlands Crust*, p. 59 (Table 4).

GEOLOGIC SETTING OF IMPACT BRECCIAS

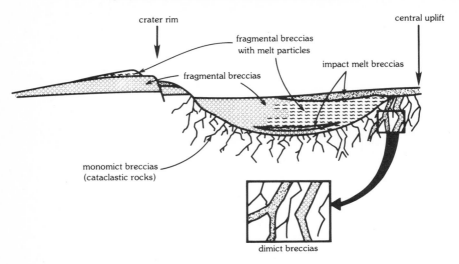

5.3 Schematic cross-section through a complex central peak crater after the crater modification stage. (Courtesy D. Stöffler, *Workshop on Apollo 16*. LPI Tech. Report No. 81-01.)

5.4a Monomict cataclastic anorthosite (lunar sample 65015,16; crossed polarizers, width of field: 0.20 mm); note intergranular brecciation [22] (Courtesy D. Stöffler.)

relation to a single impact event is shown in Fig. 5.3 [22] (see also Fig. 3.3); this work owes much to the study of terrestrial impact craters. On the Moon, the complexity of the breccias can be simulated by superimposing the situation shown in Fig. 5.3 many times at many different scales. All possible variations exist, but several distinct types may be recognized.

5.2.1 Monomict Breccias

These are composed dominantly of one rock type and are characterized by intergranular brecciation, resulting from *in situ* brecciation of single rock

5.4b Monomict cataclastic anorthosite, 15415,29 (Genesis Rock), crossed polarizers, 3 × 4 mm. (Courtesy G. Ryder.)

types, probably in the basal regions of impact events (Fig. 5.3). They are best developed in coarse grained rocks, and frequently have been referred to as cataclastic rocks (Figs. 5.4 and 5.5). These breccias probably form in craters larger than about 100 m in diameter [22] and commonly occur as clasts in breccias, lenses, and ejecta blankets. Some have been affected by later thermal metamorphism which has caused some recrystallization. This is inferred to have occurred by subsequent heating when the clasts are incorporated in hot impact melts, either in the initial, or subsequent impact events. Examples of monomict breccias include samples 65015 (anorthosite), 67955, 72415 (dunite) and 78527.

5.2.2   Dimict Breccias

As the name implies, these breccias consist dominantly of two distinct components. The most impressive examples are formed of white anorthosites and dark melt rocks. The resulting breccias, often called "black and white" rocks, are among the most striking of the large samples returned from the Moon (Figs. 5.6 and 5.7).

5.5   The Apollo 17 troctolite 76535, a monomict breccia (NASA S 73-19640).

5.6 Dimict breccia 61015, with grey, intrusive material of aphanitic, crystalline matrix texture and white brecciated anorthositic "country rock." Younger "pseudo-tachylite" veins (black and grey with black rim) occur on two sides of the sample. Width of field = 12.5 mm [22]. (Courtesy D. Stöffler.)

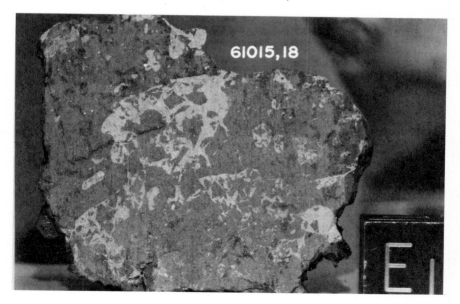

5.7 Dimict breccia 61015,18. Cube is 1 cm on side (NASA S.75 20878).

These rocks have also been referred to as "dike breccias," and probably formed by the intrusion of melt rocks into weakly shocked solid rocks. A probable location is within or near the central uplift regions of large complex craters (Fig. 5.3). The melt portions of dimict breccias appear to be rather uniform, at least for the Apollo 16 examples [23], possibly indicating formation in a single impact.

5.2.3 Feldspathic Fragmental Breccias

These lunar rocks are polymict, with several different types of clasts, set in a friable porous white feldspathic matrix. Dark melt rocks are common among the clasts, along with fragments of minerals and glass (Figs. 5.8 and 5.9). These breccias tend to be rather weakly coherent, due to the clastic matrix.

The melt-free varieties are inferred to form as shocked crater ejecta in the upper portions of the target stratigraphy, while those examples containing melted material (equivalent to the Ries crater suevite which is an impact-produced breccia containing both shock-metamorphosed rock fragments and

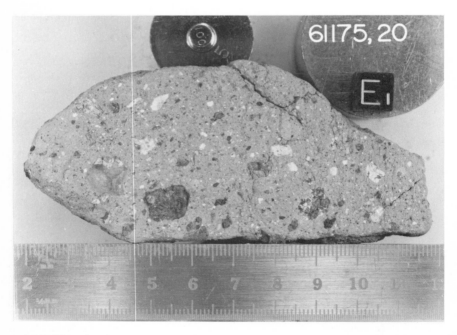

5.8 Feldspathic fragmental breccia 61175, showing a wide variety of clast types (NASA S78-31342).

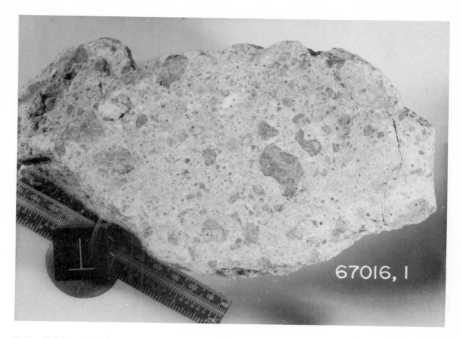

5.9 Feldspathic fragmental breccia 67016 (NASA S.81-26051).

glass bombs) come from the lower portions of the pre-crater stratigraphy. They occur in simple craters mainly as ejecta sheets outside the crater rim, or as fall-back breccia within the crater (Fig. 5.3) and form in craters larger than a few hundred meters in diameter. Examples of these include the white breccia boulders from the North Ray crater (Apollo 16: 67455, 67475). The "white rocks" at Cone crater (Apollo 14: 14063, 14082) are examples of breccias with some admixed melt material, while the Apollo 16 sample 67016 is a particularly good analogy to terrestrial suevite [24]. There is a very large compositional and textural variation in the clast populations of these breccias, indicative of rather low energy conditions during their formation.

5.2.4 Impact Melt Breccias

This class comprises the most abundant type of samples returned from the lunar highlands. Typically they consist of rock and mineral fragments set in a crystalline matrix, which has the texture of an igneous rock (Figs. 5.10 and 5.11). These rocks result from the crystallization of an impact melt laden with rock fragments resembling some of the melt-rich suevites, being transitional between the feldspathic fragmental breccias and the true impact glasses. The

5.10a, b   Highland impact melt rock 77135 showing vesicular structure (NASA S.72-56387, S.72-56391).

melt breccias are inferred to form in the deeper central part of large craters (greater than 1–5 km diameter) by shock melting, more or less below the impact point (Fig. 5.3). Various textural subtypes are recognized [25] (e.g., granular: 72215; ophitic: 61016, 62295; poikilitic: 77135).

5.11a Thin section of impact melt 68415,126. Field of view 4 × 3 mm crossed polarizers. (Courtesy G. Ryder.)

The further class of impact melts and glassy melt breccias is dealt with in Sections 3.8, 4.4.2 and 5.3. These samples comprise clast-free glasses generally occurring as clasts or coatings of breccias, in addition to their occurrence in the regolith as glasses and agglutinates (Sections 4.4.1, 4.4.2). The impact melt breccias grade into the true clast-free melt rocks (Section 5.3) and the separation of these classes is partly semantic.

5.11b This section of impact melt rock 77135, 7 showing vesicle. Field of view 3 × 4 mm, crossed polarizers. (Courtesy G. Ryder.)

5.2.5 Granulitic Breccias

Granulitic breccias are abundant in the sample collections. They have been heated to near-melting temperatures; some are partly melted and all have recrystallized, but retain rock fragments or relics to indicate their original formation as breccias of various classes. The clasts are usually

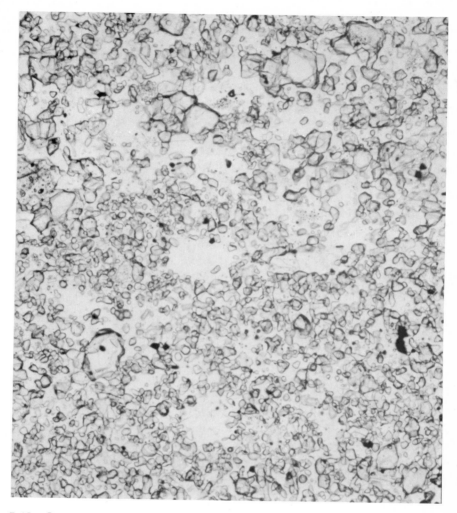

5.12 Granulitic breccia 79215,98 showing typical granoblastic or mosaic texture. Width of field 1 mm (plane polarized light). (Courtesy G. Ryder.)

recrystallized along with the matrix, which is commonly feldspathic (Fig. 5.12). These breccias have formed by thermally induced recrystallization of earlier-formed breccias at temperatures up to 900°C [26]. Many have been granulated and recrystallized several times, so that they contain breccias within breccias. Some of their compositions may approach that of the bulk composition of the crust due to this extensive reworking. Examples include

15418, 67955, 73215, 77017, 78155 and 79215. Two sources for the heat necessary to cause the metamorphic recrystallization have been suggested: (1) global metamorphism (the so-called Apollonian metamorphism) [27, 28] and (2) high temperatures from nearby lenses of impact melt trapped within the breccias [29].

Judging by terrestrial analogies, hot ejecta blankets are rare [30]. The Bunte breccia does not appear to have exceeded temperatures of 500–600° C [31]. The granulitic breccias, in contrast, appear to have been at 1000° C for an extended period, estimated by Stewart as 1–10 m.y. [27]. The implications for a crustal thermal event are considerable. It must be early, and if this hypothesis is correct, these rocks are probably the earliest fragments of crust to survive the great meteorite bombardment, and may date back to 4.2 aeons or earlier. The alternative explanation of slow cooling of local patches of impact melt avoids the global connotations of Apollonian metamorphism.

5.2.6 Basalts in Highland Breccias

Occasional fragments of basaltic composition have been returned from the highlands. Their presence has raised the question of the existence and duration of basaltic volcanism before the termination of the catastrophic bombardment at 3.8 aeons. The composition of these basalts is somewhat different from most of the common species returned from the maria proper. They are higher in Al_2O_3 (11–14%) and have lower iron contents. Their REE patterns resemble those of the VLT basalts from Luna 24 [32], and they exhibit variable Eu anomalies. Examples include 14053, 14072 and clasts from breccias 14063, 14321 and 73255 [33].

The ages of these basalts appear to be in the range of about 3.9–4.0 aeons, more or less contemporaneous with the final great meteoritic collisions (see Section 6.4.1). If volcanism had been as widespread in pre-Imbrian as in post-Imbrian time, then basaltic clasts might be expected to constitute at least one percent of the highland sample return. Their scarcity, as well as the geochemical evidence, argues against extensive pre-3.8 aeon basaltic volcanism.

Since the volcanic activity is concentrated at the surface, this estimate of the likely frequency of basaltic clasts is a minimum value. Extensive sub-crustal sill and dike formation might be expected also to contribute clasts to breccias. Scenarios which call for increased early basaltic volcanism, of which the visible mare basalts constitute the declining phase [34], should contribute even more basaltic clasts.

If the present visible amount of mare basalt were mixed by cratering into the highland crust, it would not produce a distinctive geochemical signature. The highland crust contains Mg, Cr, V and other elements which are enriched in mare basalts. One scenario would be to account entirely for these mafic components by admixture, through cratering, of appropriate amounts of mare basalt. Various constraints have been noted which make this scenario unlikely [34, 35] and a vast amount of very early (pre-4.2 aeons) basaltic volcanism, comprising about 20% of the volume of the highland crust would be needed. Isotopic evidence for this event is lacking.

5.3   Melt Rocks

A distinctive suite of igneous-appearing rocks is also present in the Apollo collections from the highlands. These may be chemically distinguished from the mare basalts, principally Al_2O_3/CaO ratios, which are a factor of two higher in the highland samples. A suite of these has been selected as a "reference suite" as part of the NASA Basaltic Volcanism Study Project [36, 37], which covers the whole range of lunar highland compositions. A debate ensued over whether these rocks were of primary igneous origin or crystallized from melts generated by meteorite impact. The presence of clasts, the frequent evidence of high concentrations of siderophile elements, and the bulk compositions all tended to favor the latter view. They are, in general, medium to dark grey in color, usually fine-grained and occasionally vesicular. They may be glassy or wholly crystalline (e.g., sample 14310). Some are clast-laden microbreccias and hence would be classified as impact melt breccias (Section 5.2.4). The fragment-laden melt rocks appear to have formed mainly between about 4.05 and 3.85 aeons, and, therefore, appear to be younger than the granulitic breccias. Examples of rocks with true igneous textures (clast-free impact melt rocks) include 14310, 68415 and 78235 [38–42]. (See Section 5.4.4 for a discussion of "KREEP basalts.")

5.4   Highland Crustal Components

The previous sections classify the physical nature of the rocks sampled by the Apollo missions. Before going on to specify the compositions and discuss petrogenesis, it is necessary to consider the broader aspects of the problem. The Moon, as a planet, is simple enough to make broad scale overviews more profitable than on complex planets, a view with much historical truth. A basic premise in this approach is that the major element chemistry is not seriously affected by the impact process and that the dominant effect is not partial melting, but rather is impact-induced melting and mixing.

A major controversy has centered around questions of the components making up the highland crust. It was clear from the time of the Apollo 14 and 15 missions that many of the breccia compositions could be resolved into mixtures of two or three components. The Apollo 16 sample return aided greatly in this interpretation. The most useful key in unravelling the composition of the battered lunar highland crust has been the fact that the chemistry of the samples has survived the impact process, whereas most information on texture and other petrological and mineralogical indexes has been destroyed by the bombardment. A trap awaited the too-ready use of chemical composition of glasses as indicative of "primary" rock types. Impacts produce glasses. Chemical fractionation or partial melting mostly does not occur, so that the frequency of glass compositions can provide an index of the relative abundance of rock types. This approach was used by the "Apollo Soil Survey." However, the impact process is an extraordinarily efficient mixer, as shown, for example, at large terrestrial impact sites such as Manicougan [43, 44]. The effect of this process on the Moon has been to produce some average crustal compositions which have been mistaken for primary rock types; e.g., highland basalt, with about 26% Al_2O_3, is close in composition to the average highland crust (Section 5.9) [45]. Thus, the compositions due to mixing must be carefully disentangled from the original or pristine highland crustal compositions. An additional artifact of this sort has been the identification of "low-K Fra Mauro basalt" (LKFM) as a primary rock type. No unambiguous rock of this composition has been returned [46]. Once this complicating factor was understood, lunar highland crustal rock types became simpler and easier to fit into a rational petrologic scheme. A primary piece of evidence comes from the REE abundances and distribution patterns in highland breccias.

The REE patterns for most lunar highland samples are sub-parallel [47–50]. La/Yb ranges from 1.3 to 3.4. Strong negative and positive Eu anomalies are superimposed on these otherwise very regular patterns. As Haskin [50] comments, "So little change suggests a relatively simple genetic relationship among highland rocks." The range in the absolute abundances of the LIL elements is very large (>4000). As with many other lunar problems, the geochemistry of the samples provides the decisive clues. The two basic REE patterns which dominate highland crustal REE geochemistry are: (a) the anorthosite pattern, with La/Yb about 3 and Eu/Eu* about 20 [51], and the so-called KREEP pattern with La/Yb about 2 and Eu/Eu* about 0.10. Mixing of these two patterns accounts for nearly all the intermediate patterns.

A third component is required to provide the concentrations of Mg, Cr, etc., observed in the highland breccias. This is variously considered to be derived from an initial frozen crust, from infalling late accreting planetesimals or from trapped liquids within the crystallizing crust. Mare basalts cannot account for it, and are "probably only a minor component" [52]. Wänke et al. [53] considered that it was a primary chondritic component, but Ryder [52] examined the evidence and stated that "There is no basis for believing that the Mg-component is a primary chondrite-related non-lunar material." Taylor and co-workers considered it to be either a frozen crust or due to infalling planetesimals of lunar composition [47]. Plots of Ti/Sm [54] serve to distin-

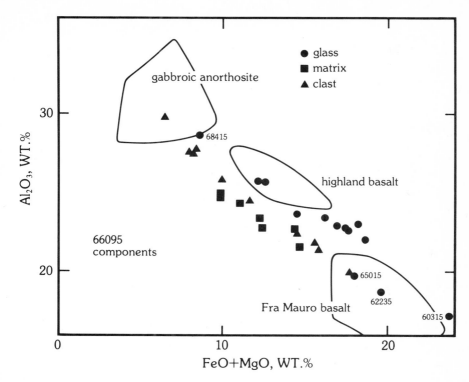

5.13 Plot of Al$_2$O$_3$ versus FeO + MgO showing the compositions of 66095 matrix, vein glass, and clasts relative to the compositions of rocks and glasses of Fra Mauro basalts, highland basalts, and gabbroic anorthosites [55]. (Courtesy L. A. Taylor.)

guish it clearly from chondritic material, thus removing the possibility of deriving the high Mg content from that source. The nickel and other sidero-phile element abundances are too low to make chondritic meteorites more than a trivial contributor.

Further insight into the difficult problem of the nature of the highland crustal components has been provided by a study of the "rusty" rock 66095, which is a polymict breccia. It provides within its mass of 1185 g a reasonable representative sampling of the highland crust. Bulk compositions approximate that of the soil at the Descartes site. On an Al$_2$O$_3$ versus FeO + MgO diagram, individual clasts, glasses, and matrix correspond to gabbroic anorthosite, "highland basalt," and Fra Mauro basalts (Fig. 5.13). The REE patterns range from less than 10 times chondritic, with Eu enrichment, through patterns at about 20 times chondritic, with no Eu anomaly, up to patterns about 200 times chondritic with deep Eu depletion. These character-

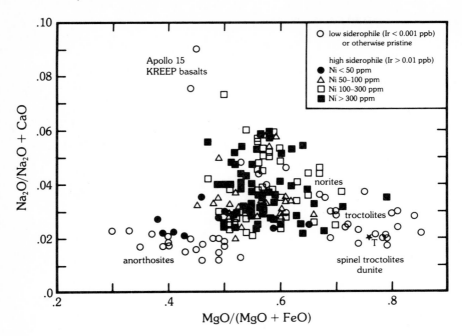

5.14 Highland rock compositions showing that high siderophile element abundances tend to fall within the boundaries created by samples with low siderophile element abundances [35]. (Courtesy G. Ryder.)

istics encompass most of the chemical variations observed in highland breccias. Garrison and Taylor [55] caution against attempts to over-refine the end-members of the highland crust by introducing additional elusive components. "Highland basalt" is almost certainly a mixture. They propose a three-component mixing model comprising ANT, KREEP and a small meteoritic component. This avoids the question of the origin of the Mg component.

Plots of MgO/(MgO + FeO) versus $Na_2O/(Na_2O + CaO)$ [49] indicate a triangular range of highland compositions, outlined by anorthosite, KREEP and troctolites, very close to the Taylor and Bence [47] components. The siderophile-rich samples cluster within the triangle, while the pristine components fall toward the outer boundaries [56] (Fig. 5.14). This result, like the REE patterns, is good evidence for mixing models. Garrison and Taylor [55] note that "recent modelings of lunar highland rocks have fallen prey to the over-interpretation of the chemical data" and continue that "we do not feel that any model can be totally quantified and realistically explain the highland breccias" [55, p. 411], particularly in view of their history of multiple brecciation.

One problem encountered in mixing models is the identification of end-members precisely defined to explain all the trace element chemistry of the breccias. Clearly the end-member components should be reasonably abundant and "The use of dunite (74215) as an end-member is unwarranted because of its scarcity" [55]. Most of the elements in the norite component used by Ryder [50] are really those of the KREEP component. Wasson et al. [56] have introduced SCCRV, an Mg-rich rock type characterized by high Sc, Cr and V contents and representing yet another addition to the collection of lunar acronyms. Garrison and Taylor [55] complain that it is "an elusive n+1 component" and that "even dunite is more realistic than SCCRV" as a possible end-member component for highlands breccias [55, p. 411]. Clearly there is little agreement on the nature of the Mg component, except that it is not chondritic, but some resolution of these questions will become apparent.

The view is taken here that too much fine-tuning of the components is not possible, but that the overall chemistry of the highland crust is consistent with a minimum of three initial components, together with the meteoritic addition. The later complexities have arisen as a consequence of the meteoritic bombardment of a crystallizing crust.

5.4.1 Anorthosites

This significant and widespread component [57] in the lunar highlands is principally responsible for the light color of the highlands, in contrast to the darker basalts of the maria, familiar as the features of the "man in the moon." Plagioclase feldspar constitutes the bulk (95%) of these pale rocks; they are very calcium-rich, with compositions typically in the range An_{95-97}. This uniformity of anorthosite composition indicates that the parent magma body was very large, probably moon-wide. The distinction between a single "magma ocean" and many smaller bodies of melt is in part semantic. The evidence for moon-wide crystal fractionation processes cannot easily distinguish between these alternatives. Low-calcium pyroxene is the next most abundant constituent, but olivine and clinopyroxenes may also be present, although the mafic minerals are only present as minor components.

Full details of the mineralogy are given by James [58] and Norman and Ryder [59]. Examples include 15415 (the genesis rock), 15437, 60025, 60055, 60215, 61015, 61016, 62255, 64475, 65035, 65315, and 67075. These rocks have been mostly granulated by impact. The chemical compositions of some selected rocks are given in Table 5.2.

The REE abundances (Fig. 5.15) are all low, with a pronounced positive Eu anomaly. The La/Yb ratios become steeper as the total REE fall, typical of plagioclase patterns. Europium is nearly constant at 0.6–0.9 ppm. The $^{87}Sr/^{86}Sr$ ratios of the anorthosites are very low. This constitutes primary evidence that they are relics of the early crust [60]. Sample 60025, for example,

Table 5.2a Major element compositions of typical highland samples.[†]

	Anorthosites				Gabbroic Anorthosites		Anorthositic Gabbros		
	15415 (1)	65315 (2)	60055 (3)	61016 (4)	68415 (5)	65055 (6)	15455 (7)	60335 (8)	66095 (9)
SiO_2	44.1	44.3	44.3	45.0	45.5	45.5	44.5	46.2	44.5
TiO_2	0.02	0.012	—	0.02	0.32	0.28	0.39	0.58	0.71
Al_2O_3	35.5	34.9	34.0	34.6	28.6	28.5	26.0	25.3	23.6
FeO	0.23	0.31	0.34	0.30	4.25	3.90	5.77	4.51	7.16
MnO	—	0.006	0.10	—	0.06	0.05	—	0.07	0.08
MgO	0.09	0.25	0.33	0.20	4.38	4.81	8.05	8.14	8.75
CaO	19.7	19.1	19.0	19.6	16.4	16.1	14.9	14.4	13.7
Na_2O	0.34	0.30	0.34	0.40	0.41	0.44	0.25	0.52	0.42
K_2O	—	0.007	0.01	0.01	0.06	0.13	0.10	0.23	0.15
Cr_2O_3	—	0.003	0.005	0.01	0.10	0.08	0.06	0.13	0.15
P_2O_5	0.01	0.001	—	0.05	0.07	0.13	—	0.19	0.24
Σ	99.99	99.2	98.4	100.2	100.1	99.8	100.0	100.3	99.4

[†]Values given in wt.%.
Sources of data (also for Tables 5.3a and 5.4):
   Apollo 15 PET NASA SP-289, 6–6 (1);
   Hubbard, N. J., et al. (1971) *EPSL*. 13: 73 (1);
   Hubbard, N. J, et al. (1974) *PLC 5*: 1234 (4);
   Taylor, S. R., et al. (1973) *PLC 4*: 1454 (7);
   Vaniman, D. T., and Papike, J. J. (1980) *Lunar Highlands Crust,* p. 276 (5,6,8,9);
   Wänke, H. (1974) *PLC 5*: 1307 (2);
   Warren, P. H., and Wasson, J. T. (1978) *PLC 9*: 188 (3);
   Wolf, R., et al. (1979) *PLC 10*: 2112 (1,4,5).

has an $^{40}Ar/^{39}Ar$ age of 4.13 aeons [61]. Some anorthosites (so-called troctolitic anorthosites) contain 15-20% olivine or pyroxene, without appreciable differences in the REE patterns. Sample 76535 is one example. It has an initial $^{87}Sr/^{86}Sr$ of 0.6990, but the Sm-Nd age is 4.25 aeons [62, 63]. (See [41, 56, 64, 65] for further data on anorthosites.)

5.4.2   The Mg Suite

Norites, troctolites, dunite, spinel troctolite, and gabbroic anorthosites are collectively designated as the Mg-rich plutonic rock suite (Fig. 5.16). Typical compositions are given in Table 5.3 [42]. Norites do not contain olivine and have flat REE patterns, with about the same abundance of Eu, and either small negative or positive anomalies (Figs. 5.17 and 5.18). Examples include 15445c, the civet cat clast from 72255 and large samples from Apollo

Table 5.2b Major element compositions of typical lunar highland samples.[†]

	Anorthositic norite	Norites		Spinel troctolite	Troctolite	Dunite	LKFM		MKFM
	78255	77075	78235	73215, 32	76535	72417	14310	65015	15386
	(1)	(2)	(3)	(4)	(5)	(6)	(7)	(8)	(9)
SiO_2	47.3	51.2	49.5	44.7	42.9	39.8	47.2	47.0	50.8
TiO_2	0.67	0.33	0.16	—	0.05	0.03	1.24	1.26	2.23
Al_2O_3	27.4	15.0	20.9	31.2	20.7	1.3	20.1	19.7	14.8
FeO	2.64	10.7	5.05	3.05	5.0	11.9	8.38	8.59	10.6
MnO	0.046	0.17	0.08	—	0.07	0.11	0.11	0.11	0.16
MgO	5.97	12.9	11.8	3.42	19.1	45.4	7.87	9.31	8.17
CaO	15.0	8.82	11.7	17.2	11.4	1.1	12.3	11.9	9.71
Na_2O	0.45	0.38	0.35	0.47	0.20	0.013	0.63	0.55	0.73
K_2O	0.084	0.18	0.061	0.075	0.03	0.002	0.49	0.36	0.67
Cr_2O_3	0.14	0.39	0.23	—	0.11	0.34	0.18	0.19	0.35
P_2O_5	—	—	0.04	—	0.03	—	0.34	0.41	0.70
Σ	99.7	100.1	99.8	100.2	99.6	100.0	98.8	99.4	99.0

[†]Values given in wt.%.

Sources of data (also for Tables 5.3b and 5.4):

Bence, A. E., et al. (1975) LS VI: 38 (4);
Higuchi, H., and Morgan, J. W. (1975) *PLC 6*: 1628 (3, 4, 6);
Haskin, L. A., et al. (1974) *PLC 5*: 1213 (5);
Keith, J. E. (1974) *PLC 5*: 2122 (3);
Laul, J. C., and Schmitt, R. A. (1975) *PLC 6*: 1234 (6);
Vaniman, D. T., and Papike, J. J. (1980) *Lunar Highlands Crust,* p. 276 (7, 8, 9);
Warren, P. H., and Wasson, J. T. (1978) *PLC 9*: 188 (1, 2);
Wolf, R., et al. (1979) *PLC 10*: 2112 (1).

17 (stations 7 and 8). Spinel troctolites (e.g., 67435c) consist of olivine and pleonaste enclosed in plagioclase and appear to be of deep-seated origin, as indicated by the presence of Al-enstatite and Mg-Al spinel. One of the most interesting samples which possibly belongs to the Mg suite is the well-known dunite 72415-8, which is mainly crushed olivine [66, 67]. There is much variation among the REE in different samples of this rock (Fig. 5.18). It contains only 160 ppm Ni, and 2500 ppm Cr, typical lunar ratios for these elements. The Rb-Sr internal isochron age is 4.45 aeons with an initial $^{87}Sr/^{86}Sr$ of 0.69900 ± 7 [60].

The plagioclase and ferromagnesian minerals in the Mg suite show wide ranges in composition, and these are correlated in a manner simply related to fractional crystallization. The relationships of this suite of rocks to the anorthosites on the one hand and the KREEP components on the other are the subject of much current debate (see also [41, 56].)

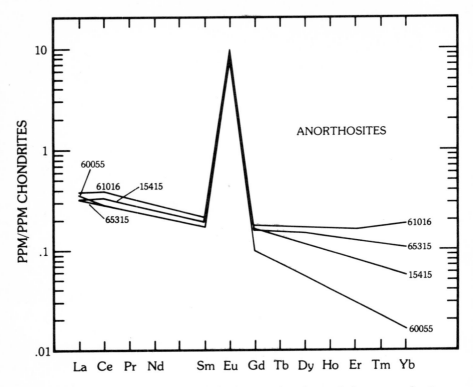

5.15 Chondrite-normalized rare-earth element abundances in lunar anorthosites. Data from Table 5.3. Note the low overall abundances and the high enrichment of europium, due to the selective uptake of Eu^{2+}. See Appendix VIII for rare-earth element chondritic normalizing factors.

5.4.3 KREEP

The existence of high concentrations of KREEP has beguiled the lunar community since the first discovery during the Apollo 12 mission of a layer in the regolith of "light gray fines" and of rock 12013, picked up by Astronaut Pete Conrad on an unscheduled traverse. The surprise in finding extreme concentrations of elements typical of the residual liquids produced by fractional crystallization, on a Moon which many considered to be a primitive unfractionated body, was compounded by the widespread occurrence of such a component. This contrasts with the so-called lunar granites, which are of trivial extent in the whole lunar picture [68–71]. Mostly they appear to be fragments of interstitial glass from mare basalts. Such interstitial glasses (up to 150 microns) are reported by many authors studying mare basalts. High

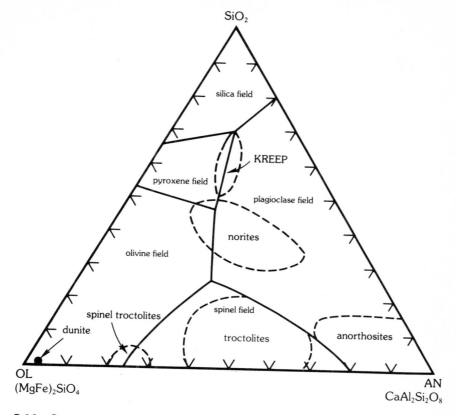

5.16 Compositions of apparently pristine lunar highland samples plotted on the olivine-anorthite-silica pseudoternary diagram. (Adapted from [42].)

silica glass fragments have not been reported from the Apollo 16 highland soils with one exception (see Section 4.4.2).

Although KREEP was thought at one time to be rather restricted in areal extent, and perhaps even confined to be the Imbrium basin [71, 72], the consensus emerges that it is in fact widely distributed on the Moon [73], although it does appear to be restricted to regions where the highland crust thickness is 60 km or less. Because KREEP is so widespread, it cannot be dismissed as a small or trivial volume of residual melt produced from crystallization of a local intrusion. In magma ocean models, it represents perhaps the final 2% residue. Studies of Sm/Nd isotopic systematics indicate that the samples measured come from a common source [73]. Sm/Nd ratios and $^{143}Nd/^{144}Nd$ ratios are very close for samples from Apollos 12, 14, 15 and 16. The idea that KREEP was derived as the final residual liquid from the

Table 5.3a Trace element compositions of lunar highland samples.[†]

	Anorthosites				Gabbroic Anorthosites		Anorthositic Gabbros		
	15415	65315	60055	61016	68415	65055	15455	60335	66095
I. Large cations									
Cs	0.02	0.015	—	0.0003	—	—	0.12	0.28	0.40
Rb	0.22	0.17	—	0.02	1.7	—	1.1	6.2	4.2
K	151	58	85	80	650	1100	830	1900	1200
Ba	6.3	5	11	7	76	80	42	190	240
Sr	173	170	—	180	182	140	220	163	154
Pb	0.27	—	—	—	0.78	—	1.0	2.0	15
K/Rb	690	340	—	4000	380	—	750	300	290
Rb/Sr	0.001	0.001	—	0.0001	0.009	—	0.005	0.038	0.027
K/Ba	24	11.6	7.7	11	8.5	14	20	10	5
II. Rare earth elements									
La	0.12	0.12	0.13	0.14	6.8	6.2	3.0	20.0	18.5
Ce	0.33	—	0.27	0.37	18.3	16.0	6.7	52	50
Nd	0.18	—	—	0.21	10.9	—	3.73	32	—
Sm	0.046	0.04	0.040	0.058	3.09	2.6	0.88	9.1	8.8
Eu	0.81	0.74	0.76	0.77	1.11	1.0	1.67	1.28	1.63
Gd	0.05	—	—	0.054	3.78	—	0.95	10.6	—
Tb	—	—	—	—	—	0.55	0.14	1.9	—
Dy	0.044	0.056	—	0.065	4.18	—	0.84	11.5	9.4
Ho	—	—	—	—	—	—	0.17	2.5	—
Er	0.019	—	—	0.040	2.57	—	0.46	6.77	—
Yb	—	0.026	0.035	0.045	2.29	2.1	0.36	6.23	4.9
Lu	—	0.004	0.004	0.01	0.34	0.29	0.06	0.68	0.90
Y	—	—	—	—	22	19	4.8	57	70
La/Yb	3.4	4.6	3.7	3.1	3.0	3.0	3.0	3.2	3.0
Eu/Eu*	51.6	51.6	64.7	45.2	0.99	1.08	5.58	0.40	0.54
III. Large high valence cations									
U	0.0017	—	—	0.0015	0.32	0.31	0.05	0.92	1.0
Th	0.0036	—	—	—	1.26	1.18	0.23	2.75	2.2
Zr	—	15	—	2.4	100	72	11	290	320
Hf	—	0.49	—	—	2.4	2.1	0.17	6.9	5.0
Nb	—	—	—	—	5.6	—	0.95	10	21
Th/U	2.1	—	—	—	3.94	3.79	4.6	3.0	2.2
Zr/Hf	—	31	—	—	42	34	65	43	64
Zr/Nb	—	—	—	—	18	—	11.5	29	16
IV. Ferromagnesian elements									
Cr	—	—	33	21	700	500	440	910	1000
V	—	—	—	—	20	35	16	26	110
Sc	—	0.39	0.55	0.5	8.2	7.2	—	8.5	6.8
Ni	1.0	1.4	1.9	1.0	180	390	12	720	710
Co	—	0.058	0.84	1.0	11	29	10	37	44
Cu	—	2.1	—	—	12	2.4	1.3	7.4	3.9
Fe, %	0.18	0.24	0.26	0.23	3.30	3.03	4.49	3.50	5.56
Mn	—	47	75	—	470	390	—	540	620
Zn	0.26	—	0.60	1.6	4.8	0.56	1.9	4.3	92
Mg, %	0.05	0.15	0.20	0.12	2.54	2.90	4.86	4.91	5.28
Ga	—	—	3.8	—	2.0	3.0	2.6	3.1	3.8
Li	—	—	—	—	5	2.2	—	13	11
V/Ni	—	—	—	—	0.11	0.09	1.3	0.04	0.15
Cr/V	—	—	—	—	35	14	28	35	9

[†]Data in ppm.

Table 5.3b Trace element composition of lunar highland samples.[†]

	Anorthositic norite	Norites		Spinel troctolite	Trocto-lite	Dunite	LKFM		MKFM
	78255	77075	78235	73215, 32	76535	72417	14310	65015	15386
I. Large cations									
Cs	—	—	0.064	0.007	0.001	0.014	0.54	0.42	—
Rb	—	—	0.92	0.30	0.24	0.045	12.8	10.2	18.5
K	700	1500	510	620	220	16	4080	3000	5600
Ba	86	160	80	61	33	4.1	630	570	840
Sr	—	—	—	—	115	8.2	250	145	190
Pb	—	—	—	1.9	—	—	6.2	4.7	—
K/Rb	—	—	550	2070	920	360	320	300	300
Rb/Sr	—	—	—	—	0.002	0.005	0.05	0.07	0.10
K/Ba	8.1	9.4	6.4	10.2	6.7	3.9	6.5	5.3	6.7
II. Rare earth elements									
La	3.3	7.2	—	4.2	1.51	0.15	56	66	84
Ce	7.8	—	9.2	12	3.8	0.37	144	185	210
Pr	—	—	—	1.54	—	—	17	26	—
Nd	5.0	8.5	5.4	6.3	2.3	—	87	120	130
Sm	1.20	3.0	1.49	1.82	0.61	0.080	24	27	38
Eu	1.21	0.98	1.03	0.50	0.73	0.061	2.15	1.97	2.72
Gd	—	—	—	2.05	0.73	—	28	36	45
Tb	0.23	0.74	—	0.42	—	0.017	5.1	6.3	7.9
Dy	—	—	2.26	2.71	0.80	0.11	33	38	46
Ho	—	—	—	0.57	—	0.023	6.5	8.3	—
Er	—	—	1.47	1.62	0.53	—	20	24	27
Yb	0.98	3.9	1.64	1.66	0.56	0.074	18	21	24
Lu	0.14	0.59	0.24	0.26	0.08	0.012	2.5	2.8	3.4
Y	—	—	—	20	—	—	175	174	—
La/Yb	3.4	1.8	—	2.5	2.7	1.9	3.1	3.1	—
Eu/Eu*	2.93	0.87	1.97	0.80	3.34	2.1	0.25	0.19	0.20
III. Large high valence cations									
U	0.19	0.5	0.19	0.23	0.020	0.006	3.10	2.2	2.8
Th	0.44	1.57	0.59	0.80	—	—	10.4	8.9	10.0
Zr	49	210	—	79	24	—	840	940	970
Hf	0.67	3.5	—	2.0	0.52	0.10	21	22	32
Nb	—	—	—	6.3	—	—	52	48	—
Th/U	2.3	3.1	3.1	3.5	—	—	3.4	4.0	3.6
Zr/Hf	73	60	—	40	46	—	40	44	31
Zr/Nb	—	—	—	12.5	—	—	16	20	—
IV. Ferromagnesian elements									
Cr	990	2650	—	—	—	—	1250	1280	2400
V	—	—	—	—	—	50	36	—	—
Sc	4.6	17	—	—	—	4.3	20	17	24
Ni	22	6	12	247	44	160	270	490	13
Co	23	33	—	—	—	55	17	43	18
Cu	—	—	—	—	—	—	5.0	5.1	—
Fe, %	2.05	8.31	3.93	2.37	3.89	9.25	6.51	6.67	8.24
Mn	335	1320	600	—	540	850	850	850	1240
Zn	0.095	3.3	1.5	6.2	1.2	2	1.8	1.5	—
Mg, %	3.60	7.78	7.12	2.06	11.5	27.4	4.75	5.62	4.93
Ga	5.1	4.0	—	—	—	—	3.2	3.8	3.5
Li	—	—	—	—	3.0	—	22	21	27
N/Ni	—	—	—	—	—	—	0.13	—	—
Cr/V	—	—	—	—	—	—	35	—	—

[†]Data in ppm.

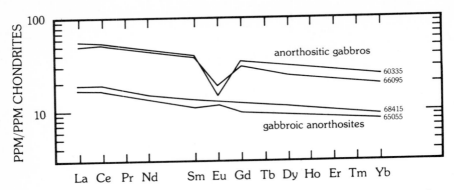

5.17 Rare-earth element abundances in lunar gabbroic anorthosites and anorthositic gabbros. Data from Table 5.3. Note that in comparison with Fig. 5.15, the Eu anomaly ranges from slightly positive to negative, and that the total REE abundances are higher.

crystallization of a magma ocean was stated by Taylor and Jakeš [74] and Taylor [75]. Major reviews have been undertaken by Warren and Wasson [76] and Meyer [77] to which the reader is referred for much valuable material.

KREEP is elusive as a rock type [46]. Some so-called pristine fragments (i.e., lacking in siderophile elements) exist (15382, 15386) and various clasts in

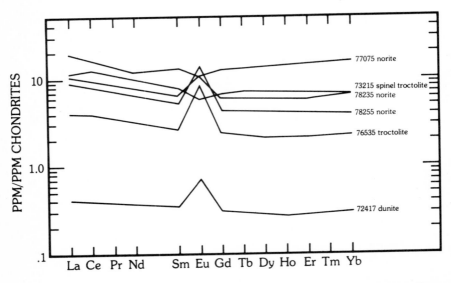

5.18 Rare-earth element abundances for norites, troctolites and the dunite. Data from Table 5.3.

Table 5.4 Volatile and siderophile trace elements in highland samples.[†]

	Anorthosites				Gabbroic anorthosites		Anorthositic gabbros		
	15415	65315	60055	61016	68415	65055	15455	60335	66095
Ir	—	—	0.013	0.01	4.6	10	0.024	—	—
Re	0.0008	—	—	0.0022	0.43	—	0.006	—	—
Ni, ppm	1.0	1.4	1.9	1.0	180	390	12	720	710
Au	—	1.0	0.014	0.02	2.7	5.0	0.042	—	18*
Sb	0.70	—	—	—	530	—	0.22	—	—
Ge	1.2	—	17	—	73	240	9.4	—	—
Se	0.23	—	—	0.4	98	—	8.3	230	—
Te	—	—	—	—	—	—	2.6	—	—
Ag	1.7	—	—	0.29	—	—	1.7	—	—
Zn, ppm	0.26	—	0.60	1.6	4.8	0.56	1.9	4.3	92*
In	0.18	—	3.6	—	—	6.4	0.05	—	680*
Cd	0.57	—	0.57	—	2.8	—	1.0	—	—
Bi	0.1	—	—	—	—	—	0.14	—	—
Tl	0.09	—	—	—	—	—	0.054	—	—
Br	2.3	—	—	—	—	—	35	—	—

[†]Data in ppb except where noted.
*Rusty rock sample.

Table 5.4 *(Continued).*

	Anorthositic norite	Norites		Spinel troctolite	Troctolite	Dunite	Fra Mauro LKFM		MKFM
	78255	77075	78235	73215, 170	76535	72417	14310	65015	15386
Os	—	—	—	—	—	—	10.5	—	—
Ir	0.43	0.25	0.14	1.0	0.005	0.005	10.5	17	0.06
Re	0.021	0.022	0.012	0.04	0.0012	0.005	1.0	1.8	—
Ni, ppm	22	6	12	247	44	160	270	490	—
Au	0.11	0.026	0.42	2.6	0.0025	0.25	4.3	11	0.22
Sb	—	—	0.08	0.71	0.01	0.47	4	14	—
Ge	58	11	19	110	1.7	30	110	—	60
Se	—	—	7.5	4	4.1	5	120	290	—
Te	—	—	—	—	0.28	—	—	—	—
Ag	—	—	0.40	130	0.12	0.3	—	—	—
Zn, ppm	0.095	3.3	1.5	6.2	1.2	2	1.8	1.5	—
In	0.05	—	—	—	—	—	30	—	—
Cd	4.2	5.4	2.9	3.1	0.6	0.4	8.4	9.3	10
Bi	—	—	0.05	0.49	0.04	0.4	—	—	—
Tl	—	—	0.02	1.0	0.012	0.05	—	—	—
Br	—	—	6	14	3.2	8.4	—	—	—

breccias have been so identified (15405c, 72275c, 77115c) (Table 5.4). It is probably fair to say that no topic has caused more confusion in our understanding of lunar petrogenesis than has KREEP; this emphasizes the hazards of attaching labels. However, some resolution of these problems appears possible. The chemical relationships are relatively simple. A dominating characteristic is the simplicity of the REE patterns (Fig. 5.19), possibly best illustrated by "Super-KREEP" (15405, 85) which has an La content 700 times the chondritic abundance [78]. The absence of unambiguous primary examples of KREEP rocks makes a discussion of their mineralogy somewhat ambiguous. The major element chemistry of KREEP-dominated samples is close to that of the olivine-plagioclase-pyroxene peritectic point in the experimental system silica-anorthosite-olivine [for $Mg/(Mg + Fe) = 0.7$] [79]. This fact has led to the popularity of traditional style igneous models, such as derivation by partial melting in the shallow lunar interior (about 120 km depth). This partial melting model is difficult to reconcile with the Rb-Sr isotopic evidence, since the Apollo 15 KREEP basalts have high initial ratios; further, it is not possible to generate Apollo 15 KREEP basalts by partial melting at 3.9 aeons, but this can be done at about 4.3 aeons, the average model age. These and other models involve large-scale lunar volcanism and the eruption of a primary magma of KREEP composition. The widespread use of the term KREEP basalt has exacerbated the problem. The basic question in KREEP genesis is how to reconcile the high concentrations of the incompatible elements with the major element chemistry and the high $Mg/(Mg + Fe)$ values. The view adopted here is that mixing of primitive and late stage liquids provides an explanation for the puzzling chemistry. The question of KREEP "volcanism" is addressed in the next section.

5.4.4   KREEP Volcanism in the Lunar Highlands?

Among the many questions which revolve around the question of KREEP, and of possible pre-mare volcanic activity in the lunar highlands, the occurrence of so-called KREEP basaltic volcanism has often been raised [80, 81, 82]. There is no doubt about the existence of high-Al basalts in the highlands breccias (see Section 5.2.6). These are similar to the familiar mare basalts and do not present a special problem. The "KREEP basalts" are different. Their high content of REE, Th, U, K, etc., together with their high $Mg/(Mg + Fe)$ ratios call for a special petrogenesis, distinct from that of other basalts of undoubted volcanic origin. Accordingly, the evidence for their existence as a "volcanic" rock needs to be carefully examined, because of the problems which they present for lunar petrology. KREEP components are widespread and LKFM appears to be moon-wide [83]. As is argued elsewhere, KREEP is not just restricted to the Imbrium basin and its surrounding terrain. In this context, we note the orbital thorium data, which shows that the

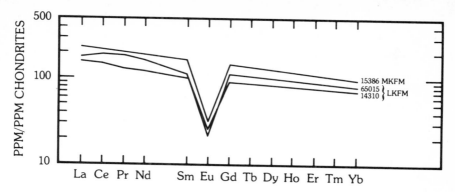

5.19 Rare-earth element abundances in Fra Mauro basalts. Note the extreme enrichment of the total REE and the deep depletion in europium, forming a pattern reciprocal to that of the anorthosites in Fig. 5.15.

Apennine Mountains, as well as the Apennine Bench Formation, have high Th concentrations (~11 ppm) [84]. Thus, popular interpretations of the Apennine Bench Formation as volcanic KREEP flows, extruded after the Imbrium collision but before the mare basalts, are not really supported by the orbital thorium data.

Great petrogenetic problems arise. Mare basalts were being erupted at 3.85 aeons (10003), and basaltic-style volcanism predated the Imbrium collision (Section 5.2.6). If the interpretation of the Apennine Bench Formation as volcanic flows is correct, then two contrasting styles of lunar volcanism were occurring at the same time. This occurs, of course, on the Earth in a complex tectonic environment, but raises many problems on the Moon. If KREEP is volcanic, it is fair to ask where all the flows are. Is the Apennine Bench the only demonstrable example? In view of the widespread nature and high concentrations of K, U, Th, etc., observed in the highland crust, it is remarkable that KREEP basalts so rarely appear. It seems more credible to account for the Apennine Bench Formation as an impact-derived melt sheet or debris flow from the Imbrium basin collision ponded against the Apennine Mountain scarp. The two primary examples of volcanic KREEP basalts usually quoted are samples 15382 and 15386. The evidence that these are "igneous" depends in part on texture and in part on their absence of siderophile elements. This recalls the argument over 14310 (Section 5.7) which was considered truly igneous on the basis of petrological evidence, until the high contents of Ni and other siderophile elements testified to its impact origin. The absence of siderophile elements does not constitute decisive evidence for pristinity (see Section 5.7). Many impact melts at terrestrial craters do not have a siderophile element signature [30]. The position is taken here that the case for KREEP volcanism is not proven, despite some suggestive evidence and many opinions

to the contrary [see for example *Basaltic Volcanism* (1981) Section 1.2.10.5].
The widespread occurrence of impact melt sheets associated with large basins
may provide terrains that are very difficult to distinguish, by photogeological
techniques, from volcanic flows. The isotopic data lend support to this interpretation. Model ages for KREEP point back to the initial differentiation of
the Moon, not to more recent volcanic events, and the view is adopted here
that large-scale fractional crystallization of the magma ocean is the ultimate
source of KREEP [50, 74, 75]. Small scale partial melting events may have
occurred locally during meteoritic bombardment of the growing crust.

An important observation is that there is a good correlation between high
abundances of the REE (KREEP component) and position on the SiO_2-Ol-
An diagram [85]. Samples with less lthan 15 ppm Sm (65 times chondritic
abundances) show scatter, consistent with an origin by mixing. Those with
higher abundances of REE tend to lie close to the cotectic line. This may be a
consequence of partial melting, or it may simply reflect the dominating
influence of the KREEP end member in a mixing model.

5.5   Volatile Components in the Highland Crust

The nature and origin of the volatile elements that are present in minor
amounts in the highland breccias are important since the bulk moon composition is highly depleted in these elements. Some rocks (e.g., 66095) have high
concentrations of Cl, Pb, Br, Zn, Rb, Ag and Tl. Various sources for these
elements may be suggested:

(a) They may be added during meteoritic or cometary impact. Either
comets or Cl chondrites possess appropriate levels of volatile elements.

(b) These elements, although low in abundance in the Moon, will be
concentrated into the residual melt following crystallization of the
magma ocean. They will thus be associated with the KREEP component. They do not enter many common silicate lattice sites and mostly
are extreme examples of incompatible elements.

(c) They will be readily remobilized during impact events, or by "impact-
induced metamorphism" and so concentrated near the surface. This
volatilization model has useful predictive capabilities. For example,
volatile trace elements in breccias may be expected to have anomalous
abundances, although the major and refractory trace elements will be
unaffected. This model, which involves migration of volatile elements
from their original source in KREEP, explains many of the effects
ascribed to "fumarolic" activity on the Moon.

(d) They may be due to "fumarolic" activity. Fumaroles imply volcanic
activity and call for complicated models of igneous activity; the alternative model of secondary mobilization is much more realistic.

For the most part, these elements in lunar samples are easily leachable [86] by water and hence probably reside along grain boundaries, in lattice dislocations and along cracks. Sample 66095 has highly water-leachable and low residual chlorine. Hence, most of the Cl_2 is probably introduced at a late stage. Garrison and Taylor [55] consider that this chlorine attacks native Fe-Ni metal grains, forming $FeCl_2$. "This delinquescent $FeCl_2$ undergoes oxyhydration [in the terrestrial atmosphere] to β FeOOH (akanéite) resulting in the notorious rust" ([55], p. 313).

Sample 66095 is a complex breccia with components ranging all the way from gabbroic anorthosite (e.g., 68415) to Fra Mauro basalts (e.g., 65015) (Fig. 5.13). It is the most volatile-rich of the lunar samples, containing Br, Cl, F, Bi, Cd, Pb, Tl and Zn. These volatiles will appear mainly as surface coatings and will easily be redistributed and flushed out during the metamorphic effects accompanying meteorite impacts. The water-bearing phase in 66095 is FeOOH (akanéite) [87] (Figs. 5.20 and 5.21).

Various sources for the water are:
(a) fumarolic activity;
(b) cometary impact;
(c) oxyhydration of $FeCl_2$ (Lawrencite) due to contamination with terrestrial water vapor.

FeOOH cannot exist for any length of time under lunar vacuum (10^{-7} torr) conditions and high lunar daytime temperatures ($140°C$), and it is concluded that the water is terrestrial [88].

Sample 66095 contains 15 ppm Pb, 85% of which is not supported by U and Th [88]. The most reasonable hypothesis is that the Pb is indigenous to the Moon. Excess Pb in 66095 was evolved in a U rich (high μ) reservoir from 4.47 to 4.01 aeons, when it migrated into 66095. Nunes and Tatsumoto [89] conclude from these data that 4.47 aeons is the best estimate for the time of the early differentiation of the Moon. From this it appears this KREEP will generate highly radiogenic lead, and will probably also concentrate most of the lunar ^{204}Pb that does not get into sulfide phases. Since the volatiles are readily moved around during impact processes, parentless lead should be fairly common in lunar breccias.

Further insights into the occurrence of lunar volatiles in lunar highlands breccias have been gained by experimental investigations [90, 91] (see Section 4.5.2). In thermal release studies, the major fractions of Pb, Zn and Cd in rusty rock 66095 were released below $1000°C$, which suggests that they are present mainly on grain surfaces. Experimental studies involving native iron chlorapatite [$Ca_5Cl(PO_4)_3$] and synthetic lunar basalts demonstrated vapor deposition including $FeCl_2$, iron phosphides (Fe_3P) and P. These, including high P metal, could form by disequilibrium thermal metamorphism during impact, which may therefore be an important vapor transport mechanism. Accordingly, there is no requirement to introduce volatiles by either volcanism or by

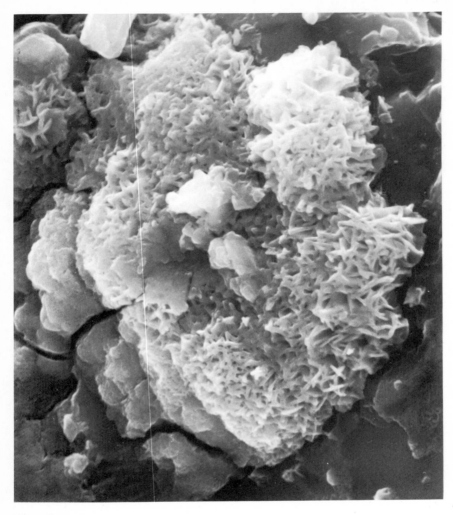

5.20   Rosettes of akaganéite in the rusty rock 66095. Magnification 6000 ×. (NASA S-73-37369. Courtesy D. S. McKay.)

addition from meteorites. Redistribution by impact processes appears adequate [55]. Note that lead, but not uranium or thorium, will be so affected; therefore, excess Pb, unsupported by Th and U, may be frequently mobilized by this mechanism.

In summary, it seems possible to account for the high concentrations of volatiles in samples such as 66095 and 61016 by two causes: (a) they contain

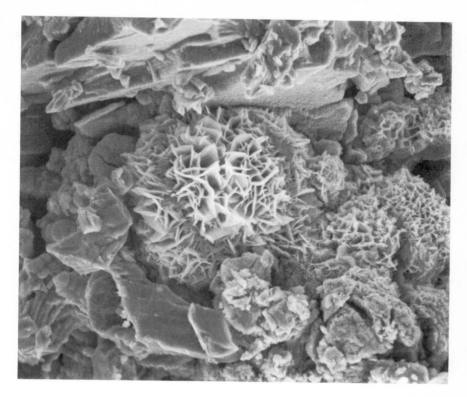

5.21 SEM photograph of one of the three morphologies of "rust" on lunar sample 66095. Compositionally the rosettes contain iron, chlorine, and occasionally up to 5% Ni. The central rosette is 7 microns in diameter. (NASA S-73-17705. Courtesy D. S. McKay.)

a large KREEP component and (b) volatiles have been added by redistribution from other KREEP-rich samples during large impacts [92].

5.6   The Ancient Meteorite Component

It was clear to many early observers that the lunar highlands had been subjected to a massive early bombardment. Once the dispute over the volcanic versus impact origin of the large craters was resolved, then a predictable consequence of the meteorite impact hypothesis was that some chemical signatures derived from the colliding bodies should be found on the surface. Experience at terrestrial impact craters, however, showed that traces of the

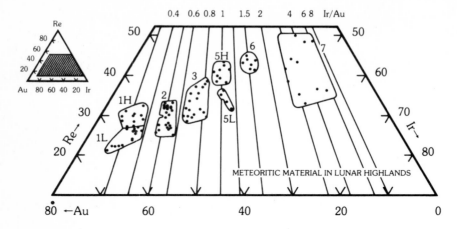

5.22 Ancient meteoritic components in lunar highland samples can be divided into about 8 groups, according to the proportions of siderophile elements Au, Re, and Ir. Reality of groups has been confirmed by objective statistical tests, such as cluster, factor, and discriminant analysis. Some groups are heavily represented at one landing site, e.g., Groups 1H and 7 at Apollo 16 and Groups 2, 3 at Apollo 17 [95]. (Courtesy E. Anders.)

impacting meteorite were often difficult to find, since the size of the crater and the volume of ejected country rock are about an order of magnitude larger than the meteorite. The most reliable chemical indexes are the siderophile trace elements. These are heavily depleted both in the terrestrial and in the lunar crusts. The determination of Ir, Re, Os, Pt, Pd, and Au at the parts per billion level is a task for neutron activation methods of analysis. These elements are abundant in the native iron of many classes of meteorites, so that they form the best index and signature of the meteoritic contribution. They can also be used in a contrary sense, so that their absence may identify "pristine" samples which have apparently escaped the bombardment. It has already been shown (Section 5.4) that the post-mare meteorite component present in the regolith is due principally to C1 carbonaceous chondrites.

Ir/Au ratios are particularly diagnostic, and have been used to identify eight groups of projectiles of differing composition, which have impacted the Moon [93–95]. From the distribution of these groups at various landing sites, it is tentatively concluded that seven of the groups are associated with basin-forming impacts rather than craters (Figs. 5.22, 5.23). One group (1H) appears to be associated with a local crater at the Apollo 16 site. There also appears to be some temporal correlation, with the more refractory "moon-like" compositions falling first. If this interpretation is correct, then it provides good evidence for heterogeneous accretion models for planetary growth (see Section 9.6).

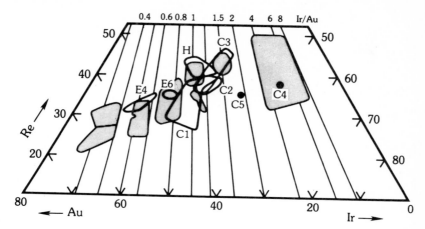

5.23 The ancient lunar highland meteorite components compared with the abundances for Au, Re and Ir in the various chondrite groups. Ancient meteoritic groups cover wider compositional range than do present-day chondrites. Except for Group 7 (where the low Au content causes scatter), they are no more distended than are some chondrite classes, e.g., C1's. However, since the Ir-Au-Re diagram does not fully resolve C1, C2, and C3 chondrites, one cannot rule out the possibility that some lunar groups are composite [94].

Many uncertainties beset this approach. If the highlands have undergone a saturation bombardment [30, 96], then the preservation of chemical evidence for a particular impact becomes more difficult and requires much selenodetective work.

5.7   Pristine Rocks in the Lunar Highlands

The widescale destruction of original textures and petrological and mineralogical relationships among the highland crustal rocks, by the raining meteoritic bombardment, has led to a search for those samples which might have escaped this fate. Pristine highland rocks are generally considered to be those produced by primary igneous activity, and which have subsequently retained their original chemical identity [97, 98]. Four characteristics may be used in assessing pristinity: (a) a low abundance of siderophile elements is essential. Since the bulk moon appears to be depleted in siderophile elements, this is a key parameter. The abundance of Ni, Au, Ge, Ir, Os and Re should be less than 3×10^{-4} times C1 chondrite compositions [99]; (b) various textural or mineral compositions that identify a monomict cumulate origin; (c) identifiable regolith components, including high concentrations of KREEP components, may indicate a polymict origin; (d) low $^{87}Sr/^{86}Sr$ initial ratios or old ages are supportive evidence, since impact will reset the radiometric clocks. It is clear that the unequivocal identification of pristine rocks is not a simple task, but calls for the highest geochemical and petrological skills. Thus, many pristine rocks are brecciated with few remaining vestiges of igneous texture. Basaltic textures can be produced in impact melts (Section 5.3). The compositional fields of the silicate minerals overlap those of the polymict rocks. The most acceptable method is chemical: the controversy over whether rock 14310 was an impact melt or a primary igneous basaltic rock [100] should be a salutary reminder of the difficulties of making these decisions on the basis of petrography.

Possibly the best case for demonstrating that the search for pristine rocks (a lunar equivalent of the holy grail) is useful is shown in Fig. 5.14 where the siderophile element distribution in pristine rocks and polymict rocks is shown. The polymict breccias show high siderophile element signatures, and cluster in the center of the diagram. The rocks with low siderophile elements cluster toward the compositions of postulated primitive highland crustal components (anorthosite, Mg suite and KREEP). This diagram constitutes probably the best evidence for the pristine rock concept [49].

However, the converse that one can calculate the indigenous siderophile element abundances on the Moon from such rocks encounters serious problems [98, 101] since the addition of small amounts of meteoritic contamination, which produce no effect on the other element concentrations, can never be excluded.

It is also clear that most highland rocks have been involved in impact events, and the definitions for pristinity are not exclusive. Thus, many impact melts at terrestrial impact sites have no detectable siderophile element component [30, 102] and so meet the criteria for pristinity [97, 99]. The difficulty of finding a chemical signature of the impacting body is a familiar problem to investigators at terrestrial impact sites. Thus, caution is needed in applying the concept of pristinity. Accordingly, the absence of a siderophile element signature is not a decisive criterion.

The use of Ni/Co ratios in metal grains as an index of pristinity has been proposed [103]. However, projectile metal is usually different in composition to the impacting meteorite and much redistribution may take place with change of composition [104]. The concept that low siderophile rocks were produced in an impact melt, with extraction of the siderophile elements by a metal phase [101], would remove any genetic significance from apparently pristine rocks. There seems however to be no evidence to support this proposal [97, 105, 106].

5.8   The Orbital Chemical Data

A fundamental problem on the Moon is how to relate the information obtained from the individual samples and landing sites with overall surface or bulk moon compositions. This problem is much less difficult than might be supposed from analogies with the Earth, on account of the simpler history of the Moon, the uniformity imposed by cratering, and the excellent photogeological coverage. The question whether the landing sites were in some way atypical was resolved by the two orbital geochemical experiments flown on Apollo 15 and 16. One measured the secondary X-rays produced at the lunar surface by the primary solar X-rays. On this basis, data for Si, Al, Fe, Mg and Ti were obtained. A second experiment (the gamma-ray spectrometer) measured the natural radioactivity of the surface, providing information on Th, K and Ti abundances.

The depth of sampling of the XRF experiment was about 10 microns, the "very surface" layer (see Section 4.2), and the data collection was limited, of course, to those areas in sunlight. The gamma-ray experiment, in contrast, probed somewhat deeper, on the order of a few tens of centimeters. Both of these experiments provided exceedingly valuable geochemical data. Although the data base of 3–4 elements might be thought sparse, it will be shown later that it is possible to extrapolate from these numbers to provide nearly complete trace and major element information about the areas surveyed.

The area of the Moon surveyed was limited to the track of the orbiting spacecraft. This was at a high inclination for the Apollo 15 mission (26°), but nearly equatorial (9°) for the Apollo 16 flight. The agreement, where overlap occurred, between data from the two missions was generally within 10%. Correlation with surface sample chemistry was established by measurements over the landing sites. Full details of the analytical methods and the experimental details are given by Adler [107] and Arnold [108]. (For reviews, see [109, 110].)

The spatial resolution of the data is about 50 km for the XRF data and about 70 km for the gamma-ray values. An important initial observation from the Al/Si data is that the ratio is greater in the highlands than in the maria, consistent with the more aluminous nature of the highlands and reflected by the high content of plagioclase feldspar. The first-order observation from the gamma-ray data is that the distribution is exceedingly inhomogeneous, with large values in the western maria regions around Mare Imbrium. This striking evidence for the lateral heterogeneity of the distribution of thorium across the lunar surface exceeds that predicted from the first-order highlands-maria dichotomy. The absolute abundance of thorium in the highlands is also a matter of great consequence for geochemical theories of lunar evolution.

Refinements of the data set have produced an average highland surface crustal value of 0.9 ppm Th [111, 112].

A basic question is the degree to which this surface sampling is representative of deeper layers in the highland crust. This question is addressed in the next section, but some direct evidence comes from the orbital data itself. Data collected over the Aristarchus Plateau reveal that the crater Aristarchus and its ejecta blanket have Th values as high as any on the Moon. The crater has sampled material to a depth of at least 3 km and the "thorium concentration of 12–13 ppm clearly demonstrates that the plateau material "has high concentrations of the radioactive elements" to at least that depth; it does not consist of gabbroic or anorthositic highland material lightly mantled with radioactive volcanic material" [112, 113], although opinions still persist about the "skin-deep" nature of the Th concentrations [114].

Other high abundances of K and Th are located generally around the Imbrium basin, with abundances of about 2000 ppm K and 10–12 ppm Th being common [113]. The Apennine Bench Formation has about 11 ppm Th; although this is sometimes interpreted as evidence for KREEP volcanism, the occurrence of similar values around the Imbrium basin and at the Aristarchus impact crater are interpreted here to indicate that the Imbrium impact sampled a KREEP-rich stratum in the lunar highland crust.

In another context, the presence of high-Al ejecta blankets around Mare Crisium, the Smythii basin and the crater Langrenus indicate that ejecta are coming from a deep Al-rich (hence, plagioclase-rich) layer, contrary to notions that the anorthosites form also a thin surficial layer.

The Kant Plateau, an elevated region to the east of the Apollo 16 landing site, has higher Al and lower Mg concentrations than the rest of the central highlands, leading to suggestions that it might constitute "primordial terra crust" [115, 116] (Fig. 5.24). The Kant Plateau forms part of the rim of the Nectaris basin and must be comprised of primary ejecta from that basin. If this anorthositic material was sampled at the Apollo 16 Descartes site, it could provide an age for the Nectaris event (see Sections 5.10.4). The existence of any remnants of primordial crust is exceedingly doubtful (see Sections 3.1 and 3.6).

The sharpness of the boundaries between maria and highlands is instructive. The Al/Si values in both highlands and maria reflect rather closely the values in the bedrock samples from these areas (Fig. 5.25). Thus, this primary difference in composition persists to the "very surface." This observation places severe constraints on the lateral movement of fine-grained material from the highlands to the maria by dust transfer mechanisms.

A further consequence of the data has been noted with reference to the problem of the origin of the smooth highland plains of Cayley Formation type. The lack of correlation between the gamma-ray data and the areas mapped as Cayley Formation might preclude their derivation as a uniform

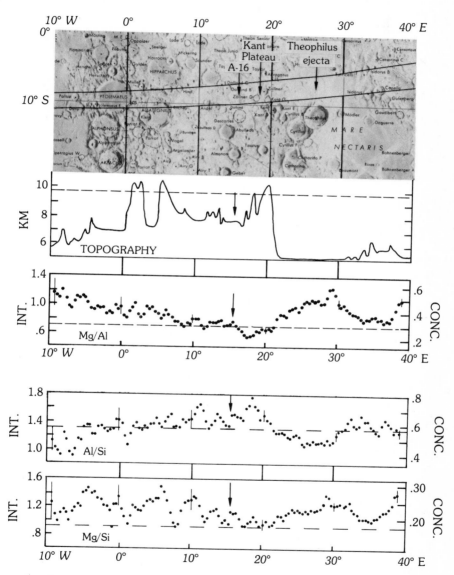

5.24 The variation in topography, Mg/Al, Al/Si and Mg/Si ratios in the vicinity of the Apollo 16 site [115]. (Courtesy C. Andre.)

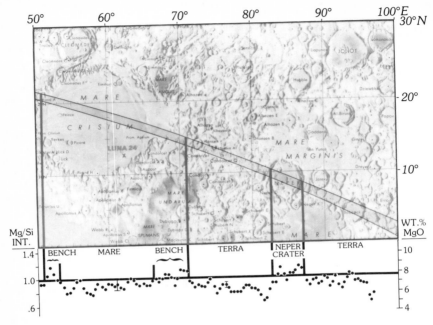

5.25 Highlands-mare contrasts in Mg/Si ratios from orbital data. Note the high-Mg basalts at the edges of Mare Crisium and in the Crater Neper [115]. (Courtesy C. Andre.)

ejecta blanket from a single source but secondary cratering may complicate this picture by excavating older ejecta blankets. The lateral variations in composition indicate that the cratering and ringed-basin formation have not completely homogenized the highland rocks. The belief that the intense cratering might produce an exceedingly uniform highland crustal composition appears unfounded, although the common occurrence of "highland basalt" compositions indicates that mixing is an important process. If terrestrial experience is a guide, this occurs mainly during the production of melt rocks during impacts, and so is limited to a few percent of the target rocks.

Regional geochemical anomalies are probably not due to lateral heterogeneities produced during global differentiation of the magma ocean. This event appears to have been uniform and moon-wide, as is evidenced by the uniform isotopic and chemical characteristics of KREEP on a moon-wide basis. Other factors account for the various geochemical anomalies: (a) basin-forming processes, (b) early reworked mare lavas, (c) buried mare lavas as shown by the presence of dark-halo craters. Thinly buried mare units may make an important contribution to the existence of geochemical anomalies

[117]. Values for Fe [118], Mg [119], and Ti [120] have also been recovered from the gamma-ray data.

5.9   The Chemical Composition of the Highland Crust

Initial observation with low-powered telescopes revealed that the highlands were elevated, relative to the maria [121]. This led to analogies, both for density and composition, to the terrestrial continents. A crucial step was to resolve the nature of the large craters. If these were of volcanic origin, then lunar magmas would possess different properties than those on Earth. The resolution of the crater problem was accomplished effectively by Gilbert and Baldwin (Section 3.2), but the concept of widespread volcanism on the lunar highlands continued, finally influencing the selection of the Apollo 16 landing site. The terrestrial continent analogy suggested the possibility that the lunar highlands were of granitic composition [122, 123]. This concept was reinforced by two factors: (a) Detailed mapping of the highlands revealed the presence of light plains units of varying stratigraphic age, which would be consistent with deposition by a mechanism analogous to terrestrial ignimbrite eruptions. The highly viscous lavas required to generate such ash-flow deposits would necessarily be acidic, analogous to terrestrial dacites or rhyolites in composition [124]. (b) This concept was agreeable to those workers who wished to derive tektites from the Moon, since it provided an acidic source material. The distinctions between the composition of terrestrial rhyolites and tektites were ignored.

The first direct analysis of the lunar highland surface by Surveyor VII, on the rim of Tycho, revealed a totally different composition, similar to terrestrial high-alumina basalt [125], and widely different from granitic compositions. The first clue to the mineralogical nature of the highlands came from the discovery of anorthosite fragments in the Apollo 11 soil samples [126]. This immediately suggested a possible resolution to a major geochemical dilemma by providing a sink for europium. Thus began the "magma ocean" concept.

The next surprise was the discovery at the Apollo 12 site of the light gray fines (e.g., 12033) and rock sample 12013, both enriched to an extraordinary degree in large ion lithophile elements [127]. The evidence for an extreme degree of element fractionation, reminiscent of the concentrations observed in terrestrial pegmatites, was dramatically reinforced by the Apollo 14 sample return from Fra Mauro [128]. The existence of large areas of K-rich material (KREEPUTH) at the lunar surface, indicated by the gamma-ray orbital data, led to new concepts of the origin of the highland crust. Since the Fra Mauro site was on the Imbrium ejecta blanket, these unusual compositions focused attention on possible sources, either local or from within the Imbrium basin.

Attempts to explain the chemical composition of the crust with its high content of refractory elements led quickly to the development of heterogeneous accretion models [129]. The late addition of a layer rich in refractory elements at first appeared to pose fewer problems than those caused by melting and cooling of large volumes of the Moon, since this was required to happen at a very early stage of lunar history. The geochemical debate over homogeneous versus heterogeneous accretion, which led to some extremely high-alumina moons [130], was eventually resolved in favor of homogeneous models (or homogenization after accretion) [131]. The key evidence was the recognition that the ratios of volatile/refractory elements were similar in mare basalts (derived from the deep interior) and in the highland crust [132]. This linking of the highland chemistry with that of the source regions of the mare basalts provided a ready explanation for the europium anomalies and many other geochemical problems [74].

Meanwhile, the sampling problems persisted. How representative of the crust were the surface samples, and how might they be related to the Al/Si and Th values derived from the orbital geochemistry experiments? A solution to this problem was proposed by the Apollo soil survey [133], in which the proposition was advanced that the glass spherules formed during impact would be representative of the principal components of the highland crust. (Composition of the spherules could be rapidly determined by microprobe analysis.) The amount of mixing which has occurred and the tendency of impact melts to homogenize differing source rocks [30] render this approach less useful for distinguishing "primordial" rock types [45]. If these conclusions are valid, then there should be a prominent grouping of glass compositions, representing the average crust. The "highland basalt" composition (26% Al_2O_3) [133] probably meets this criterion. Other samples of granulitic breccias (e.g., 67955, 76230, 77017, 78155, 79215) [134] have bulk compositions close to that of the average highland crust.

5.9.1 Element Correlations

The observation that many of the refractory trace elements were correlated in lunar highland samples of differing composition, and from different landing sites [135, 136], is a key which enables estimates to be made for the abundances of many trace elements. These can be related via the gamma-ray thorium abundances to moon-wide compositions. By this means, estimates for average lunar highland surface compositions can be made.

These correlations among many distinctive elements in the highland samples (Fig. 5.26) have been instructive to geochemists [137]. The correlations depend on various factors such as similarity of geochemical behavior due to resemblances in ionic radius, valency or bond type which lead to conventional geochemical coherence [e.g., K/Rb, Th/U, Zr/Hf, REE (except

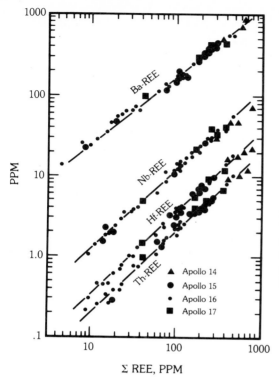

5.26 The close correlation observed between many involatile elements in the highland samples. In this diagram, the abundances of Ba, Nb, Hf, and Th are plotted against total REE abundances. Correlations with individual REE would be similar. Many other similar correlations exist. These shown here are selected to emphasize the correlations between elements of dissimilar geochemical behavior.

Eu), Ba/Rb, Rb/Cs, Fe/Mn], and concentration of incompatible elements into residual melts (e.g., K/U, Zr/Nb, K/Zr, K/La). This type of correlation needs to be treated with the understanding that it is not due to fundamental geochemical properties, but is accidental to some degree. The fact that these correlations are highly significant statistically indicates the degree of uniformity in crustal processes on the Moon. The close correlations between volatile and refractory elements (K/U, Ba/Rb, K/Zr, K/La) both in mare and highland samples indicates that both regions were originally homogeneous with respect to volatile and involatile elements. This is a primary piece of evidence for homogeneous accretion of the Moon, or homogenization following accretion.

Correlations resulting from mixing of different rock types during intense cratering of the highlands is a further process that will contribute to the observed close inter-element ratios in highland samples. The fact that the chemistry of the highland samples appears to be controlled by three end-member components (anorthosite, Mg-component and KREEP), and that the large ion lithophile element abundances are dominated by the KREEP component, lead us to expect simple element correlations.

Table 5.5 Highland crustal composition, compared with data from granulitic breccia 78155.

Oxide	wt.%	78155	Element	ppm	78155	Element	ppm	78155
SiO_2	45	45.6	La	5.3	4.0	U	0.24	0.28
TiO_2	0.56	0.3	Ce	12	10.2	Th	0.9	1.0
Al_2O_3	24.6	25.9	Pr	1.6	1.5	Th/U	3.8	3.6
FeO	6.6	5.8	Nd	7.4	6.3	K/U	2500	2320
MgO	6.8	6.3	Sm	2.0	1.81	Zr	63	54
CaO	15.8	15.2	Eu	1.0	0.87	Hf	1.4	1.49
Na_2O	0.45	0.3	Gd	2.3	2.3	Zr/Hf	45	36
K_2O	0.075	0.08	Tb	0.41	0.39	Nb	4.5	—
Cr_2O_3	0.10	0.10	Dy	2.6	2.6	Zr/Nb	14	—
			Ho	0.53	0.61	Ti	3350	—
Element	ppm		Er	1.51	1.69			
Cs	0.07	0.11	Tm	0.22	—	Cr	680	680
Rb	1.7	2.1	Yb	1.4	1.73	V	24	—
K	600	650	Lu	0.21	0.26	Sc	10	13
Ba	66	59				Ni	100	80
Sr	120	147	ΣREE	39.3	34.3	Co	15	14
			Eu/Eu*	1.4	1.3	Fe%	5.13	4.51
			Y	13.4		Mg%	4.1	3.8

Data sources for 78155:
   Warner, J. L., et al. (1977) *PLC 8:* 2057.
   Wänke, H., et al. (1976) *PLC 7:* 3482.
   Hubbard, N. J., et al. (1974) *PLC 5:* 1235.

The preservation of so many element correlations in highland rocks is thus a predictable consequence of mixing of a few components. The parallel nature of the REE patterns is only the most striking of these, since any igneous fractionation would change the slopes of the patterns. These correlations greatly facilitate the calculation of highland crustal abundances.

5.9.2 Highland Crustal Abundances

In this section a table of element abundances (Table 5.5) in the lunar highlands is presented. The method used is similar to that described previously ([137], pp. 249–253) and the details are not repeated here. The technique is to employ the orbital data to provide moon-wide averages, at least over the orbital tracks. Thus, the averages are not dominated by a specific region [138].

Al/Si and Mg/Si ratios from the orbital data for the highlands are 0.62 (± 0.10) and 0.24 (± 0.05), respectively. Concentrations of SiO_2 are relatively uniform at 45% in highland samples, yielding values of 24.6% Al_2O_3 and 8.6% MgO. A value of 6.6% FeO is obtained from the MgO/FeO relationship observed in the highland soils and breccias. This value compares with 6.9%

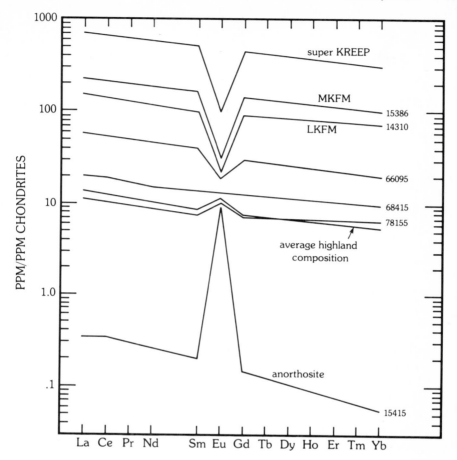

5.27 Rare-earth element abundances in various typical highland rock samples. Data from Table 5.3. Note that the average highland composition proposed in this book shows enrichment in europium, and is close in compostion to the granulitic breccia 78155.

FeO from the orbital data [118]. MgO orbital values average 6.8% [119], being lowered by the far-side highland values in comparison with the sample data. The orbital value is adopted here. From the Fe/Cr relationship, a figure of 0.10% Cr_2O_3 results. Allowing a typical lunar Na_2O abundance of 0.45%, the remaining major constituent is CaO, which yields, by difference, a value of 15.8%. This value is consistent with that observed in samples of this approximate composition and with the Ca/Al relationship. The Th value is 0.9 ppm [111, 112]. Using a Th/U value of 3.8, then U = 0.24 ppm; from the lunar K/U ratio of 2500, then the K abundance is 600 ppm [137] which is in

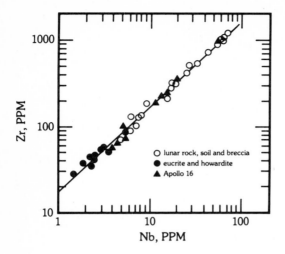

5.28 The correlation between zirconium and niobium for lunar samples and the howardite and eucrite groups of meteorites [140].

agreement with a recent orbital average estimate of 500 ppm [139]. Rubidium and cesium values are calculated from $K/Rb = 350$ and $Rb/Cs = 24$.

From the Th/REE correlation (Fig. 5.26) a total REE value of 39 ppm is obtained. Assuming that the La/Yb ratio is the same as for other highland rocks, REE values can be obtained from the data illustrated in Fig. 5.27, where the average pattern is compared with the other highland REE data. The pattern is rather close to that of "highland basalt" [137], consistent with the derivation of that well-known composition by overall mixing during impact events. A notable consequence is that the europium enrichment in the highland crust is pronounced (Eu/Eu* = 1.4). Since the overall lunar REE pattern has no Eu anomaly (Chapter 8), the origin of this europium enrichment in the highlands is a major factor to be explained in theories of highland crustal formation.

Values for the other trace elements (Ba, Hf, Nb) may be derived from the data illustrated in Fig. 5.26. Values for Zr come from the well-marked Zr/Nb correlation (Fig. 5.28)[140]. Yttrium values are derived from the similarity of chondrite-normalized Y and Ho values. Cr/V and Cr/Sc ratios (Table 5.5) provide V and Sc values while Sr values are derived from Sr/Eu and Rb/Sr ratios.

The composition in Table 5.5 is considered to be representative of the overall highland crustal surface (as noted above), in the same manner as estimates of continental crustal composition are representative of the terrestrial crust. For comparison, the composition of the granulitic breccia 78155 is given in Table 5.5. The abundances of both major and trace elements in this sample are remarkably close to the average highland composition, derived from a differing set of criteria. The next step, extending these results to draw conclusions about the composition of the total highland crust, is less readily taken. How representative is this surface composition to that of the bulk highland crust to depths of 60–80 km? How reliable are the geophysical estimates of crustal thickness? Does the crust vary laterally in chemistry and/or thickness? From what depth are samples brought to the surface by the large impacts which form the ringed basins? How much mixing and homogenization has resulted from these and smaller collisions? These questions have been debated throughout the text (see Chapters 2 and 3, especially), and they will be addressed again in Chapter 8. The overall evidence indicates that this composition is representative of the upper 30 km of crust at least. Probably it extends to the base of the crust, although this may be irregular due to mantle uplift during multi-ring basin formation.

The consequences of a more aluminous composition, which will lower the crustal density (Section 7.2), is that a thinner crust will accommodate the geophysical requirements; conversely, a less aluminous crust needs to be thicker. The crustal composition discussed here forms about ten percent of the lunar volume, so that the elemental concentrations comprise a significant fraction of their whole moon abundances (Section 8.4), in contrast to those of the mare basalts which comprise perhaps only 0.1% of the lunar volume.

5.10   Age and Isotopic Characteristics of the Highland Crust

5.10.1   The Oldest Ages

Few samples of the highland crust have avoided having their isotopic systems reset by the meteoritic bombardment. These include the black and white breccia 15455, of which the white portion yields an Rb-Sr age of 4.52 aeons, and the dunite (72417) (Fig. 5.29). How reliable are these old ages? All show some signs of disturbed isotopic systematics, with data points falling off the isochrons.

Redistribution of volatile elements such as Rb (Section 4.5.2) is particularly likely and may affect KREEP ages. Some debate exists over the effects of impact in resetting ages. Studies at the Ries Crater show that granite clasts shocked to 450 kbar have lost most (99%) of their radiogenic ^{40}Ar [141], hence the view that only melted rocks which have reset ages need reconsideration. Rb-Sr isochron ages of greater than 4.5 aeons have been obtained for troctolite 76535. Other methods give younger ages. For 76535, both Sm-Nd and Ar-Ar methods yield ages near 4.3 aeons, distinctly younger than the Rb-Sr age, but these may represent excavation ages from a deep hot crust. However, the initial ^{87}Sr/^{86}Sr ratios are close to BABI, and no pre-4.6 AE ages were obtained. Accordingly, these old ages are probably meaningful [142]. A summary of those samples which yield very old ages is given in Table 5.6 and

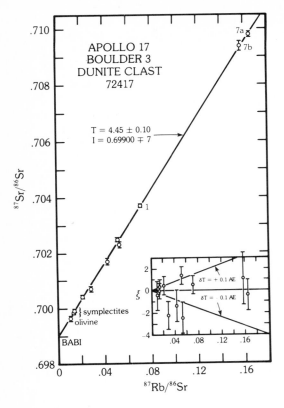

5.29a The oldest lunar rocks. Rb-Sr evolution diagram for mechanically separated samples of the dunite. The insert shows deviations ξ (in parts in 10^4) of the measured $^{87}Sr/^{86}Sr$ from the best-fit line. The present measured enrichment in $^{87}Sr/^{86}Sr$ is 1.4%. The initial $^{87}Sr/^{86}Sr$ of the dunite is essentially equal to BABI = 0.69898.[Papanastassiou, D. A., and Wasserburg, G. J. (1977) *PLC* 6: 1473.]

Fig. 5.30. Their significance is considerable. They provide additional evidence for crustal formation at times around 4.4 aeons. The principal conclusion is that the highlands crust is very old, dating back almost to the origin of the solar system.

The existence of such ancient lunar ages places very tight time constraints not only on the accretion of both Earth and Moon, but also on their possible relationship, fission origins and time of melting of the outer portions of the Moon. By 4.4 aeons, solid-liquid equilibria must have already acted to establish both the isotopic and chemical characteristics of the mare basalt source regions (Section 6.4) and the reciprocal nature of mantle and crustal geochemistry. Nevertheless, a solid crust in the sense that we observe it today may not have been in evidence. The existence of formed bodies of crustal rocks prior to 4.2 aeons is conjectural on account of the "stonewall" effect [143].

The age of the Earth and meteorites is usually given as 4.56 aeons. This age dates the time of fractionation of the U-Pb system from low values of

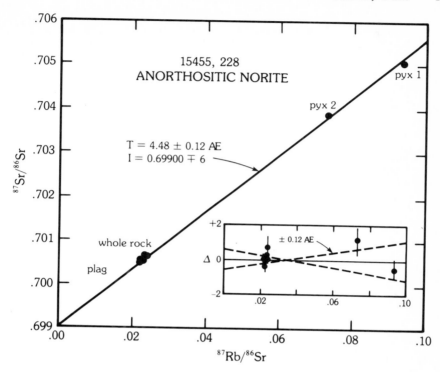

5.29b The oldest lunar rocks. An isochron age of 4.48 aeons for the anorthositic norite 15455, 228. (Courtesy L. E. Nyquist.)

Table 5.6 Old ages on highland crustal rocks.

Rock Sample		Age	Reference[†]
15455	Anorthositic gabbro (Rb-Sr)	4.48 ± 0.10	1
72417	Dunite (Rb-Sr)	4.45 ± 0.10	2
76535	Troctolite (Rb-Sr)	4.61 ± 0.07	3
77215	Norite (Rb-Sr, Sm-Nd)	4.4	4
78236	Norite (Sm-Nd)	4.49	7
Various anorthosites ($^{40}Ar/^{39}Ar$)		4.4 – 4.5	5,6

[†]References:
1. Nyquist, L. E., et al. (1979) Lunar Highlands Conf. Abstracts, 122.
2. Papanastassiou, D. A., and Wasserburg, G. J. (1975) *PLC 6:* 1467.
3. Papanastassiou, D. A., and Wasserburg, G. J. (1976) *PLC 7:* 2035.
4. Nakamura, N., et al. (1976) *PLC 7:* 2309.
5. Dominik, B., and Jessberger, E. (1978) *EPSL.* 38: 407.
6. Huneke, J. C., and Wasserburg, G. J. (1979) LPS X: 598.
7. Carlson, R. W., and Lugmair, G. W. (1979) Lunar Highlands Conf. Abstracts, 9.

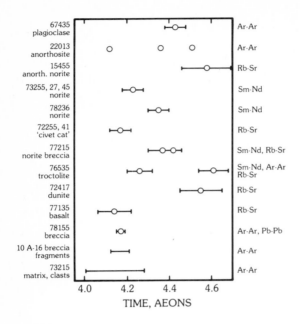

5.30   List of lunar highland samples with ages greater than 4.1 aeons. Multiple ages listed for single samples indicate discrepancies between data obtained from different radiometric systems, or multiple plateau ages from ^{40}Ar–^{39}Ar studies. The samples listed belong to the Mg-rich suite with the exception of 22013 (anorthosite), 77135 and the 73215 matrix clasts. [Carlson, R. W., and Lugmair, G. W. (1981) *EPSL*. 52: 235. Courtesy G. W. Lugmair.]

μ (^{238}U/^{204}Pb) of the average solar system material (0.3) to primitive terrestrial values of about 7.5. Whether or not this fractionation occurred before, during, or after the accretion of the Earth, it *is* clear that it occurred close to the accretion event (within 20 ± 10 million years).

The following sequence of events thus had to occur within 150–200 million years of the general accretion of the solar system:

(a) Accretion of the Moon (from previously fractionated material not of solar nebula composition).

(b) Melting of at least the outer half and probably the whole Moon.

(c) Crystallization of the magma ocean to provide the old anorthosites, troctolites and dunites.

(d) Formation and isotopic closure of the source regions of the mare basalts.

(e) Invasion of the highland crust by the residual liquids from the crystallization of the magma ocean.

The next question to be addressed is the time at which early differentiation of the crust occurred. An age of 4.46 aeons for the time of the crustal differentiation is derived from the U-Pb-Th systematics [144] (Fig. 5.31). The Rb-Sr and Sm-Nd systematics combine to yield somewhat younger ages; however, crystallization of the magma ocean was apparently complete by 4.35 aeons. We note again the problems associated with determining this age from

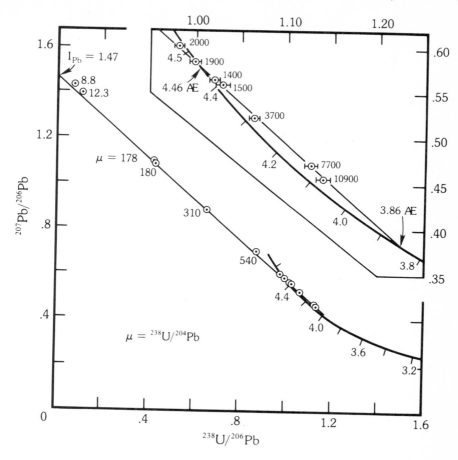

5.31 U-Pb evolution diagram for highland breccias. Inset shows data points near concordia curve with error bars. The measured value of $\mu = ^{238}U/^{204}Pb$ is given for each sample. Of 13 data points, 10 lie precisely on a line intersecting concordia curve at 4460 m.y. and 3860 m.y. The upper intersection represents the time of early lunar differentiation and crust formation. The lower intersection corresponds to a time of intense bombardment, which has reset the ages. Samples whose data points do not plot precisely on the line may have impact ages slightly different from 3860 m.y. The bombardment affected almost all lunar highland rocks, causing severe U/Pb fractionation in the breccias. This is reflected by the linear distribution of the data points on the graph and the strong correlation of μ values with position of the data points [144]. (Courtesy G. J. Wasserburg.)

highland crustal rocks, in which the systematics may be reset down to 4.2 aeons or to younger times by the meteoritic bombardment.

5.10.2 The 4.2 Aeon Ages

A number of highland samples have ages of about 4.2 aeons. An extensive investigation of Apollo 16 breccias carried out by Maurer et al. [145] established three age groups based on $^{40}Ar/^{39}Ar$ using new decay constants; 48 soil samples from the 2–4 mm fines collected near North Ray crater (50 million years old) were investigated. The oldest group had ages from 4.06–4.15 aeons and were all anorthositic. These are regarded as coming from impact craters up to a few hundred kilometers in diameter, which excavated an upper anorthositic layer. This might, however, represent the Kant Plateau anorthositic layer recognized by high Al/Si ratios (Section 5.8) (most probably primary Nectaris ejecta) and so date the Nectaris event itself. The second group had ages from 3.83 to 3.96 aeons and were noritic anorthosite and anorthositic norite breccias. These showed a higher degree of annealing (melted matrix) and contained one totally melted KREEP basalt. This group was considered to be connected with the Nectaris basin-forming event, which in this model had sampled deeply enough to excavate deep-seated KREEP. Most ages in this group are from 3.83 to 3.96 aeons. Their interpretation gives an age for the Nectaris event of 3.92 aeons, supporting the multi-impact cataclysm model in which Nectaris and other major basins were formed within a 0.1 aeon time span. This interpretation is model dependent. The older ages might well be due to the Nectaris event. If they are not, it might reasonably be inquired which event is in fact represented by them. The third group was composed of much younger breccias (< 2.5 aeons). Only one of the 48 samples fell outside the three groups; its age was 3.55 aeons and it is thought possibly to have come from Theophilus (100 km diameter), which is within three crater diameters of the Apollo 16 site. Rb-Sr dates on a suite of pristine anorthosites yielded ages of 4.17 ± 0.20 aeons (2σ) with an initial ratio of 0.69906 ± 2, which is within the error limits of LUNI (0.69903), as defined by the purest anorthosites [146]. These samples, which contain high Sr and low Rb concentrations, are the best candidates for preserving primitive $^{87}Sr/^{86}Sr$ ratios. A "pristine" lherzolite yields an Sm-Nd isochron of 4.18 aeons [147]. Several questions are raised by these and similar data and various interpretations are possible. For example:

(a) Crystallization of the lunar highland crust lasted until about 4.2 aeons.

(b) Crystallization of the main magma ocean was complete at 4.4 aeons, but some of these samples are the products of later intrusions into the crust.

(c) The isotopic systems remained open until about 4.2 aeons due to high temperatures within the crust. Thus, the 4.2 aeon ages could either represent excavation ages, or the time at which the temperature became low enough to close the isotopic systems.

(d) The ages are reset by the bombardment.

Many of the ages are $^{40}Ar/^{39}Ar$ ages, and it is generally considered that these were reset during large cratering or basin events, with opinions differing as to whether the age distribution is dominated by the large basin impacts [148, 149, 150] or during formation of medium-sized craters [151]. Argon loss or retention may occur as a result of (a) heating or melting during crater formation, (b) annealing in a hot ejecta blanket, either in a short time at high temperatures or over a long period at lower temperatures, (c) uplift and cooling following a major basin impact (this only works for large impact basins where the depth of excavation is many kilometers) and (d) crystallization from either a primary igneous or meteorite impact melt [148]. Few highland samples appear to belong to this igneous category. Both 14310 and 68415 are impact melts.

There are several arguments in favor of the resetting of argon ages by big basin impacts rather than by small craters. There are many examples of rocks which are identified as coming from young lunar impact craters, such as North Ray, Camelot or Cone. There are no examples where any significant resetting of the ^{40}Ar-^{39}Ar ages has occurred as a result of the formation of these kilometer-sized craters [148].

Examples exist of a few highland rocks which have been reset at times younger than 3.8 aeons. These include 14318, 61016 and 63335 which yield ages of 3.7 aeons. The anorthosite 60015, which has a glass coating, gives a well-defined age of 3.50 aeons [150], which cannot be due to a basin-forming event. Since the number of large craters younger than 3.0 aeons is small (about 15), the chances of finding rocks reset by these events (e.g., by Copernicus or Tycho) are likewise small.

Table 5.7   Calculated temperatures and heating times required to produce 70% argon loss from lunar anorthosites.

Temperature θ (°C)	Heating time[†] τ (years)	Conduction length[*] a	Depth[§] d (km)
720	1	4(m)	24
560	10^2	40	19
450	10^4	400	15
370	10^6	4(km)	12
300	10^8	40	10
270	$5 \cdot 10^8$	90	9

[†] The time required to produce 70% argon loss is calculated by extrapolating laboratory diffusion data assuming an activation energy of 50 (kcal mole^{-1}K^{-1}).
[*] $a = (2\kappa\tau)^{1/2}$ is the depth from which significant heat loss can occur (e.g., from an ejecta blanket in time τ). Calculated using $\kappa = 10^{-5}$ km^2 year $^{-1}$.
[§] $d = \theta/(d\theta/dz)$ is the depth below the surface at which time the ambient temperature equals θ. Calculated assuming $(d\theta/dz) = 30°C$ km^{-1}.
Turner, G. (1977) *PCE* 10:181.

The high frequency of $^{40}Ar/^{39}Ar$ ages of 3.8 aeons in the lunar highlands may be due to resetting by excavation from depths below the 350°C isotherm. Although heating during the basin-forming process may be minor and the ejecta blankets are cool, rather than hot, nevertheless, much melt is produced in these events.

Table 5.7 shows the time needed to lose 70% of argon for various temperatures and inferred depths of burial. It must be recognized, however, in contrast to terrestrial examples, that multiple impact histories are involved. Even though the amount of melt, and thus igneous appearing rocks with reset ages may be small in each event, the cumulative total due to the extended bombardment history explains the rarity of ages older than about 4.2 aeons.

5.10.3   Basin Ages and the Lunar Cataclysm

This topic has many implications for lunar history, with diverse conclusions being reached from isotopic and cratering studies. A principal difficulty is to identify samples as resulting from particular basin-forming collisions. This is becoming more difficult as we understand the complexities of large collisions, the likely temperatures in ejecta blankets, and the relative roles of secondary and primary ejecta. A recent survey of basin ages has been given in *Basaltic Volcanism on the Terrestrial Planets* (Chapter 7 [152]).

Nectaris Basin

There is considerable controversy over the age of this basin. The Descartes site, sampled at Apollo 16, is only 60 km from the outer rim of the Nectaris basin, and accordingly, primary ejecta from that basin must be present at depth. The Cayley plains are too young to be Nectarian, but the hilly Descartes Formation might represent primary ejecta. If so, two possibilities exist for sampling. North Ray crater may have excavated deeply enough to sample it. Alternatively, samples from stations 4 and 5 on Stone Mountain may be of Nectaris ejecta (see Section 5.8 on Kant Plateau). A Nectaris age of 4.2 aeons was assigned by Schaeffer and Husain [150] from analyses of Apollo 16 site material. The work reported earlier by Maurer et al. [145] is relevant here. They considered that the older (> 4.06 aeons) ages represented medium-sized impacts. The large number of 3.83–3.96 aeon ages were thought to come from a basin-sized impact which dug deeply enough to penetrate a conjectured upper crustal anorthosite layer, and excavate KREEP rich material. This was considered to be the Nectaris event, and hence these data provide an age of 3.92 aeons for Nectaris. This interpretation is dependent on our models of the highland crust. A reasonable case can be made that the anorthositic Kant Plateau is primary Nectaris ejecta. Accordingly, the anorthositic samples with 4.06 aeon ages might be derived from this region and so date the Nectaris event. At least two other possibilities exist.

Imbrium secondaries might have plowed up Nectarian ejecta, or the site is dominated by primary Imbrium ejecta [20]. An unknown contribution from Serenitatis primaries or secondaries may be present.

Serenitatis Basin

Various age estimates from the Apollo 17 site converge on 3.86 aeons as a probable age for this basin excavation. However, this estimate has not gone unchallenged. The highland samples collected at the Apollo 17 site are dominated by poikilitic, impact-produced melt breccias, not necessarily Serenitatis ejecta, but possibly deeper uplifted material.

A separate class of rocks are identified as aphanitic (crypto-crystalline— too fine grained to be readily visible). These light-colored rocks occur at Boulder 1 at Station 2 (73215, 73235, 73255), and have different chemistry, clast populations and petrography, compared to the more common impact melts, and are thought to come from shallower depths. Wood [153] suggested that they came from deep within the basin and were average crustal samples. Dence [154] suggested that the average crust was of low-K Fra Mauro composition, which is unlikely, particularly since granulitic breccias such as 78155, matching estimates of highland crustal compositions from the Apollo 16 site, are present. Spudis and Ryder [155] suggest that samples from smaller impacts such as Littrow (30-km diameter, 50 km from the landing site) and Vitruvius (30-km diameter, 80 km from the landing site) are possible candidates. They suggest that the Taurus-Littrow highlands are not dominated solely by Serenitatis ejecta, but rather have a complex multiple-impact history involving other basins.

Imbrium Basin

This is usually dated from the Apollo 14 site, where most of the KREEP-rich melt rocks have ages of 3.82 aeons (e.g., 14310, 14073, 14276, 14001). The general interpretation of 14310 as an impact melt has lent credence to this as dating the age of the Imbrium event. Basaltic clasts within the Apollo 14 breccias have ages of 3.86–3.88 aeons. There appears to be a general consensus about this age for a number of reasons. The oldest ages from the mare basalts are only just younger (and indeed 10003 is older!) so that there is some agreement between the geochronologists and the geologists on this interpretation. The possibility that the Apollo 14 samples are local material, merely disturbed by the Imbrium event [30], seems less likely. The Apollo 14 site at Fra Mauro is only 550 km from the main Imbrium rim and the material there must have a high component of primary ejecta from Imbrium. How much of this material was melted and reset, and how much of the ejecta blanket was cold is a matter of debate. The presence of thermal effects and of local magnetic fields is taken to indicate the presence of local pools of melt within the ejecta blanket. The problems of Cayley-type plains and the importance of secondary ejecta should not be over-emphasised at the Apollo 14 site, close to the Imbrium basin rim.

The Lunar Cataclysm

One interpretation of the age data given here could support the concept that Nectaris and the post-Nectarian basins all formed within about 100 million years. There are, as noted earlier (Section 3.14), many pre-Nectarian basins, and the occurrence of several large collisions in such a period is not suggestive of a spike in the cratering flux and depends heavily on interpretation of individual samples. The principal philosophical support is the agreement about the Imbrium age, dated at 3.82 aeons. This is based on the analyses of the melt rocks at the Apollo 14 site and the assumption that these date the Imbrium event. The South Serenitatis basin event has a possible age of 3.86 aeons. The Nectaris basin may have an age of 3.92 aeons, although the older ages of 4.1 aeons seem equally likely to date the Nectaris event [150]. There are sufficient uncertainties in the interpretation of the age data and in the assignments to specific basins to make the cataclysm a non-unique interpretation. Even on the most conservative assumption, the cataclysm does not appear to be remarkable. Between Imbrium and Nectaris, only the formation of twelve multi-ring basins occurred. Imbrium was followed by Orientale, but Nectaris was preceded by 29 basins (Appendix IV, Table 3.1).

5.11   Evolution of the Highland Crust

The processes which formed the highland crust are in principle rather simple, but the details are complex, as is typical of most geological processes, recalling the statement of Urey that "Nature has a great capacity to produce most surprising results within the limitations of the basic laws of physical science." The anorthosite component is derived from flotation of plagioclase during crystallization of the magma ocean. The Mg-rich component is derived principally from trapped cumulus liquids with some probable later intrusions, while the KREEP component represents the final residual melt from the magma ocean which invades the highland crust. This scenario is complicated by the details of crystallization of the magma ocean and by the repeated infalling of large projectiles, which continued well beyond the crystallization of the magma ocean at about 4.4 aeons. The detailed processes responsible for the evolution of the highland crust are now considered.

5.11.1   The Magma Ocean

There is a considerable amount of evidence in support of the concept of large scale lunar differentiation. Geochemical balance problems provide one argument. The lunar highland crust is generally considered to be at least 60 km thick and contains perhaps 25% Al_2O_3. Reasonable lunar compositions do not exceed 6% Al_2O_3 so that the amount of aluminum in the highland crust is about 40% of the total lunar budget. Potassium, uranium and thorium are concentrated near the lunar surface by two orders of magnitude in excess of their lunar abundances. Europium is concentrated in the highland feldspathic crust to about the same degree as aluminum, and is depleted in the source regions of the mare basalts. Upper limits for the U content are set by heat flow as well as by overall cosmic abundance arguments. The isotopic data (Pb, Sm-Nd, and Rb-Sr) all point toward large-scale early differentiation. The Sm-Nd systematics of KREEP, which are uniform from all landing sites, imply a moon-wide event. In all these debates, an integrated approach is required involving geophysical models for crustal structure and thickness, and geochemical models which integrate orbital and sample data. The mere existence of a single differentiated rock sample does not, of course, imply moon-wide differentiation nor justify the existence of magma oceans. No one proposes that lunar "granites" are widespread or demand more than trivial amounts of parent material. It is the combination of the whole set of evidence which requires moon-wide differentiation.

The geochemical evidence does not specify the physical state of the initial differentiation. It does necessitate the operation of crystal-liquid fractionation on a moon-wide scale so that, for example, 50% of the europium and a similar amount of the potassium content of the bulk moon now reside in the lunar highland crust, which comprises about ten percent of the lunar volume. This may be carried out either by an extended sequence of small melting episodes, or by a magma ocean. For simplicity, the concept of a single magma ocean is adopted throughout the text since the geochemical evidence demands that most of the Moon was involved in a crystal-liquid fractionation sequence.

This differentiation must be completed quickly, in planetary terms. The time constraints on this process are discussed in Section 5.11.2. Less than 200 million years are available, and even less time (100 million years) if the accretion of the Moon takes 10^8 years. It is thus tempting to associate the differentiation and the thermal energy required with the lunar accretionary process. In contrast, the melting and eruption of the mare basalts does not require a massive energy source. They comprise about 0.1% of the lunar volume so that the amount of heat necessary to form them by partial melting in the lunar interior is over three orders of magnitude less than that required for the initial differentiation. Although various questions have been raised about the possible existence of the magma ocean, the alternatives are even less attractive and "the one time existence of a magma ocean is a reasonable conclusion"[156].

The question of the initial depth of the magma ocean is relevant here. Estimates have ranged from 200 km [157] to whole-moon melting [158].

Estimates of 800 km (85% of the lunar volume) [74], consistent with models of internal structure, were proposed at an early stage. The possible presence of a discontinuity at 400–480 km is not uniquely specified by the seismic data (Section 7.6.1) and is not considered to be a reasonable constraint. The requirement for concentration of elements into the lunar crust requires minimal depths of 500 km [159, 160], but more realistic depths exceed 800–1000 km [161]. If a lunar core is present, as is argued in Sections 7.6.2 and 8.4, then whole moon melting is effectively demanded. The geochemical necessity for massive near-surface concentration of elements by crystal-liquid fractionation is most readily accommodated by whole moon melting. It should be noted here that effectively all mare basalts have negative europium anomalies (Section 6.3.3) so that a massive removal of that element into the crust has occurred.

The possibility that mare basalts are derived by partial melting mainly from depths of 400–500 km lends support to the concept of whole-moon melting. The mare basalt source regions crystallize following prior crystallization of olivine, orthopyroxene and plagioclase (as is shown by the ubiquitous Eu depletion in the basalts). No evidence of the presence of garnet appears in the REE patterns of the mare basalts, indicating that Ca and Al were not present at depth in sufficient amounts to crystallize garnet as would be the case for undifferentiated lunar material. Probable convective overturning and sinking of dense cumulates forming late in the crystallization sequence has occurred. Very large volumes of olivine and orthopyroxene, which crystallized earlier, must be present at greater depths. It is thus a reasonable assumption that the Moon was melted to depths considerably in excess of 400–500 km. In summary, the evidence which requires effectively whole-moon differentiation and hence a "magma ocean" includes (a) the presence of a feldspathic crust comprising 10% of lunar volume, (b) complementary highland and mare basalt Eu anomalies and general geochemical characteristics, (c) enrichment of incompatible elements in the crust, (d) the isotopic uniformity of KREEP, and (e) the isotopic evidence for early differentiation of mare basalt source regions, completed by about 4.4 aeons.

5.11.2 How Long did the Early Highland Crust Take to Evolve?

There is a considerable body of evidence which suggests that lunar differentiation occurred as early as 4.47 aeons [144]. One interpretation of the isotopic data is that early lunar magmatism involving highly fractionated sources continued for at least 200 million years [156]. As long as the bombardment of the highlands continued, with projectiles capable of excavating craters and basins greater than 150 km in diameter, it is reasonable to expect continued reworking of the highland crust.

Further insights into this question are gained from calculations about the time taken for the magma ocean to solidify [162]. The time required depends on the thickness of the crust which develops, since heat loss is then controlled by conduction through the solid crust. Early estimates forbade a molten Moon by giving timescales for conductive cooling that exceeded the age of the solar system. Later estimates swung to the other extreme when the importance of loss of heat by convective processes began to be realized. Estimates for the solidification time of a 200-km thick magma ocean are now about 10^7–10^8 years, one or two orders of magnitude lower than the original estimates. Various unknown factors beset these calculations. The thickness and rate of growth of the crust is uncertain, and since this is occurring (from the isotopic evidence) at about 4.4 aeons, a heavy bombardment was continually breaking up the crust. The diameter of planetesimals following planetary condensation

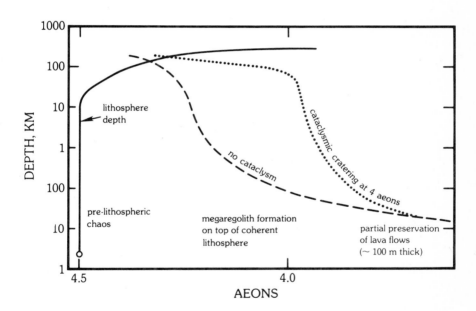

5.32   The relationship between cratering history and the growth of the lunar crust, according to Hartmann [96]. The competition between formation of coherent lithosphere (magma ocean solidification in absence of cratering, solid curve) and pulverization of lithospheric layers (dotted and dashed curves). In first 10^8 yr. or so (before cross-over of curves) cratering would be so intense that any lithospheric crust would be 100% penetrated on a timescale short compared to the solidification time. Later, solid lithosphere layers could form at the bottom of a megaregolith tens of kilometers (10^2 km?) deep, probably partially welded into coherent breccia. Plausibly thick lava flows on the surface would avoid complete pulverization only after about 3.8 to 4.2 aeons ago.

has been estimated to be about 10 km [163]. These will punch through the growing crust, and hence promote radiative cooling. If no crust is present, heat loss by radiation will cause the magma ocean to crystallize in a few decades [164, 165]! Accordingly, the meteoritic flux rate is an important controlling factor on cooling rates. The isotopic data (see Section 6.5) show that the source regions of the mare basalts were closed by 4.4 aeons (i.e., crystallized and cooled to temperatures below which no isotopic redistribution occurs). At this time, the surface crust was still being rapidly destroyed by the meteorite bombardment (the so-called stonewall effect [166, 167]) (Fig. 5.32). Any rock which crystallized at depths of 100 km at 4.5 aeons would be destroyed rapidly. The intense early cratering would inhibit formation of any coherent igneous rock, and so the highlands probably began as a megaregolith tens of kilometers deep [166]. Plagioclase would have to comprise ~80% of the crust if it were to float [157] (melt density = 2.83 gm/cm^3); accordingly, plagioclase would dominate the early crust. Before plagioclase can crystallize, at least 50% of the olivine must have crystallized and the temperature must have dropped by 400°C [165]. Thus, a solid conducting crust will not grow until the temperature drops below about 1150°C. This will be continually broken up by the raining planetesimals down to ages of about 4.2 aeons, long after the deeper parts have frozen. Accordingly, there is little problem with heat loss, since a solid crust, through which heat can only be lost by conduction, will not be established until after much of the crystallization of the magma ocean is completed.

The evidence is consistent with models which call for rapid crystallization of the crust before 4.3 aeons, breakup of this crust by projectiles to about 4.2 aeons, and resetting of ages down to 3.80 aeons by the large basin-forming impacts.

5.11.3 Crystallization History of the Lunar Crust

A possible sequence of events is as follows: Crystallization of bulk moon composition begins in the magma ocean. Olivine crystallizes until the olivine-plagioclase peritectic line is reached at which point plagioclase begins to crystallize. Calcium-rich plagioclase (anorthosite) will float in the dry lunar magma [168] although it would sink in a terrestrial wet magma. Thus, plagioclase floats to begin the formation of the highland crust. It is very uniform in composition (Section 5.4.1). Some intercumulus liquid is trapped at this point, and differentiates to low Mg/(Mg + Fe) values, typical of the anorthosites. The composition of the liquid where the plagioclase appears is that of low-K Fra Mauro basalt (with about 10–15 ppm Sm = 40–60 times chondritic REE values or 18–26 times bulk moon compositions). Crystallization continues with clinopyroxene joining the crystallization sequence. Feldspar cumulates continue to accrete. The difference between this well-ordered

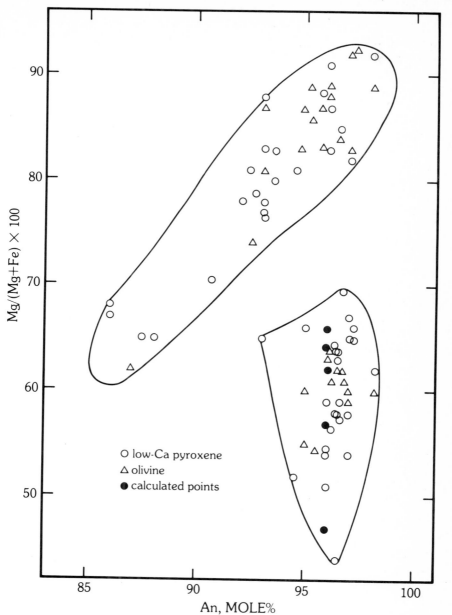

5.33 The relation between the anorthite content of plagioclase and the Mg number in co-existing mafic minerals in lunar highland samples. Similar trends are shown in the banded zone of the terrestrial Stillwater intrusion, except that the gap between the lunar trends is filled [169]. (Courtesy Linda Raedeke and Stuart McCallum.)

crystallization sequence and the actual situation is that there is a continuing infall of planetesimals of whole moon composition since the siderophile element content is low, excluding a meteoritic component. These remix, remelt, add heat, and generally create an extremely complex picture in detail. In some areas, the already solidified crust is remelted, and recrystallizes. Longhi [156] has described such scenarios. No rocks survive in the near-surface environment until the crust reaches a thickness of 1–10 km, with this stage being as late as 4.2 aeons or even younger [166]. The so-called stonewall effect operates until that time. Although the surface of the Moon is in turmoil, crystallization can proceed more quietly at depth and the zoned cumulate source regions, from which the mare basalts would subsequently be derived, escape the bombardment. Thus, the isotopic systematics of the mare basalt source regions are likely to exhibit more regularities than those in the battered and reworked highland crust. The impacting bodies may induce convective overturn in the magma ocean, and cause rafts of dense zones of accumulating crystals (e.g., ilmenite-rich zones later to be the source regions for the high-Ti basalts) to sink to the bottom of the magma ocean.

The most direct terrestrial analogue to the lunar highlands appears to be the Stillwater Intrusion [169], although useful constraints come from the study of other layered intrusions such as the Skaergaard [170]. Particular insights have come from the concept of cumulus and intercumulus models for crystallization involving trapped liquids. None of the terrestrial examples remotely approaches the scale of the lunar magma ocean, and differences between lunar and terrestrial oxidation states affect the behavior of elements such as Fe and Cr. The absence of water on the Moon is another major difference affecting many properties. The most significant of these appears to be the demonstration that plagioclase (anorthite) will float in anhydrous magmas parental to the lunar crust [168]. An important contribution from the Stillwater studies has been the demonstration that two apparently diverse fractionation trends can develop during crystallization of a single magma. Such trends were observed in the lunar highland samples (Fig. 5.33)[171, 172], which draws attention to a steep decrease in $Mg/(Mg + Fe)$ while the An content of the plagioclase remained effectively constant. This behavior is the reverse of that observed in terrestrial fractional crystallization, where the An content of feldspar should decrease with decreasing $Mg/(Mg + Fe)$. The same two trends have been observed in the Stillwater crystallization sequence [169]. In this example, only one magma is involved, since widely separated samples possess the same initial Nd isotopic ratios [173]. Sm-Nd data are lacking in the anorthosites at present. The explanation for the two trends in the Stillwater is as follows: the vertical trends of $Mg/(Mg + Fe)$ with relatively constant An content are ascribed to the crystallization of intercumulus liquid trapped in a plagioclase rich crystal mush. The abundance of plagioclase buffers the An composition (in the "middle" BZ). The other conventional trend, in the upper and lower "banded" zones of the Stillwater, is the result of normal fractional crystallization, with resulting decrease of Mg/(Mg + Fe) and An content (Fig. 5.33).

The same effects on the Moon are shown by the vertical trend exhibited by anorthosites, noritic anorthosites and troctolitic anorthosites (the Mg suite) from the Apollo 15 and 16 missions. The noritic troctolites of Apollo 15 and 17 show the effect of conventional fractional crystallization. Various granulitic impactites fall between the two trends [174]. The major remaining questions focus on whether one magma or two are involved on the Moon. In the case of the Stillwater Intrusion, these trends have developed during the crystallization of one magma. The question is partly semantic, partly on whether the "gap" in the lunar plot (Fig. 5.33) is real or will be filled in by later data, and partly depends on trace element evidence.

The dunite (74215), which plots on the evolved trend, resembles the anorthositic suite. The complexity of magma ocean crystallization, the continuing bombardment, the trapping of intercumulus liquids all contribute to masking the overall simplicity of the crystallization patterns. The initial $^{87}Sr/^{86}Sr$ ratios of the anorthosites may be slightly older. All this constitutes evidence for prior crystallization of anorthosites.

Insights can be gained from Sc/Sm and Ti/Sm relationships among the anorthosite and the Mg rich suites (Fig. 5.34). The former rocks are products of direct crystallization of the magma ocean, involving removal of olivine and pyroxene crystallization before reaching plagioclase saturation. The anorthosites and Mg-suite are clearly separated on Ti/Sm and Sc/Sm plots. In both cases, the anorthosites plot nearer to chondritic (and whole moon) ratios. Titanium, Sm (= REE) and Sc are all refractory elements and have chondritic ratios in the bulk Moon. The depletion in Ti in the Mg suite indicates that ilmenite has been removed, while the Sc depletion indicates removal of pyroxene. Ilmenite is not, however, a liquidus phase in the Mg suite. This paradox can be explained by the mixing of a primitive component, with high Mg/(Mg + Fe) values, and a differentiated component, from which ilmenite has already been removed. Mafic minerals are not abundant in the anorthosites, but there are some mafic members of the Mg suite (61224, 67667) which have Ti/Sm and Sc/Sm ratios more like the anorthosites. The dunite has near chondritic values for these ratios so that it is an equivocal member of the Mg suite.

Trace element evidence in olivines and plagioclases is consistent with differing crystallization histories for the two groups. Thus, the plagioclases from the anorthosites show a very restricted range in composition [175], but those from other rock types have a wide range in values, which is indicative of an origin either as separate events, perhaps in small intrusions, or by mixing. Strontium shows a positive correlation with sodium, the reverse of that expected, but explicable if other minerals (e.g., olivine or orthopyroxene) are

5.34a The relationship between Mg number and Ti/Sm ratios in lunar highland rocks. Figure shows the difference between the anorthosites which have Ti/Sm in the chondritic range, and the rocks of the Mg suite (troctolites and norites). The latter are characterized by high Mg numbers (indicative of a primitive undifferentiated composition) and low Ti/Sm ratios (Sm serves as an index for the other rare-earth and incompatible elements). The

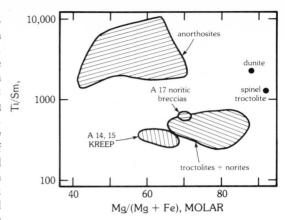

Mg suite rocks and KREEP have similar Ti/Sm ratios, indicative of a highly fractionated source. This combination of primitive and fractionated chemical characteristics is typical of KREEP and the Mg suite. It is here interpreted as evidence for mixing of primitive and late stage material during the raining bombardment of the highland crust (cf. Fig. 5.32). (Courtesy G. A. McKay.)

crystallizing simultaneously. All lunar plagioclases, including those from mare basalts, lie on this trend. The uniformity of the trace element contents of plagioclase from the anorthosites is consistent with a common origin, and hence supportive of the magma ocean concept.

5.34b The relationship between Sm abundances (representative of REE and incompatible elements generally) and the Ti/Sm ratio (cf. Fig. 34a). Note that the dunite plots among the anorthosites, and both have C1 type abundances. The troctolites and norites of the Mg suite show the highly fractionated geochemical characteristics of KREEP, with marked depletion in Ti interpreted as being due to prior removal of ilmenite. This is not petrologically feasible

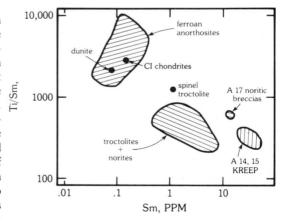

because of the primitive unfractionated major element compositions and Mg numbers. A mixing scenario is is preferred to explain the geochemical paradoxes in these two diagrams. (Courtesy G. A. McKay.)

Similar relationships are noted for the trace element content of olivine. Olivines from anorthosites have consistently lower contents of minor elements than those from other rock types [176], except for Ca and Mn, although such olivines have high Fe contents, with Fo values ranging from 66 down to 45. Olivines from other rocks show a higher and more variable concentration of these elements (Al, P, Ti, Cr). Chromium is notably low in olivines from anorthosites and rarely exceeds 50 ppm. In the Mg-rich rocks, Cr is typically about five times higher, consistent with the more primitive nature of portions of the Mg suite.

The complexity of these crystallization sequences in detail have resulted in many models, in particular to account for the high Mg suite. Remelting of early basic cumulates would provide high $Mg/(Mg + Fe)$ [177], but not the highly fractionated REE patterns. Contamination of solid rock by KREEP does not explain why the anorthosites are not contaminated. The Mg suite shows many of the characteristics of a differentiated suite, as is well displayed in the An versus $Mg/(Mg + Fe)$ diagram.

A viable but complex hypothesis appears to be magma mixing [178]. The unfractionated nature of the major elements ($Mg/Mg + Fe$) is due to a primitive magma, while the addition of a relatively small amount of highly fractionated residual liquid from the magma ocean dominates the trace element characteristics (Fig. 5.35). One problem is that very little primitive magma is available when the residual liquid stage is reached. Rare-earth element patterns will be flatter than KREEP due to dilution with "chondritic" patterns, although the KREEP abundances are large enough to swamp most of this effect.

Another scenario is that many of the rocks of the Mg suite have crystallized subsequent to the main magma ocean crystallization as small intrusions within the crust, thus explaining many of their trace element characteristics [26, 56, 97, 99]. The data are not yet capable of distinguishing between these alternatives, but it is judged here that the mixing scenarios, induced by the effects both of the bombardment and the pervasive infiltration of late differentiated liquids into the crust is the more reasonable scenario.

The origin of KREEP has been commented upon several times in this chapter. There is an extensive debate on this subject (e.g., [179–183]). Two important facts need to be considered. First, extremely high concentrations of the incompatible elements are present (Section 5.4.3) with the concentrations of the REE up to 700 times chondritic abundances [78]. The second is that the Sm-Nd isotopic systematics [73] are uniform on a moon-wide basis (Fig. 5.36). The Nd isotopic evolution in KREEP is complementary to that of the high-Ti basalts (Section 6.4) (Fig. 5.37). Variations in Rb-Sr isotopic systematics are probably due to differential movement of volatile Rb in KREEP breccias during impacts (Section 5.5).

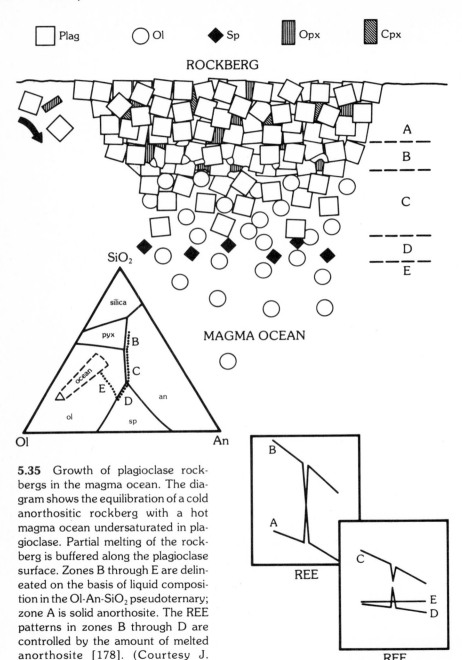

□ Plag   ◯ Ol   ◆ Sp   ▥ Opx   ▨ Cpx

5.35 Growth of plagioclase rockbergs in the magma ocean. The diagram shows the equilibration of a cold anorthositic rockberg with a hot magma ocean undersaturated in plagioclase. Partial melting of the rockberg is buffered along the plagioclase surface. Zones B through E are delineated on the basis of liquid composition in the Ol-An-SiO$_2$ pseudoternary; zone A is solid anorthosite. The REE patterns in zones B through D are controlled by the amount of melted anorthosite [178]. (Courtesy J. Longhi.)

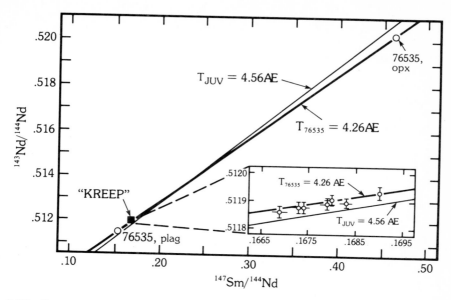

5.36 Samarium-neodymium systematics of "KREEP" samples from different Apollo sites. The Sm-Nd evolution diagram indicates the remarkably close clustering of Sm-Nd data of KREEP samples. Resolution of the points from the 4.56 AE reference isochron for Juvinas and the rough alignment along the 4.26 AE reference isochron obtained from troctolite 76535 suggests a close relationship between the Sm/Nd system of KREEP and other lower crustal materials [73]. (Courtesy G. W. Lugmair.)

A basic premise of the model adopted here is that KREEP originates as the final melt residuum following crystallization of the magma ocean. The volume may be about one or two percent. This low density iron-rich and trace-element-rich residual melt (4.4 aeons) invades the crust [184] where it is mixed in by the continuing bombardment with the later products of magma ocean crystallization.

5.12   The Crust of the Earth

There is a primary division into oceanic and continental crusts. The oceanic crust is young (< 200 m.y.), comprised primarily of basalt and is recycled into the mantle. It may serve as a model for conditions in the earliest Archean. The continental crust, in contrast, is old and complex, but has grown throughout geological time. The oldest rocks have ages of 3.8 aeons, postdating the decline in the massive planetesimal bombardment. Is there any sign of a primitive crust analogous to that of the moon? There is no isotopic or

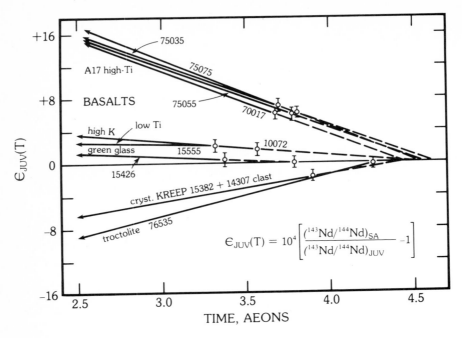

5.37 Differentiated ^{143}Nd evolution for highland samples (KREEP and troctolite) and mare basalts [73]. This diagram clearly shows the early differentiation of the Moon and the development of separate source reservoirs for mare basalts. The time of the early differentiation is not well resolved by the Sm-Nd systematics, except that it is early.

chemical evidence of the existence of such a crust. The principal information comes from Rb-Sr and Sm-Nd isotopic systematics, which indicate derivation of the present crust from primitive mantle, and the absence of recycled material of continental composition [185].

The composition of the upper continental crust, is approximately that of granodiorite [186]. There is too high a content of the radioactive heat producing elements, K, U, and Th in the accessible upper crust for this composition to persist to depths in excess of 10–15 km. The composition of the whole crust, down to the Mohorovičić discontinuity at 40 km is thus model dependent, as is that of the inaccessible lower crust. A reasonably close approximation to the bulk composition is provided by that of the voluminous silica-rich island arc volcanic rocks. These are derived, at subducting plate boundaries, from the mantle and represent the only reasonable present-day source of igneous rocks, both voluminous and silica-rich, to be viable candidates for additions to the continental crust. There is a considerable body of geological evidence indicating that the latter stages of continental growth involved lateral accretion of

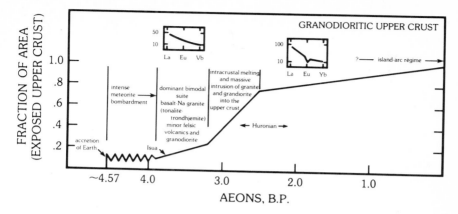

5.38 Model for the evolution of the terrestrial continental crust during geological time. Principal events of crustal evolution are indicated, as well as average REE patterns for the *upper* crust, as derived from sedimentary rock data [186].

island-arcs. However this model is only valid so long as the present-day plate-tectonic regime was operating. There is a growing body of isotopic evidence, principally from the Sm-Nd system, which indicates that a massive increase in the volume of continental material occurred in the period between 3.0 and 2.5 aeons, at the close of the Archean [187].

Figure 5.38 shows a recent attempt to model the growth rate of the crust. The largely episodic nature of this event in the late Archean raises several questions. Although the island-arc model is valid back to beyond 1.0 aeon, it is probably not an appropriate model for the massive late Archean additions to the continents. The Archean crustal composition is not very different to that of the present day total crust, except that it may be higher in Ni and Cr. Within about 200 m.y. of the massive episodic additions from the mantle, large scale intra-crustal partial melting produced the granodioritic upper crust, leaving a depleted lower crust. The ubiquitous depletion in the rare earth element, Eu, observed in upper crustal sedimentary rocks, was produced by this process. This relative concentration is due to differences in ionic size and valency from those of common mantle cations, so that the most "incompatible" elements (e.g., Cs, Rb, K) are the most strongly concentrated in the crust. About 30% of the total mantle budget of these elements have been so concentrated in the continental crust. This accordingly sets a lower limit on the volume of the mantle which has been involved. The extraction of these elements has led to the formation of "depleted" mantle. The geochemical characteristics of Mid-Ocean-Ridge basalts (MORB), which exhibit relative depletion of light REE, K, Rb, etc. as well as the isotopic systematics (particularly Sm-Nd) indicate that they are derived from such regions.

The thermal regime responsible for the evolution of the terrestrial crust may be briefly commented upon. Heat production in the early Archean was at least three times its present value. Whether the surface heat loss was substantially greater is unknown. At present 50% of the surface heat loss occurs at mid-ocean ridge (spreading plate boundaries) [188]. Either faster plate movement or many more plates are needed in the Archean. The continental growth model outlined here is consistent with a steady temperature buildup in the mantle (heat production exceeding heat loss) until about 3.0 aeons, when massive mantle melting and production of continental crust ensues. This event transfers substantial quantities of K, U, and Th into the crust. Intra-crustal melting, producing granodiorite, occurs within a few hundred million years of this event. The massive increase in continental volume changes the tectonic style, and initates the present regime of linear style tectonics and subduction zones. About 80% of continental volume is thus emplaced by about 2.5 aeons, and the upper continental crust has not changed in composition since that time.

It can thus be seen that there are great differences in the origin and evolution of the two planetary crusts with which we are most familiar. The lunar crust, although greatly complicated by the details of its crystallization history and the raining bombardment, results essentially from an early planetary-wide melting episode. The *upper* continental crust of the Earth is the product of at least three successive partial melting events from the mantle and lower crust.

5.13   Other Planetary Crusts

No chemical data are available for the Mercurian crust (Section 2.5), but the preservation of the record of an early intense bombardment indicates that crustal formation was completed well before 4.0 aeons. The reflectance spectra indicate a silicate surface much resembling that of the Apollo 16 soils [189], possibly indicating a composition containing less than about 5% FeO. The evidence for early differentiation and lack of global expansion reinforce the similar situation observed on the Moon.

The resemblance between the reflectance spectra of the lunar highlands and of Mercury may mean that the crust of the planet is of anorthositic gabbro composition on the average. The remote-sensing data are integrated over the whole planetary surface, and resolution of individual areas, for example the smooth plains, is not possible. No spectra typical of mare basalts have been observed [189]. Both the similarities and differences between Mercury and and the Moon are important constraints for our understanding of early solar system history.

Most of the details of the Martian crust have been discussed in Section 2.6. The evolution of the crust appears to have continued, with the eruption of basaltic lavas, until relatively recently, as indicated by the young age of Olympus Mons. The question of the elevation of the Tharsis plateau is dealt with in Section 7.6.3; whether this plateau is due to uplift or to superposition of a series of lavas is uncertain.

A comprehensive review of the Venusian crust is given by Phillips et al. [190]. The geology of Venus has been described in Section 2.7 and little can be added here. The Venera gamma-ray data indicate K/U ratios of 10^4 and abundances comparable with those in terrestrial granites (see Section 4.13).

Many details of the surface are not yet clear, e.g., the circular features on the Median Plains may be impact craters or volcanic features [191]. The age of the surface features (Section 2.7) is uncertain, but if these circular features are craters, then the plains are very old. Accordingly, the operation of plate tectonic processes at present is unlikely. The continental masses (Ishtar and Aphrodite) present major puzzles, possibly indicative of an early period of crustal evolution (see Section 7.6.3). A principal conclusion is that in many respects Venus is dissimilar to the Earth, a fact of great significance for planetary formation (Chapter 9).

Little compositional evidence is available for the surface crusts of the satellites of the outer planets, and debate, for example over the surface of Io, is too speculative for much discussion here. The grooved terrains, exhibited by Ganymede and Callisto, provide some new insights into planetary crustal evolution. The former provides an excellent example.

Ganymede, the third of the four Galilean satellites of Jupiter, with a radius of 2638 km, is a little larger than Mercury. Areas of an older cratered crust are separated by younger grooved terrain. The most reasonable interpretation is that the older crust has been split, and new material injected from below (192).

Ganymede has a density of 1.99 gm/cm^3, consistent with a silicate-ice mixture. Expansion on such a planet can occur by the following mechanism. A mixture of silicate and ice accretes. The ice deep within the planet undergoes polymorphic change to higher density forms. These have densities ranging from 1.16 to 1.66 gm/cm^3. The initial planetary crust is heavily cratered at times earlier than 4000 million years. Heating on a longer time scale due to the radioactive elements K, U and Th present in the silicates causes melting in the ice in the interior. The water migrates outward.

The change from high density ice to lower density water provides an expansion of the planet of up to 5–7%. The water refreezes near the surface in the low density Ice I polymorph, after disrupting the older frozen crust. Thus we find some evidence for minor expansion in Ganymede. This is explicable on the known properties of ice and it is not necessary to invoke any more mysterious mechanism (192).

References and Notes

1. Lyell, C. (1830) *Principles of Geology*, Vol. 1, John Murray. Lyell commented (p. 82) that a race of intelligent amphibia would have arrived much sooner at a proper understanding of geological processes than land-dwelling species. The truth of this comment was fully appreciated with the development of the plate tectonic hypothesis.
2. Windley, B. W. (1977) *The Evolving Continents*, Wiley.
3. *New Yorker*. May 11, 1981, p. 2.
4. Haines, E. L., and Metzger, A. E. (1980) *PLC 11*: 689; Bielefeld, M. J., et al. (1976) *PLC 7*: 2661.
5. Metzger, A. E., et al. (1977) *PLC 8*: 978.
6. Davis, P. A. (1980) *JGR*. 85: 3209.
7. Bills, B.G., and Ferrari, A. J. (1977) *JGR*. 82: 1306.
8. Thurber, C. H., and Solomon, S. C. (1978) *PLC 9*: 3481.
9. Kaula, W. M., et al. (1972) *PLC 3*: 2189.
10. Phillips, R. J., and Lambeck, K. (1980) *Rev. Geophys. Space Phys*. 18: 27.
11. Airy, G. B. (1855) *Phil. Trans. Roy. Soc. London* B145: 101.
12. Pratt, J. H. (1855) *Phil. Trans. Roy. Soc. London* B145: 53.
13. Lingenfelter, R. E., and Schubert, G. (1973) *Moon*. 7: 172.
14. Aggarwal, H. R., and Oberbeck, V. R. (1979) *PLC*. 10: 2689.
15. Thompson, T. W., et al. (1979) *Moon Planets*. 21: 319; (1980) *Lunar Highlands Crust*, p. 175; (1981) *Icarus*. 46: 201.
16. See Section 5.10.3.
17. Wolfe, E. W., and Bailey, N. G. (1972) *PLC 3*: 15.
18. Mattingly, T. K., and El-Baz, F. (1973) *PCL 4*: 55.
19. Spudis, P. D., and Ryder, G. (1980) *Multi-ring Basins*, p. 86.
20. Wilhelms, D. E. (1980) Apollo 16 Workshop, 10; (1979) LPS X: 1251.
21. Hawke, B. R., and Head, J. W. (1980) *Multi-ring Basins*, p. 36.
22. Stöffler, D., et al. (1980) *Lunar Highlands Crust*, p. 51; Stöffler, D., et al. (1979) *PLC 10*: 639.
23. McKinley, J. P., et al. (1981) LPS XII: 691.
24. Norman, M. D. (1981) LPS XII: 776.
25. a) Poikilitic textures comprise large host crystals enclosing many smaller crystals of other phases.
b) Ophitic textures comprise elongated feldspar crystals embedded in pyroxene or olivine.
Both terms are good examples of jargon.
26. James, O. B. (1980) *PLC 11*: 365.
27. Stewart, D. B. (1975) LS VI: 774.
28. Warner, J. L., et al. (1977) *PLC 8*: 2051.
29. Stöffler, D., et al. (1979) LPS X: 1177.
30. Grieve, R. A. F. (1980) *Lunar Highlands Crust*, p. 187.
31. Miller, D. S., and Wagner, G. A. (1979) *EPSL*. 43: 351.
32. Blanchard, D. P., and Budahn, J. R. (1979) *PLC 10*: 803.
33. James, O. B., and McGee, J. J. (1980) *PLC 11*: 67.
34. Ryder, G., and Taylor, G. J. (1976) *PLC 7*: 1741.
35. Ryder, G. (1980) *PLC 10*: 575.
36. Vaniman, D. T., and Papike, J. J. (1979) *GRL*. 5: 429.
37. *Basaltic Volcanism* (1981) Section 1.2.10.
38. Irving, A. J. (1975) *PLC 7*: 363.
39. Prinz, M., and Keil, K. (1977) PCE 10: 215.

40. Meyer, C. E. (1977) PCE 10: 506.
41. Warren, P. H., and Wasson, J. T. (1977) *PLC 8*: 2215; (1978) *PLC 9*: 185.
42. Norman, M. D., and Ryder, G. (1979) *PLC 10*: 531.
43. Phinney, W. C., and Simonds, C. H. (1977) *Impact Cratering*, p. 771.
44. Grieve, R. A. F., et al. (1977) *Impact Cratering*, p. 791; Floran, R. J., et al. (1981) *JGR*. In press.
45. Meyer, C. (1978) *PLC 9*: 1551.
46. Reid, A. M., et al. (1977) *PLC 8*: 2321.
47. Taylor, S. R., and Bence, A. E. (1975) *PLC 6*: 1121; Taylor, S. R. (1976) *PLC 6*: 3461.
48. Wänke, H., et al. (1977) *PLC 8*: 2237.
49. Ryder, G. (1979) *PLC 10*: 561.
50. Haskin, L. A. (1979) Lunar Highlands Crust Abstracts, 49.
51. The ratio Eu/Eu* is a measure of the depletion or enrichment of europium relative to the neighboring rare-earth elements, samarium and gadolinium. Eu is the measured elemental abundance in the sample; Eu* is the theoretical concentration for no Eu anomaly, and is calculated by assuming a smooth REE pattern in the region Sm-Eu-Gd. Values of Eu/Eu* less then 0.95 indicate depletion, and values of greater than 1.05, enrichment of europium relative to the neighboring REE.
52. Ryder, G. (1979) *PLC 10*: 560.
53. Wänke, H., et al. (1977) *PLC 8*: 2191.
54. McKay, G. A., et al. (1978) *PLC 9*: 661.
55. Garrison, J. R., and Taylor, L. A. (1980) *Lunar Highlands Crust*, p. 395.
56. Wasson, J. T., et al. (1977) *PLC 8*: 2237.
57. The simpler term "Anorthosite" is preferred to usages such as Ferroan Anorthosite, with its slightly horsey overtones. Although it is useful to note that the accompanying minerals have high Fe/Mg ratios consistent with crystallization from a magma ocean, the prefix appears to be unnecessary. Smith, J. V., and Steele, I. M. (1976) *Amer. Mineral.* 61: 1074.
58. James, O. B. (1980) *PLC 11*: 365.
59. Norman, M D., and Ryder, G. (1979) *PLC 10*: 531.
60. Papanastassiou, D. A., and Wasserburg, G. J. (1972) *EPSL*. 17: 52; (1975) *PLC 6*: 1467.
61. Schaeffer, O. A., and Husain, L. (1974) *PLC 5*: 1541.
62. Lugmair, G. W., et al. (1976) *PLC 7*: 2009.
63. Papanastassiou, D. A., and Wasserburg, G. J. (1976) *PLC 7*: 2035.
64. Dowty, E., et al. (1974) *EPSL*. 24: 15.
65. Prinz, M., and Keil, K. (1977) PCE 10: 215.
66. Laul, J. C., and Schmitt, R. A. (1975) *PLC 6*: 1231.
67. Dymek, R. F., et al. (1975) *PLC 6*:301.
68. Taylor, G. J., et al. (1979) *Lunar Highlands Crust*, p. 166.
69. Roedder, E., and Weiblen, P. W. (1970) *PLC 1*: 801.
70. Glass, B. P. (1976) LS VII: 296.
71. Arnold, J. R. (1972) *Apollo 15 PSR*, p. 16-1.
72. Hawke, B. R., and Head, J. W. (1978) *PLC 9*: 3285.
73. Lugmair, G. W., and Carlson, R. W. (1978) *PLC 9*: 689; (1978) *EPSL*. 39: 349; (1979) *EPSL*. 45: 123.
74. Taylor, S. R., and Jakes, P. (1974) *PLC 5*: 1287.
75. Taylor, S. R. (1975) *Lunar Science: A Post-Apollo View*, Pergamon, Fig. 7.2.
76. Warren, P. H., and Wasson, J. T. (1979) *Rev. Geophys. Space Phys.* 17: 73.
77. Meyer, C. (1977) PCE 10: 239.

78. Nyquist, L. E., et al. (1977) LS VIII: 738.
79. Walker, D., et al. (1972) *PLC 3*: 797; Hess, P. C., et al. (1977) *PLC 8*: 2357.
80. Irving, A. J. (1975) *PLC 6*: p. 363.
81. Ryder, G. J., and Spudis, P. D. (1980) *Lunar Highlands Crust*, p. 353.
82. Ryder, G., and Taylor, G. J. (1976) *PLC 7*: 1741.
83. Charette, M. P., et al. (1977) *PLC 8*: 1049.
84. Metzger, A. E., et al. (1979) *PLC 10*: 1701.
85. McKay, G. A., and Weill, D. F. (1977) *PLC 8*: 2339.
86. Jovanovic, S., and Reed, G. W. (1974) *PLC 5*: 1685.
87. Taylor, L. A., et al. (1974) *PLC 5*: 743.
88. Nunes, P. D., and Tatsumoto, M. (1973) *Science*. 182: 916.
89. Taylor, L. A., and Burton, J. C. (1976) *Meteoritics*. 11: 225.
90. Cirlin, E. H., and Housley, R. M. (1980) *PLC 11*: 349.
91. Cirlin, E. H., and Housley, R. M. (1981) *PLC 12*: 529.
92. Hunter, R. H., and Taylor, L. A. (1981) LPS XII: 488.
93. Gros, J., et al. (1976) *PLC 7*: 2403.
94. Hertogen, J., et al. (1977) *PLC 8*: 17.
95. Janssens, M. J., et al. (1978) *PLC 9*: 1537.
96. Hartmann, W. K. (1980) *Lunar Highlands Crust,* p. 155.
97. Warren, P. H., and Wasson, J. T. (1977) *PLC 8*: 2215; (1978) *PLC 9*: 185; (1979) *PLC 10*: 583; (1980) *PLC 11*: 431.
98. Norman, M. D., and Ryder, G. (1979) *PLC 10*: 531.
99. Warren, P. H., and Wasson, J. T. (1980) *PLC 11*: 431.
100. James, O. B. (1973) USGS Prof. Paper 841.
101. Delano, J. W., and Ringwood, A. E. (1978) *PLC 9*: 111.
102. Grieve, R. A. F. (1980) *Lunar Highlands Crust*, p. 189.
103. Ryder, G., et al. (1980) *PLC 11*: 471.
104. Taylor, L. A., et al. (1976) *PLC 7*: 837.
105. Norman, M. D., and Ryder, G. (1979) *PLC 11*: 531.
106. Wolf, R., et al. (1979) *PLC 10*: 2107.
107. Adler, I., et al. (1972) NASA SP-289, 17-1; (1972) NASA SP-315, 19-1.
108. Arnold, J. R., et al. (1972) NASA SP-289, 16-1; (1972) NASA SP-315, 18-1.
109. Adler, I., and Trombka, J. I. (1977) PCE 10: 17.
110. Haines, E. L., et al. (1976) IEEE Trans. Geoscience Electronics GE 15: 141.
111. Metzger, A. E., et al. (1977) *PLC 8*: 969.
112. Haines, E. L., et al. (1978) *PLC 9*: 2985.
113. Hawke, B. R., et al. (1981) LPS XII: 415.
114. Haskin, L. A. (1979) Lunar Highlands Crust Abstracts, 49.
115. Andre, C. G., and El-Baz, F. (1981) *PLC 12*: 767; Andre, C. G., et al. (1979) *PLC*: 1739.
116. Andre, C. G. (1981) *Multi-ring Basins*, p. 1.
117. Hawke, B. R., and Spudis, P. D. (1980) *Lunar Highlands Crust*, p. 467.
118. Haines, E. L., and Metzger, A. E. (1980) *PLC 11*: 689.
119. Bielefield, M. J., et al. (1976) *PLC 7*: 2661.
120. Metzger, A. E., and Parker, R. E. (1979) *EPSL*. 45: 155. The average Ti value quoted (0.8% Ti) demands a very large Fra Mauro or KREEP component and is probably about twice the correct value. They suspect an "unidentified systematic offset."
121. Galilei, G. (1610) *Nuncius Siderus*, Padua.
122. O'Keefe, J. A., and Cameron, W. S. (1962) *Icarus*. 1: 271.
123. Walter, L. S. (1965) *Ann. N.Y. Acad. Sci.* 123: 367.
124. O'Keefe, J. A., et al. (1967) *Science*. 155: 77.
125. Jackson, E. D., and Wilshire, H. (1968) *JGR*. 73: 7621.

126. Wood, J. A. (1970) *PLC 1*: 965.
127. LSPET (1970) *Science:* 2697.
128. LSPET (1971) *Science.* 173: 681.
129. Gast, P. W. (1972) *Moon.* 5: 121.
130. E.g., Anderson, D. L. (1973) *EPSL.* 18: 301.
131. Brett, P. R. (1973) *GCA.* 37: 2697.
132. Duncan, A. R., et al. (1973) *PLC 4*: 1097.
133. Reid, A. M., et al. (1974) *Moon.* 9: 141.
134. Warner, J. L., et al. (1977) *PLC 8*: 2051.
135. Wänke, H., et al. (1975) *PLC 4*: 1461.
136. Taylor, S. R., et al. (1973) *GCA.* 37: 2665.
137. See Taylor, S. R. (1975) *Lunar Science: A Post-Apollo View*, p. 245–249 for an extended discussion of the significance of inter-element correlations.
138. Spudis, P. D., and Hawke, B. R. (1981) LPS XII: 1028.
139. Parker, R. E., et al. (1980) LPS XII: 811.
140. Duncan, A. R., et al. (1973) *PLC 4*: 1108.
141. Bogard, D. D., et al. (1981) LPS XII: 92.
142. James, O. B., and Hörz, F. (1980) Apollo 16 Workshop, 20.
143. Hartmann, W. K. (1975) *Icarus.* 24: 181; (1980) *Lunar Highlands Crust,* p. 155.
144. Oberli, F., et al. (1978) LPS IX: 832; LPS X: 940.
145. Maurer, P., et al. (1978) *GCA.* 42: 1687.
146. Nyquist, L. E., pers. comm., 1981.
147. Carlson, R. W., et al. (1976) LPS XII: 126. The lherzolite has 58% olivine, 21% plagioclase, 15% orthopyroxene and 5% diopside.
148. Turner, G. (1977) PCE 10: 179.
149. Jessberger, E. K., et al. (1974) *PLC 5*: 1419.
150. Schaeffer, O. A., and Husain, L. (1974) *PLC 5*: 1541.
151. Kirsten, T., and Horn, P. (1974) *PLC 5*: 1451.
152. *Basaltic Volcanism* (1981) Chapter 7.
153. Wood, J. A. (1975) *Moon.* 14: 505.
154. Dence, M. R., et al. (1976) *PLC 7*: 1821.
155. Spudis, P. D., and Ryder, G. (1980) *Multi-ring Basins*, p. 86.
156. Longhi, J. (1980) *PLC 11*: 289.
157. Solomon, S. C., and Chaiken, J. (1976) *PLC 7*: 3229.
158. Binder, A. B. (1976) *Moon.* 16: 159.
159. Taylor, S. R. (1978) *PLC 9*: 15.
160. Herbert, F., et al. (1977) *PLC 8*: 573; (1978) *PLC 9*: 249.
161. Brown, G. M. (1978) in *Origin of the Solar System* (ed., S. F. Dermott), Wiley, p. 897.
162. Minear, J. W. (1980) *PLC 11*: 1941.
163. Goldreich, P., and Ward, W. R. (1973) *Astrophys. J.* 183: 1051. See also Section 9.2. See also Wetherill, G. W. (1981) *Ann. Rev. Astron. Astrophys.* 18: 77.
164. Herbert, F., et al. (1977) *PLC 8*: 573.
165. Solomon, S. C., and Longhi, J. (1977) *PLC 8*: 583.
166. Hartmann, W. K. (1980) *Lunar Highlands Crust*, p. 166.
167. The term is appropriate enough, with its Civil War connotations. Formed bodies of troops were destroyed by the Stonewall tactics of Jackson just as the crystallizing lunar crust was destroyed by the meteorite bombardment.
168. Walker, D., and Hays, J. F. (1977) *Geology.* 5: 425.
169. Raedeke, L., and McCallum, I. S. (1980) *Lunar Highlands Crust*, p. 133; McCallum, I. S., et al. (1981) LPS XII: 676; Steele, I. M., et al. (1981) LPS XII: 1034.
170. Wager, L. R., and Brown, G. M. (1967) *Layered Igneous Intrusions*, Oliver and Boyd.

171. Steele, I. M., and Smith, J. V. (1973) *PLC 4*: 419.
172. Roedder, E., and Weiblen, P. W. (1974) *PLC 5*: 303.
173. DePaolo, D. J., and Wasserburg, G. J. (1979) *GCA*. 43: 999.
174. Bickel, C. E., and Warner, J. L. (1978) *PLC 9*: 1629.
175. Steele, I. M., et al. (1980) *PLC 11*: 571.
176. Smith, J. V., et al. (1980) *PLC 11*: 555.
177. Norman, M. D., and Ryder, G. (1980) *PLC 11*: 317.
178. Longhi, J., and Boudreau, A. E. (1979) *PLC 10*: 2085.
179. Weill, D. F., and McKay, G. A. (1975) *PLC 6*: 2427.
180. McKay, G. A. (1978) *PLC 9*: 661.
181. Warren, P. H., and Wasson, J. T. (1979) *Rev. Geophys. Space Phys.* 17: 73.
182. Gromet, L. P., et al. (1981) LPS XII: 368.
183. Hess, P. C., et al. (1981) LPS XII: 442.
184. Palme, H., and Wänke, H. (1975) *PLC 6*: 1179; Shirley, D. N. (1981) LPS XII: 979.
185. Moorbath, S. (1978) *Phil. Trans. Roy. Soc.* A288: 401.
186. Taylor, S. R., and McLennan, S. M. (1981) *Phil. Trans. Roy. Soc.* A301: 381.
187. McCulloch, M. T., and Wasserburg, G. J. (1978) *Science.* 200: 1003.
188. Pollack, H. N. (1981) pers. comm.
189. McCord, T. B., and Clark, R. N. (1979) *JGR*. 84: 7664; *Basaltic Volcanism* (1981) Section 2.2.3.
190. Phillips, R. J., et al. (1981) *Science.* 212: 879.
191. Campbell, D. B., and Burns, B. A. (1980) *JGR*. 85: 8271.
192. Squyres, S. W. (1980) *GRL*. 7: 593.

Chapter 6

BASALTIC VOLCANISM

Basaltic volcanic activity is very common in the solid planets in the solar system, appearing at the planetary surface as a consequence of partial melting deep within the silicate mantles. It is a manifestation of the internal heat budget of a planet, and may continue for billions of years, contrasting sharply with the briefer episodes of planetary evolution triggered by accretionary melting.

The evidence for extraterrestrial volcanism was first detected on the Moon by observers with vision acute enough to identify the dark patches on the lunar surface as lava flows [1], despite many contrary suggestions including asphalt, dust and sedimentary rocks [2]. In this chapter, most emphasis is placed on the well-studied lunar lavas. Like the highland crustal materials described in Chapter 5, these samples have provided enough scientific debate to engender caution about the age and composition of apparent volcanic landforms on unvisited or little explored planets. For this reason, most of the discussion in this chapter is concerned with the lunar mare basalts. These are well characterized by petrographical and geochemical studies. In contrast, the evidence for Martian and possible Mercurian lavas is based mainly on photo-geological interpretation, so that the description of these inferred basalts is addressed more properly in Chapter 2.

6.1 Floods of Basaltic Lava

The vast plains, which cover 17% of the surface of the Moon (Fig. 6.1), contrast with many terrestrial volcanic landforms, a difference which delayed

6.1 Far-side lunar maria and highlands. The large circular crater, filled with mare basalt, is Thomson (112 km diameter) in the northeast sector of Mare Ingenii (370 km diameter, 34°S, 164°E). The large crater in the right foreground is Zelinskiy (54 km diameter). The stratigraphic sequence, from oldest to youngest, is (1) formation of highland crust, (2) excavation of Ingenii basin, (3) formation of Thomson crater, (4) formation of Zelinskiy, (5) flooding of Ingenii basin and Thomson crater with mare basalt, (6) production of small craters on mare surface including a probable chain of secondary craters (NASA AS15-87-11724).

their acceptance as lava flows. The relief on the mare surfaces is nowhere great, and they are, in fact, exceedingly smooth. Slopes of 1:500 to 1:2000, and differences in elevation of less than 150 m over distances of 500 km are common. This smoothness of the maria is due to the low viscosity of the mare lavas. The mare basins are not all at the same level. The surfaces of Mare Crisium and Mare Smythii are about 4 km below the surrounding highlands, but Oceanus Procellarum averages about 1 km below the highlands. The distribution of mare basalts on the lunar surface is shown in Fig. 6.2.

An outstanding feature of the maria [3] is their tendency toward circular form caused by the flooding of the large ringed basins (Fig. 6.1). A clear distinction exists between the craters or multi-ring basins excavated by meteorite impacts and the maria formed by lava flooding. The Imbrium basin was formed by impact; Mare Imbrium is the "sea" of lava which filled it later. Mare basalts are much more common on the earth-facing side of the Moon [4] (Fig. 6.2). This dichotomy has been remarked upon frequently. It is probably due to the presence of a thinner (60–70-km-thick) highland crust on the near side compared with possible 80–90 km thicknesses on the far side. The paucity of far-side lavas is thus explained by their failure to reach the surface, except in deep basins, the height to which lavas can rise being governed by the density difference between the melt and that of the overlying column of rock.

In the southern hemisphere of the Moon, the far side contains more mare basalt than the near side, which is the reverse of the normal situation. This concentration of mare basalt is due to flooding of the deep depressions in the Australe and South Pole-Aitken basins.

Mare basalts may be more extensive than their surface exposure indicates and extend under some of the highland light plains units, covered by debris from basin-forming collisions. The principal evidence for this comes from the presence of dark halo craters (Section 3.11). A summary of the areal distribution of mare basalts has been given by Whitford-Stark [5].

6.1.1 Thickness of Mare Fill

A basic question is how thick are the basaltic lavas. In many areas, of which Oceanus Procellarum is a prime example, the flows are clearly rather thin, embaying the rugged highland topography. Around the edges of the circular maria, flooding of craters provides a measure of depth of mare fill. Early estimates of great thicknesses of mare basalts have not been substantiated and they form a comparatively thin veneer on the surface of the highlands. Thickness of mare basalts in the irregular maria (e.g., over much of Oceanus Procellarum) are probably only a few hundred meters [6–9]. These values can be estimated from the partially buried "ghost" craters on the mare surface.

The depth of mare basalt fill in the centers of large basins cannot be so readily identified, although an estimate of their thickness is significant for the mascon problem (Section 7.4.2). Although the mare basalt fill in the circular basins is a component in providing the positive gravity anomaly, mantle uplift beneath the basin appears to be the major factor.

It should be noted that the crater Kepler did not penetrate through the lava fill in Mare Imbrium [10]. In Mare Crisium, no basin peak ring remnants are observed and hence the mare basalt fill must be 2–4 km thick [11]. The central peak, if present, is also covered [12], although the probable basin diameter is so large that a central peak is unlikely (Sections 3.5, 3.6). Six large craters ranging in diameter from 6.2 to 22.5 km, with depths from 1.3 to 1.8 km, occur within the basalt fill of Mare Crisium. The two large craters, Peirce

NEAR SIDE

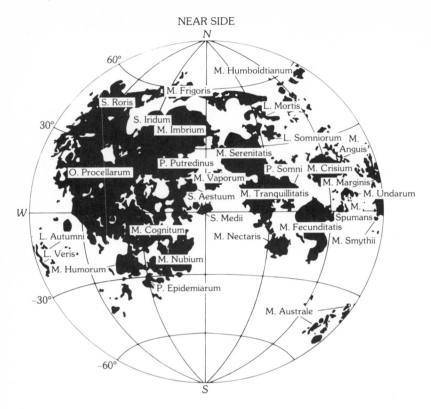

(19 km diameter, 1.49 km deep) and Picard (22.5 km diameter, 1.9 km deep) appear to bottom in highland material, and hence give a maximum thickness (near the edge of Mare Crisium) of 1.5–2.0 km. Minimum thicknesses of 1.6 km for Mare Serenitatis and 1.4 km for Crisium are indicated from the Lunar Sounder experiment [13].

A recent estimate for the thickness of the basaltic fill in Mare Imbrium is 1.5–2.5 km [14]. No indications of buried central peaks appear on the surfaces of the lava plains of the large circular maria, but probably no central peaks were formed, from analogy with Mare Orientale.

6.1.2   Age of the Oldest Mare Surface

The oldest established crystallization age for a mare basalt is for sample 10003 which has an ^{40}Ar-^{39}Ar age of 3.85 ± 0.03 aeons. This age overlaps ages of some of the highland samples and is older than the generally accepted date of 3.82 aeons for the Imbrium event. The rock is an Apollo 11 low-K mare

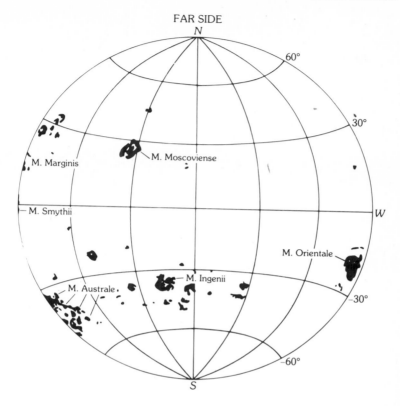

FAR SIDE

6.2   Distribution of mare deposits on the Moon. [M = Mare, O = Oceanus, S = Sinus, P = Palus, L = Lacus. From *Basaltic Volcanism* (1981) Fig. 5.4.1.]

basalt, but its relationship to the geology of Mare Tranquillitatis is unknown. The samples from the high-K Apollo 11 suite have well established dates of 3.55 aeons, so that the surface of the mare may be of that age [15].

The basalts at the Apollo 17 site were carefully sampled and yield ages from 3.69 to 3.79 aeons (with one at 3.84 ± 0.02), slightly but significantly younger than the Imbrium basin ages. This age provides, in turn, an absolute younger limit for the time of formation, both of the multi-ringed basins and of most of the highland craters. Baldwin [16] records only Hausen, Tsiolkovsky, Schrödinger, Antoniadi, Compton and Humboldt as younger than Mare Orientale, which in turn, is younger than Imbrium. Six large craters occur between the formation of Imbrium and Orientale (Section 3.17). High-Ti basalts dated from the upper surfaces of the maria are accordingly close in age to the final great cratering events.

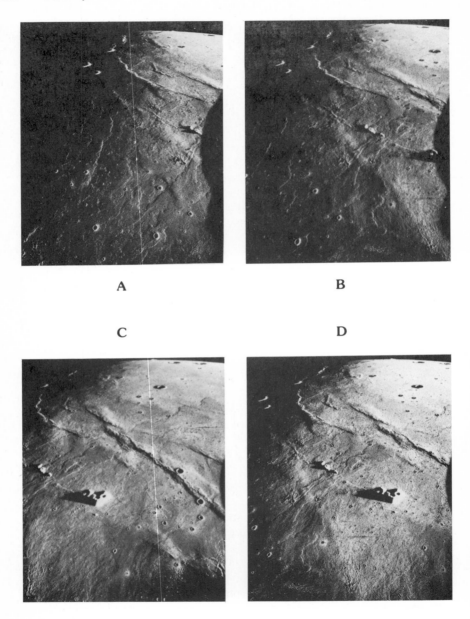

A

B

C

D

6.3   Mare basalt flow fronts illuminated under varying low sun angles in southwestern Mare Imbrium. Source of the flows is about 200 km southwest of La Hire crater (5 km diameter) seen right center in **d.** Flow thicknesses are in the range 10–30 m (Apollo pan photos 0269-0272).

6.1.3 The Lunar Lava Flows

Individual lava flows were noted by early observers [1], and some were observed in the Orbiter photographs [17], but it was not until the low sun angle photographs were taken on the Apollo 15 and 17 missions that the extent and nature of the flows were fully appreciated (Figs. 6.3 and 6.4). There were channels with levee banks along the centers of some flows larger than in terrestrial examples and typically 20–50 m high. Three stages of very late flows can be distinguished at Mare Imbrium. Figure 6.5 shows their distribution. They are of Eratosthenian age [18], and based on crater counting, dates as late as 2.5 aeons have been suggested. Before such suggestions can be confirmed, the crater-counting age techniques need further validation in view of the many past discrepancies with the radiometric ages.

The extent and volume of the flows are remarkable. The three phases of lava extrusion reached distances of 1200, 600, and 400 km. The area covered exceeds $2 \times 10^5 km^2$, an area equal to that of the Columbia River basalts. The flows ran down slopes as gentle as 1:1000. The Columbia River basalts flowed over similar slopes [19]. Some flows observed in the walls of Hadley Rille are 10–20 m thick [20] (Fig. 6.6). The average thicknesses are probably

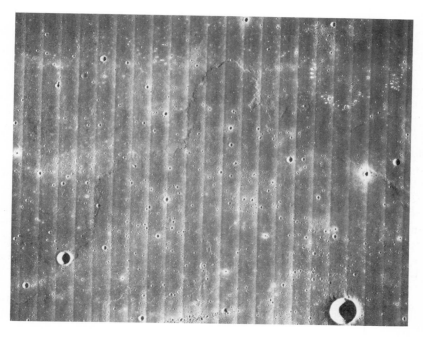

6.4 Lobate lava flow front in Mare Imbrium (NASA Orbiter V.159 M).

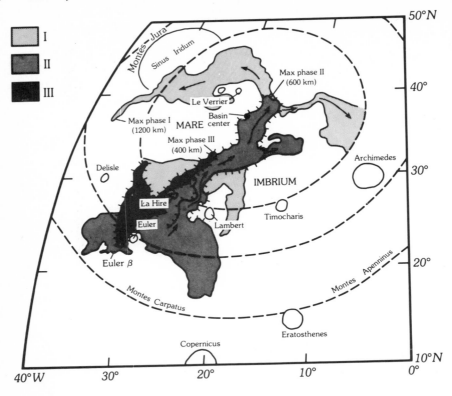

6.5 Areal distribution of the younger flows in Mare Imbrium showing three successive flows (I, II, III) and the source region near the crater Euler (28 km diameter) [18].

somewhat less than 10 m [21]. The heights of measured flow fronts range from 1 to 96 m [22], but most are less than 15 m. The flows in Mare Imbrium, although providing agreeable evidence of the volcanic nature of the mare fill, are not generally observed in the other maria. Probably this represents a difference in eruptive style, with the filling of the other maria in a manner similar to the voluminous flood basalts of Earth. In contrast, the late flows on Imbrium, derived from a discrete center, resemble some basaltic plains units on Earth formed by a series of coalescing lava shields [23]. Most mare filling must be of the first type, to account for the scarcity of discrete flows.

The length of lava flows depends as much on rates of eruption as on viscosity. The shield-building Hawaiian basalts are more fluid than the Columbia River flood basalts [21] but build domes rather than lava plateaus. The extreme lineal extent of the lunar lavas is thus due to a combination of rapid rates of extrusion coupled with low viscosity. The apparent absence of

6.6 The west wall of Hadley Rille, from Station 10 Apollo 15, showing lava flows *in situ*. This outcrop shows massive and thin-bedded units, the former with well-developed columnar jointing. The thin-bedded units are less than 1 m thick. The largest boulders on the western slope are 10 m in diameter (AS15-89-12116).

evidence for deep lava lakes, and the absence of evidence of strong fractional crystallization in deep or shallow reservoirs (Section 6.6), indicates that extrusion took place as a widely spaced series of rapid effusions.

The scarcity of traditional volcanic landforms on the Moon lends great interest to the nature and location of the volcanic vents. The youngest series of flows appear to come from a fissure about 20 km long near the crater Euler in southwestern Mare Imbrium (Fig. 6.5). This site appears to be at the intersection of two concentric ring systems, one due to the partly buried second Imbrium ring and the other to the now buried outer ring of a possibly very old "south Imbrium" basin, 900 km in diameter, centered near Copernicus. The intersection of these deep circum-basin fractures may be a significant factor in channeling the lava toward the surface.

The Herigonius region of the Moon contains particularly well exposed and mapped mare vents. Herigonius is a crater about 15 km in diameter northeast of the 110 km diameter crater Gassendi, on the north rim of the

Humorum basin [24]. Lavas from this region have flowed north to Oceanus Procellarum and south into the Humorum basin. Multi-stage eruption of lavas of a variety of compositions has occurred.

The total number of volcanic vents now recognized on the Moon is 1296 and this figure is probably an underestimate by an order of magnitude [5]. The significance of this large number of vents is that mare basalts were derived by many separate eruptive events, and were not derived from a few massive outpourings of lava [5]. This observation is critical for thermal histories requiring many individual partial melting events in the lunar interior, rather than a few massive moon-wide pulses of activity. This is also consistent with the many different compositions of mare basalts. The size of the eruptive fissures is not known, but they are probably up to 10 m in width. This figure is deduced from the observation that the vent widths for the Columbia River basalts are about 4 m wide. Effusion rates about ten times those of the Columbia River basalts as appear probable for the lunar lavas would require fissures perhaps 10 m wide [19]. The general absence of shield volcanoes on the Moon implies overall high rates of lava effusion. Presumably there are dikes and sills in the lunar crust, and much magma injection into the brecciated zones beneath the large multi-ring basins and craters may be surmised, making an unknown but probably small contribution to the mascon loadings.

6.1.4   Dark Mantle Deposits

The identification of small volcanic landforms, e.g., cinder cones, elongate spatter ridges, vents aligned along fissures and so on, has proven elusive. A potential volcanic crater at Apollo 17 (Shorty) turned out to be of impact origin. The dark mantle deposits appear more promising as a potential indicator of fire fountaining along volcanic vents. The best documented examples occur around the southern edge of Serenitatis (Fig. 6.7), notably in the Sulpicius Gallus area and in the Taurus-Littrow Valley at the Apollo 17 site. The Taurus-Littrow dark mantle is bluish dark gray compared to "very dark tan" at Sulpicius Gallus, the "blue gray" outer ring or annulus of mare lava fill in Serenitatis and the "tan gray" younger lava filling the center of the mare [25]. Figure 6.8 shows the general distribution of dark mantle deposits. Samples from the Apollo 17 site have been identified as black and orange glasses of mare basalt composition for which a fire fountain origin seems most reasonable, particularly if the rates of eruption were as rapid as those of the last flows in Mare Imbrium. Thus, the dark mantle areas may be a clue to the final centers of eruption of the mare basalts [26, 27] (see Section 6.2.3).

The low albedo area forming the outer ring of lava fill in Serenitatis has the same color and albedo as most of the lava filling Mare Tranquillitatis,

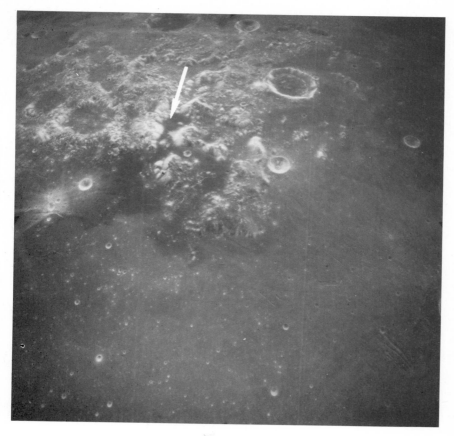

6.7 Oblique view (looking east) of the Taurus-Littrow region of southeastern Serenitatis. Apollo 17 landing site is indicated by arrow. The dark mantle deposits are
conspicuous by their extremely low albedo (Apollo metric photo AS 15-1404.) [34].

which accordingly is older than the basalt fill in the center of Serenitatis ([28],
Figs. 6.9, 6.10). The dark mantle deposits are thought to be the source region,
pyroclastic equivalents to these low albedo units. If the Apollo 11 ages for the
mare basalts reflect the last stages of filling of Mare Tranquillitatis, then the
high-K basalts thought to be at the surface at the Apollo 11 site [29] provide a
date for that event of 3.55 aeons (Section 6.4).

Dark mantle deposits have been identified in the Apollo 15 Apennine
region. These are of probable pyroclastic origin, and include regions near
Rima Fresnel (28° 30 N', 4° 15 E) and also near the Apollo 15 landing site and in
other areas (Fig. 6.8). This implies that fire fountaining has been common
along the base of the Apennine Mountains. Parts of the Apennine Bench

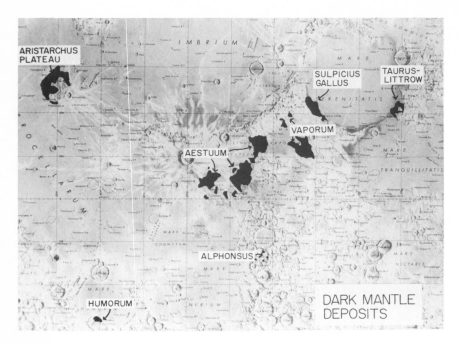

6.8   The major lunar dark mantle deposits. (Courtesy J. W. Head.)

Formation near the crater Beer are covered by pyroclastic deposits [30] (Fig. 2.13).

6.1.5   Domes and Cones

Small scale examples of Hawaiian style lava domes with central pit craters occur on the Moon. Those in the Marius Hills form the classic example, but about 200 have been mapped [31, 32]. Large shield-type volcanoes like Olympus Mons or Mauna Loa do not occur. The small lunar domes resemble small terrestrial lava shields and appear to be related to the surrounding lava plains. Small cones resembling terrestrial cinder cones have also been reported [32] in the Marius Hills area and also in Mare Serenitatis. A useful summary of these forms on the inner planets is given by Wood [33]. Forms identified as cinder cones on the Moon and Mars are much smaller than those on the Earth and the ejection velocities are calculated to be between one-third and one-tenth those of terrestrial eruptions [33]. These values imply low volatile contents both for lunar and Martian magmas.

6.9a An Earth-based photo of Mare Serenitatis, 680 km in diameter. The dark outer ring of basalt extends south-eastwards into Mare Tranquillitatis. The Apennine Mountains form the wall between Serenitatis and Imbrium to the west. The highland ridges (Haemus Mountains) radiating from the Imbrium basin are shown in the lower left portion of the photograph. A light-colored ray from Tycho, 2,000 km to the southwest, crosses the region, passing through the crater Bessel (16 km diameter) located in the south-central portion of the basin. Sulpicius Gallus (12 km diameter) is the young crater west-southwest from Bessel, inside the mare. (Lick Observatory photograph.)

6.9b Mare Crisium. The lava-flooded area is about 590 km east-west and 460 km north-south. Wrinkle ridges are well developed. The fresh crater in the south-west quadrant is Picard, 23 km in diameter. Peirce (19 km diameter) lies slightly west of north of Picard. The Luna 24 site is in the southeastern corner of the mare (Lunar Orbiter IV, 191H3).

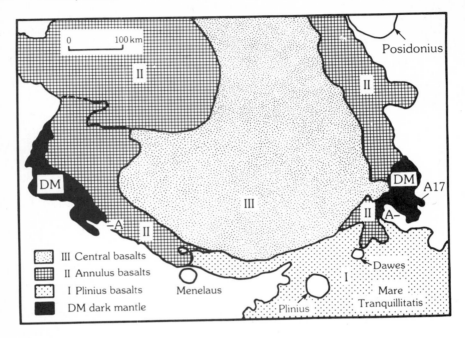

6.10   Map of Mare Serenitatis showing the distribution and relative ages of the mare basalt units. [From *Basaltic Volcanism* (1981) Fig. 5.4.7.]

6.1.6   Ridges

Wrinkle ridges commonly form concentric rings in the mare fill in the ringed basins. Good examples of concentric patterns of wrinkle ridges occur in Serenitatis, Nubium, Fecunditatis, and Tranquillitatis. Many origins have been proposed, which can be divided into three classes: (a) the ridges are due to extrusion of lavas along radial fractures; (b) they are formed when lava sheets are draped over buried highland or other structural ridges following subsidence; and (c) the ridges are produced by compressional forces due to loading of the lithosphere by the lava flows.

The morphology of the ridges supports the view that they are compressional features. They consist, not of squeezed-out lava, but of preexisting mare surface, deformed and folded [34], as is apparent from many photographs (Figs. 6.9 and 6.11). The intense crumbling that occurs along their crests, with some local examples of overthrusting, attests to their compressional origin. The absence of tension cracks seems to rule out intrusion of lava from beneath, which would stretch the crust. The ridges do not resemble

6.11 The western edge of Mare Serenitatis showing rilles (center bottom) and wrinkle ridges. View looking north. Mare Imbrium on top left horizon (NASA AS 17-0953).

terrestrial lava domes. They are thus most probably related to compression attendant on mare subsidence, which results in crustal shortening.

Clear evidence of lowering of mare basalt levels has been observed in most of the western maria where highland surfaces protrude through the basalt flows [35]. Benches interpreted as being due to former higher levels of mare basalts are a ubiquitous feature in these areas. Such subsidence will cause compression, thus providing one mechanism for producing the wrinkle ridges. The evidence for compressive stresses in the mare wrinkle ridges has wider significance for the tectonic history of the Moon: horizontal compressive stresses were active down to 3 aeons at least [36].

6.12 An oblique view of Schröters Valley, a classic example of a sinuous rille, about 200 km long and 15 km across at its widest point. The large bright ray crater is Aristarchus (40 km diameter). The subdued crater just to the right of the head of Schröters Valley is Herodotus (35 km diameter).

6.1.7 Sinuous Rilles

The resemblance of sinuous and meandering rilles (Figs. 6.12 and 6.13) to the familiar meander patterns of terrestrial river channels long encouraged speculation that liquid water had existed, at least briefly, on the lunar surface. The view that the maria were filled with dust or other incoherent material encouraged the possibility of erosion by water and it was not until the initial sample return that these ideas evaporated. As the basaltic nature of the mare basin fill was established, detailed examinations and comparisons with terrestrial lava channels, tubes and tunnels proved instructive.

6.13   Hadley Rille (about 1 km wide) at the Apollo 15 landing site, close against the Apennine Mountain scarp.

A typical example, Hadley Rille (Figs. 6.13, 6.14 and 6.15), investigated during the Apollo 15 landing, is 135 km long and sinuous, averages 1.2 km in width and 370 m in depth, and greatly resembles a collapsed terrestrial lava tube. The sinuous bends are not structurally controlled by simple fracturing, for the two sides do not match up [37]. The rille originates in a cleft adjacent to the highland scarp. Many others originate in similar fissures or craters, and

6.14 View along Hadley Rille taken during the Apollo 15 mission (NASA AS 15-85-11451).

the association with the edges of the mare basins recalls, on a much larger scale, the tendency of terrestrial basaltic lava lakes to fill from the sides as well as from beneath. The rille is deepest where it is widest, in contrast to river channels, a feature consistent with the collapsed lava tube hypothesis. Where it is shallow and flat bottomed it may have been an open channel. The physical problems of roofing across kilometer-wide channels of very fluid lavas in an environment of one-sixth terrestrial gravity are not well understood. The very fluid nature of the lava encourages turbulent rather than laminar flow, increasing the erosive power of the fluid. Elevation profiles show that the floors of the rilles do indeed slope downstream [38].

6.15   A view of the bottom of Hadley Rille, looking north from Station 2 Apollo 15 mission. The material in the rille comprises fine-grained regolith and large blocks which have rolled down the slope. The largest block is 15 m, about the size of a bus (NASA AS 15-11287).

The general absence of deltas where the rilles fade out onto the broad mare surfaces has often been noted. Explanations include: (a) the fluid nature of the lavas, as shown by the low flow fronts, and (b) the covering of rille ends by later mare flows. There is occasional evidence of ponding at the downstream end.

The consensus from the Apollo studies is that Hadley Rille is a lava channel that was partly roofed. Meteorite bombardment will tend to collapse roofs after the emptying of the channel. There is clear evidence at the Hadley site that the surface of the mare (Palus Putredinis, or the Swamp of Decay) has subsided differentially by about 100 m, after partial solidification [37]. Hadley Rille provides a suitable drainage channel.

Much insight into the subsurface of mare basins was gained from the examination of the wall of Hadley Rille. The regolith is about 5 m thick, underlain by exposed massive bedrock for about 55 m (Fig. 6.6). The remaining 300 m, to the bottom of the rille, is blanketed with talus, including many massive blocks (Fig. 6.15) up to 15 m across. The rock units in the walls are

10–20 m thick, massive, and little jointed. Many small lava ponds are associated with sinuous rilles [39]. In general, the sinuous rilles have probably been important feeder routes for filling the mare basins [23].

6.1.8   Straight and Curved Rilles

Unlike the sinuous rilles discussed above, which are clearly related to mare lava flows, the straight and curved rilles are tectonic graben features. They occasionally extend into the highlands but are more abundant on mare basalt surfaces and so are considered here.

The straight rilles are commonly more than 5 km wide, hundreds of kilometers long, and cut across craters and maria alike, indifferent to the surficial topography. Curved, or arcuate, rilles are a variant of this type. Particularly fine exmples of these, concentric to Mare Humorum, are shown in Fig. 3.17. These rilles are the lunar equivalent of terrestrial grabens, and they appear to exhibit seismic activity at present [40]. Lunar rilles of all species cluster around the edges of the maria, a fact noted by many observers (Fig. 6.11).

An important observation is that these rilles appear to cut only the older mare lavas [41]. This indicates that extensional stress in the outer regions of the Moon decreased about 3.6 aeons ago, although compressional stress, as indicated by the wrinkle ridges, continued to younger periods [36].

6.2   Mare Basalt Rock Types

More than twenty distinct basaltic types have been identified in the Apollo collection, although some exist only as very small samples. This rather fine subdivision has led to comments that the dispersion in composition among some suites of lunar basalts (e.g., Apollo 11-A, Apollo 17-A and B, and Apollo 12 pigeonite basalts) "are scarcely more variable in composition

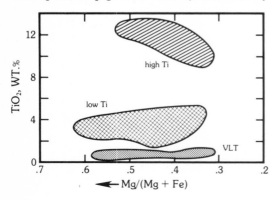

6.16 The variation between TiO_2 (wt.%) and the Mg/(Mg + Fe) ratio in mare basalts, showing the clear separation into high, low and very low titanium basalts. Adapted from *Basaltic Volcanism*, Fig. 1.2.9.4.

Table 6.1 Classification of lunar basaltic rock types.

Group	Basalt class	Mission	Sample Numbers
Low Titanium Basalts	Emerald Green Glass	Apollo 15	15426
	Very low Titanium (VLT)	Luna 24 Apollo 17	24174 70007, 70008, 78526
	Olivine	Apollo 12	12002, 12004, 12006, 12009, 12012, 12014, 12015, 12018, 12020, 12040, 12075, 12076
		Apollo 15	15016, 15445, 15555
	Pigeonite	Apollo 12	12007, 12011, 12017, 12021, 12039, 12043, 12052, 12053, 12055, 12064, 12065
		Apollo 15	15058, 15499, 15597
	Ilmenite	Apollo 12	12005, 12008, 12016, 12022, 12036, 12045, 12047, 12051, 12054, 12056
High Titanium Basalts	Low-K	Apollo 11	10003, 10020, 10029, 10044, 10047, 10050, 10058, 10062, 10092
		Apollo 17 A B C U	75055, 76136 70215, 70275 74245, 74255, 74275 70017, 70035
	Orange glass	Apollo 17	74220
	High-K	Apollo 11	10017, 10022, 10024, 10032, 10049, 10057, 10069, 10071, 10072, 10085
High Aluminum Basalts		Apollo 12 Luna 16 Apollo 14	12031, 12038, 12072 B-1 14053, 14072, 14321

Sources: Rhodes, J. M., and Blanchard, D. P. (1980) *PLC 11:* 49 (Apollo 11).
Beaty, D. W., and Albee, A. L., (1978) *PLC 9:* 359 (Apollo 11).
Rhodes, J. M., et al. (1977) *PLC 8:* 1305 (Apollo 12).
Rhodes, J. M., et al. (1976) *PLC 7:* 1467 (Apollo 17).

than vertical sections of single flows of several terrestrial basalts" [42]. Accordingly, caution should be exercised in making too fine a subdivision, or concluding that many individual flows were sampled, or drawing conclusions about proposed variations in source regions. The problems associated with the identification of terrestrial primary basaltic magmas, in a much less random sampling situation, should engender caution.

The lunar mare basalts show extreme variations in titanium content (Fig. 6.16), from about 0.5% to over 13% TiO_2. This range in titanium content has

Table 6.2 Modal mineralogy of lunar basalts.[†]

Mission	Type	Pyroxene	Feldspar	Olivine	Opaque Minerals
Luna 24	VLT	60	34	4	2
Apollo 17	VLT	62	32	5	1
Apollo 12	olivine	54	19	20	7
Apollo 15	olivine	63	24	7	6
Apollo 12	pigeonite	69	21	1	9
Apollo 15	pigeonite	62	34	—	4
Apollo 12	ilmenite	61	26	4	9
Apollo 11	low-K	51	32	2	15
Apollo 17	high-Ti	52	33	—	15
Apollo 17	VHT	48	23	5	24
Luna 16	high-Al	52	41	—	7
Apollo 14	high-Al	54	43	—	3

[†]Adapted from *Basaltic Volcanism* (1981) Table 1.2.9.6.

been used to make a broad subdivision, while a separate class is distinguished by high-Al contents. The classification is thus basically chemical, with some finer subdivisions based on mineralogy. Detailed descriptions of the mare basalts appear in *Basaltic Volcanism on the Terrestrial Planets* [43] and only a brief outline is presented here.

The classification adopted here is shown in Table 6.1. The types which are distinguished are broadly divided into low-Ti and high-Ti classes. The low-Ti basalts include very-low titanium (VLT) basalts, olivine basalts, pigeonite basalts, and ilmenite basalts. The high-Ti basalts include low-K, high-K and very-high-Ti (VHT) basalts. A third class of aluminous or feldspathic mare basalts comprises the high-Al basalts. The typical modal mineralogical composition of these basalts is given in Table 6.2. Their areal distribution is shown in Fig. 6.17.

The major element chemical compositions of representative samples of these different types are given in Table 6.3. An attempt has been made to select especially significant samples, particularly those thought to represent primary or undifferentiated samples. Representative mare basalts are shown in Figs. 6.18–6.28. A number of detailed studies since 1974 have been carried out on the Apollo basaltic sample collections. These have clarified our ideas on the interrelationships of the various basalt types. Details are given in references [44–50]. The significant findings are discussed in the succeeding sections of this chapter.

6.2.1 Mineralogy

A general survey of the whole field has been carried out by Frondel [51]. Extensive reviews have also been carried out by Smith [52, 53]. These exhaustive accounts have been updated in *Basaltic Volcanism on the Terrestrial*

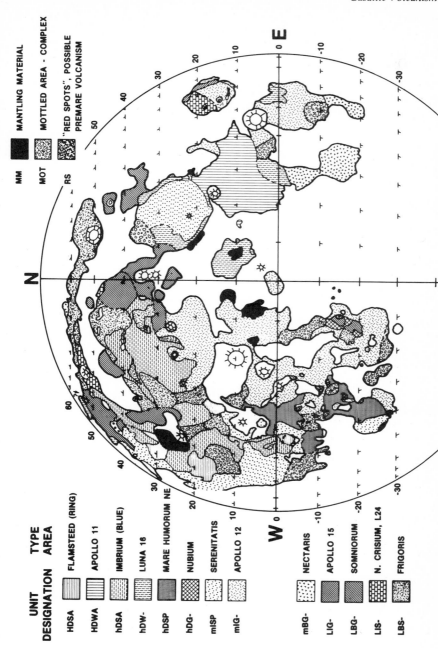

6.17 Major basaltic types on the visible face of the Moon, as derived from spectral reflectance data. The unit designations represent values for four measurable parameters (UV/VIS ratio, albedo, strength of the 1 μm band, and strength of the 2 μm band). Each of these parameters depends upon the composition of the surface. From *Basaltic Volcanism*, Fig. 2.2.9.

286

Table 6.3a Major element compositions of low-Ti lunar basalts.

	Green Glass	Luna 24 VLT	Apollo 17 VLT	Apollo 12 Olivine		Apollo 15 Olivine		Apollo 12 pigeonite	Apollo 15 pigeonite	Apollo 12 Ilmenite
	15426	24109	78526	12002	12009	15016	15545	12064	15597	12051
SiO_2	45.2	45.2	46.7	43.6	45.0	44.1	45.2	46.3	48.0	45.3
TiO_2	0.38	0.89	0.92	2.60	2.90	2.28	2.41	3.99	1.80	4.68
Al_2O_3	7.5	13.8	10.0	7.87	8.59	8.38	8.59	10.7	9.44	9.95
FeO	20.0	20.5	18.6	21.7	21.0	22.7	22.2	19.9	20.2	20.2
MnO	0.26	0.27	0.24	0.28	0.28	0.32	0.30	0.27	0.30	0.28
MgO	17.5	6.35	12.2	14.9	11.6	11.3	10.3	6.49	8.74	7.01
CaO	8.5	12.7	10.0	8.26	9.42	9.27	9.82	11.8	10.4	11.4
Na_2O	0.13	0.24	0.12	0.23	0.23	0.27	0.31	0.28	0.32	0.29
K_2O	0.03	0.01	—	0.05	0.06	0.04	0.04	0.07	0.06	0.06
Cr_2O_3	0.53	0.19	0.74	0.96	0.55	0.85	0.68	0.37	0.48	0.31
Σ	100.0	100.2	99.5	100.4	99.6	99.5	99.8	100.2	99.8	99.5
$Mg/(Mg+Fe)$	0.59	0.36	0.54	0.55	0.50	0.47	0.45	0.37	0.44	0.38

Sources: (also for 6.4a) *Basaltic Volcanism* (1981) Tables 1.2.9.1–5.
Ma, M. -S., et al. (1981) *PLC 12*: 915.
Refs. [6.44–6.50].

Table 6.3b Major element compositions of high-Ti and high-Al lunar basalts.

	Apollo 11 Low-K (10003)	Apollo 17 A (75055)	Apollo 17 B (70215)	High-Ti Apollo 17 C (74275)	U (70017)	Orange Glass (74220)	Apollo 11 High-K (10049)	Apollo 12 (12038)	High-Al Luna 16 (B1)	Apollo 14 (14053)
SiO_2	39.8	40.6	37.8	38.6	38.5	38.6	41.0	46.9	43.8	46.4
TiO_2	10.5	10.8	13.0	11.9	13.0	8.81	11.3	3.25	4.90	2.64
Al_2O_3	10.4	9.67	8.85	8.72	8.65	6.32	9.5	12.7	13.7	13.6
FeO	19.8	18.0	19.7	18.1	18.3	22.0	18.7	17.7	19.4	16.8
MnO	0.30	0.29	0.27	0.27	0.25	—	0.25	0.25	0.20	0.26
MgO	6.69	7.05	8.44	9.65	9.98	14.4	7.03	6.74	7.05	8.48
CaO	11.1	12.4	10.7	10.6	10.3	7.68	11.0	11.5	10.4	11.2
Na_2O	0.40	0.43	0.36	0.36	0.39	0.36	0.51	0.67	0.33	—
K_2O	0.06	0.08	0.05	0.06	0.05	0.09	0.36	0.076	0.015	0.10
Cr_2O_3	0.25	0.27	0.41	0.54	0.50	0.75	0.32	0.07	0.28	—
Σ	99.3	99.5	99.5	98.8	99.9	99.1	100.0	100.0	100.1	99.5
$Mg/(Mg+Fe)$	0.38	0.41	0.43	0.49	0.49	0.54	0.40	0.40	0.39	0.48

Sources: (same for 6.4b) *Basaltic Volcanism* (1981) Tables 1.2.9.1–5.
Beaty, D. W., et al. (1979) *PLC 10*: 132.
Refs. [6.44–6.50].
Hubbard, N. J., et al. (1972) *EPSL* 13: 426.

6.18   Apollo 17 high-Ti basalt 71559 (wt. 82 g) (NASA S 73-16456).

6.19   Highly vesicular olivine basalt 15016 (wt. 924 g, length 13.5 cm) (NASA S 71-46632).

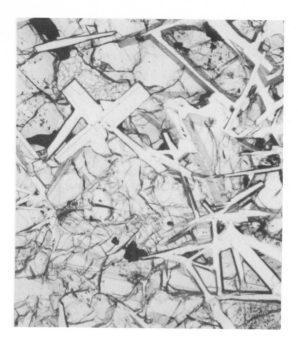

6.20 Thin section of a particle of VLT basalt 78502,24, 1.0 × 0.75 mm. Plane polarized light. (Courtesy G. Ryder.)

6.21 VLT basalt 78502,24. Crossed polarizers. (Courtesy G. Ryder.)

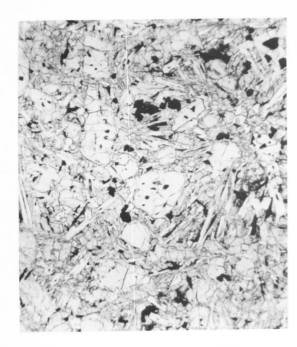

6.22a Apollo 12 olivine basalt 12002,161. Plane polarized light 3 × 4 mm. (Courtesy G. Ryder.)

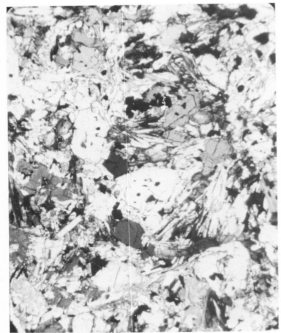

6.22b Apollo 12 olivine basalt 12002,161. Crossed polarizers. (Courtesy G. Ryder.)

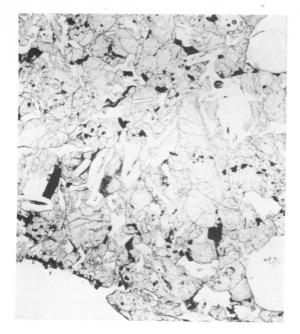

6.23a Apollo 15 vesicular olivine basalt 15016,146. Plane polarized light, 3 × 4 mm. (Courtesy G. Ryder.)

6.23b Apollo 15 vesicular olivine basalt 15016,146. Crossed polarizers. (Courtesy G. Ryder.)

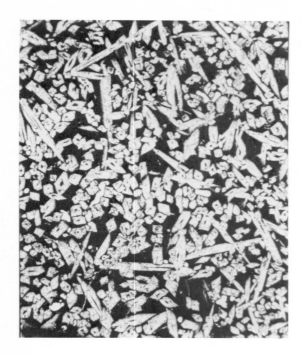

6.24 Apollo 15 pigeonite vitrophyric basalt 15597,15. Plane polarized light 3 × 4 mm. (Courtesy G. Ryder).

Planets for the lunar basaltic rocks [54] and the interested reader is referred to these sources for complete mineralogical details. Typical modal compositions for the basalts are given in Table 6.2. Pyroxene is the most abundant mineral in lunar basalts and provides useful information on petrogenesis, since it forms over much of the crystallization interval. Olivine compositions are mainly magnesium-rich, the most magnesian being Fo_{75}–Fo_{80}. These should be in equilibrium with $Mg/(Mg + Fe)$ ratios of 0.47–0.57 in the melt.

Feldspars are commonly An-rich, the range in the basalts being from An_{60} to almost pure anorthite. Reviews of spinel chemistry have been provided by Papike [55] and Haggerty [56].

6.2.2 Reduced Nature of Mare Basalts

Vesicular basalts are not uncommon in the returned lunar samples, indicating that a gas phase was present during eruption. Bubble sizes up to several centimeters in diameter have been observed (sample 15016). Many of the bubble pits or vesicles are plated with late-crystallizing minerals and accordingly show strong reflections from the crystal faces. The gas phase was most probably carbon monoxide. An abundance level in the lunar lavas of 500 ± 250 ppm has been suggested [57–59].

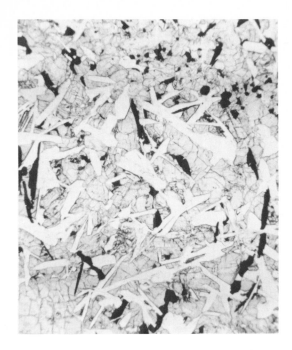

6.25a Apollo 12 ilmenite basalt 12051,56. Plane polarized light 4 × 3 mm. (Courtesy G. Ryder.)

6.25b Apollo 12 ilmenite basalt 12051,56. Crossed polarizers. (Courtesy G. Ryder.)

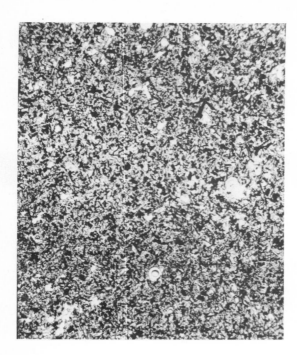

6.26 Apollo 11 high-K basalt 10049,32. Plane polarized light 3 × 4 mm. (Courtesy G. Ryder.)

This is produced by reaction of oxidized iron with carbon or carbides in the magma, leading to the production of CO and metallic iron. Such reactions could occur at depths of about 3 km in the Moon, as the magma rose toward the surface [57], rather than by reduction involving loss of sulphur [60]. The measured abundance of elemental carbon in lunar rocks is very low (10–80 ppm; Section 6.3.8), indicating that up to an order of magnitude carbon has been lost, if this process is responsible for reduction on the Moon. In contrast, H_2O and CO_2 are the dominant volcanic gases on Earth and probably on Mars.

The presence of metallic iron, FeS, and the low amount of ferric iron in the lunar lavas indicate that the mare basalts are strongly reduced with oxygen fugacity values of about 10^{-14} at 1100°C. Figure 6.29 gives the measured total rock oxygen fugacity values for lunar basalts. At a given temperature, the values appear to be independent of the texture, mineralogy, or chemistry of the basalts. The values lie about midway between the iron-wüstite and the iron-rutile-ferropseudobrookite buffers and are several orders of magnitude lower than oxygen fugacity values for terrestrial basalts, which are typically about 10^{-8} at about 1100°C.

The lunar basalts appear to have been strongly reduced at the time of extrusion, judging from the similarity in oxygen fugacity values in olivine and

6.27a Apollo 12 aluminous basalt 12038,4. Plane polarized light 4 × 3 mm. (Courtesy G. Ryder.)

6.27b Apollo 12 aluminous basalt 12038,4. Crossed polarizers. (Courtesy G. Ryder.)

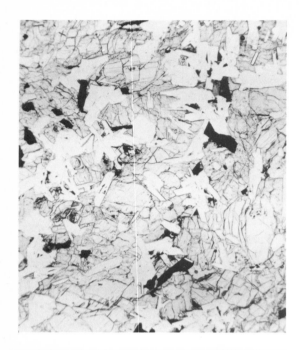

6.28a Apollo 14 high-Al basalt 14053,6. Plane polarized light 3 × 4 mm. (Courtesy G. Ryder.)

6.28b Apollo 14 high-Al basalt 14053,6. Crossed polarizers   3 × 4 mm. (Courtesy G. Ryder.)

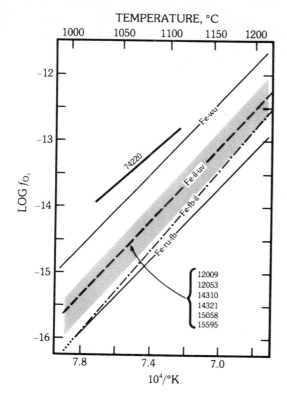

6.29 The measured bulk oxygen fugacity (f_{O_2}) value for mare basalts falls in the shaded zone, superimposed on the buffer curves of the Fe-FeO-TiO_2 system. The orange glass sample 74220 appears to be considerably more oxidized. Whether this is a local effect produced during surface eruption or of more fundamental significance is unclear [57, 61].

groundmass for the Apollo 12 basalt 12009, which reached the surface as a liquid, before crystallization set in. The experimental studies do not favor the loss of the alkali elements by volatilization. At the low oxygen fugacities of lunar magmas, sodium would volatilize as the metal, rather than as the oxide, oxidizing the magma and producing Fe^{3+}. That this has not occurred is good evidence against extensive alkali loss by volatilization [61].

The highly reduced state of lunar magmas changes the oxidation state of several elements, with interesting geochemical consequences; thus, 90% of Cr is present as Cr^{2+}, 70% of Eu as Eu^{2+}, 4% of Ti as Ti^{3+} and possibly 1% of Fe is present as Fe^{3+}. Cerium is present as Ce^{3+}, with no evidence of Ce^{4+} [62].

6.2.3 Volcanic Glasses

In addition to the wide variety of impact glasses found in the lunar regolith, examples of glass spheres of probable volcanic origin have been identified. The original and most celebrated example is the Emerald Green Glass from Apollo 15 (15425, 15426, 15427). Spheres were most abundant in

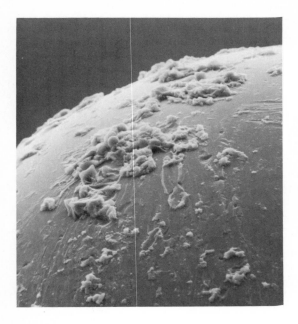

6.30 The surface of an orange glass spherule (74220) from Apollo 17, coated with spatter. The view bears some resemblance to the Apennine Mts. (cf. Fig. 5.1b). Magnification × 5000 (NASA S.73.17687). (Courtesy D. S. McKay.)

the 0.1–0.3 mm diameter size range. In the soil samples around Spur crater they contribute up to 20% of the soil by volume. Their exposure age is 300 million years, which is taken as the age of Spur crater [63]. The age of the green glass is 3.3 aeons, similar to that of the Apollo 15 basalts [64]. The significance of this age is that the origin of the glass is probably related to the extrusion of the mare basalts in the Swamp of Decay (Palus Putredenis) than to either younger volcanism, older highland crust, or impact events.

The discovery of orange soil (74220) by the Apollo 17 astronauts was a spectacular part of the mission, since it carried with it hopes of water, oxidation, young volcanic activity, and the possibility that "Shorty" crater, on whose rim the glass was found, might be the young lunar volcanic vent so eagerly awaited by some astrogeologists. All these hopes were dampened by the examination of the glass. The soil consisted almost entirely of fine glass spheres, averaging about 0.1–0.2 mm diameter (Fig. 6.30). The color is mainly orange, but ranges from yellow to black. The glass spheres have a high-Ti content (9.3% TiO_2), which causes the orange color, and they are similar in composition both to the Apollo 11 orange glasses and to the Apollo 11 and 17 basalts, although somewhat richer in MgO, zinc, chlorine, copper, lead, and other volatile components [65, 66]. The carbon contents are low (down to 4 ppm) and variable. The carbon (and many of the other volatiles) is present as a surface condensate [67]. The surface of the spheres is coated with adhering droplets of similar composition, indicating low-velocity spattering. The major element composition is close, but not identical to that of the Apollo 17 basalt. There is no evidence of the presence of ferric iron.

The age of the glass spheres is 3.48 aeons, which links them directly to that of the Apollo 17 basaltic rocks [68]. Their lead content is high (2.5 ppm), but 80% of it is leachable. This lead is among the most nonradiogenic on the Moon. The $^{238}U/^{204}Pb$ ratio for the source was 34.5 [69], compared to values of 100–1000 for some of the mare basalts. Similar parentless lead appears in the rusty rock 66095. Other glasses, probably due to volcanic action rather than to impact, appear at most sites. Thus, red and yellow glasses at the Apollo 15 site are attributed to a volcanic origin [70].

The glasses do not show a wide range of size, have a more restricted range in composition by comparison with most glasses of impact origin, and contain high concentrations of volatile elements. The orange and green glasses were liquid close to the time at which the associated mare basalts were extruded. They tend to occur along the edges of the maria. The similarities in volatile element, the occurrence in areas where lavas may be ponded, the lead isotope evidence, and the absence of a demonstrated meteoritic component in the glasses all point to an internal, rather than an external process of origin. The absence of shock effects is strong evidence against an impact origin, although this hypothesis has recently been reproposed [71].

The samples from the drive tube (75001) which sampled the soil on the rim of Shorty crater, have an identifiable meteoritic component, which does not resemble the common C1 chondritic patterns typical of average lunar soils, but resembles that of IIB iron meteorites. This observation has revived the suggestion that the orange glass was formed by impact into a lava lake, although the meteorite component might have been due to the Shorty crater event itself [71]. Most of the criteria can be explained as resulting from fire fountaining [26, 27]. In such a process, multiple droplets are built up by low velocity impacts of material with the same composition. The observation that the last flows flooding across the broad plains of Mare Imbrium came from a 20-km-long fissure near one edge is relevant, as is the distribution of the sinuous rilles, interpreted here as lava cut channels.

All these phenomena are consistent with lava effusions flooding out from fissures near the edges of the basins. During this massive outpouring, some fire fountaining occurs, spraying the homogeneous glass spherules in layers over the nearby surfaces. Often, they will be buried by subsequent flows and later excavated by impacts, as at Shorty crater, to produce locally a glass-rich regolith.

The minor differences with the products of terrestrial fire fountains, such as lack of vesicles, can be ascribed to the lower viscosity of the lunar basalts and to the much lower content of gaseous species. The green and orange glasses bear a morphological resemblance to glasses from Hawaiian lava fountains [72], but a detailed study of glass ejecta from terrestrial fire foun-

---

tains needs to be carried out. Detailed study of the individual glass spherules [70, 73] reveals large variations in major element contents and many bizarre compositional characteristics. For example, nickel decreases as Mg/(Mg + Fe) increases (Fig. 6.46), contrary to normal behavior. Accordingly, very complex magmatic processes are required to account for these trends [73]. Such processes may occur during recycling of material in fire fountains, mixing in the source regions, or may of course be produced by mixing in impact conditions. Although the fire fountain origin is judged likely, fundamental petrological information about the source regions of the glasses may not survive unchanged during the eruption and caution should be exercised in such interpretations [70]. Nevertheless, the trace element chemistry reveals that the green glasses were derived from cumulate source regions which had experienced olivine and plagioclase fractionation. They possess a distinct negative Eu anomaly [73] (see Section 6.5.2).

The high concentrations of volatile elements, and of parentless lead on their surfaces, and the ease with which these elements are leached, is consistent with droplets forming or passing through a vapor cloud. Volatiles may be widely dispersed over the Moon by such mechanisms [74]. The broader significance of the volatile content is speculative (see Sections 4.5.2 and 5.5). Volatile elements and parentless lead are widely dispersed in the regolith and in the highland crust through which the lavas pass. Accordingly, the coatings on the glass spheres probably have no wider significance, and do not demand a volatile rich interior. The geochemical evidence for a bulk depletion of volatile elements in the Moon is overwhelming.

6.3   Chemistry of Mare Basalts

6.3.1   Major elements

Twelve distinct compositions are listed in Table 6.3, although many finer-scale subdivisions have been proposed. Thus, in a recent restudy of the Apollo 11 low-K suite [75], three distinct low-K basalt types have been recognized. In contrast, the chemistry and isotopic data suggest that the high-K basalts could have come from a single flow [76, 77].

The detailed relationships among the low-Ti basalts are given in Section 1.2 of *Basaltic Volcanism on the Terrestrial Planets* [43]. The olivine and pigeonite basalts at Apollo 12 are related by olivine fractionation, but the ilmenite basalts are a separate major type. In contrast, at the Apollo 15 site the olivine and pigeonite basalts cannot be related by low-pressure fractionation. The high-Al basalts (e.g., 12038) are firmly established as a class and samples of them continue to be discovered in the sample collection [78]. Although the newly found samples are small (often <100 mg), thus raising the question of sampling size, the major element compositions are all quite characteristic of high-Al basalts (Al_2O_3 ranges from 12.0–13.3%). Nickel is low and the Ni/Co relationships in the metal resemble those of high-Ti basalts, placing them among those basalts derived from the late differentiates of the magma ocean. The REE patterns of these basalts resemble those of high-Al basalts generally, although they show so much difference in detail that they may come from separate source regions. The Luna 16 basalts from Mare Fecunditatis are characterized by high Al_2O_3 (13%), moderate TiO_2 (5%) and high Mg/Fe ratios. They have low abundances of Cr_2O_3 (0.20–0.24%) for lunar basalts and are not related to other basalt types. "No combination of fractionation of major mineral phases produces the observed compositional characteristics of Luna 16 basalts" [79].

The variations among the major elements and the great variety of mare basalts argue for heterogeneous source regions within the lunar interior. Variations in degrees of partial melting of a uniform or primitive unfractionated source region could produce the observed diversity. The Mg numbers Mg/(Mg + Fe) range in lunar basalts from 65 down to 35, in contrast with a range from 75 to 45 in typical terrestrial basalts. The Mg number for the source regions of the lunar basalts is about 80–82 compared with values of 92 for the terrestrial mantle.

In the next sections, the trace element abundances in the mare basalts are discussed. These data provide important constraints on the nature of the source regions of the mare basalts, on their fractionation history and on the evolutionary history and bulk composition of the Moon. Many of the first order conclusions, such as the depletion of the basalts in volatile and siderophile elements, and the enrichment in refractory elements, were discovered following the first sample return during the analysis of the Apollo 11 high-Ti basalts.

6.3.2 Large Cations

The large cations (Table 6.4) comprise the elements that mainly accompany potassium. In contrast to most terrestrial surface rocks, the concentration of potassium is so low in the lunar rocks that it constitutes a minor or trace element and resembles the concentration levels in low-K oceanic tholeiites or in chondritic meteorites. In the lunar basalts, potassium and other large cations are principally contained in the interstitial material, or mesostasis, and enter the lattice sites in the main rock-forming minerals only to a minor degree. Nevertheless, potassium may concentrate in the residual phases (up to 9% K has been observed) with formation of K-feldspar.

The high concentrations of barium in mare basalts is due both to the high liquid/crystal distribution coefficients which cause it to enter partial melts, but also to its inherently high lunar abundance. Barium is a refractory

Table 6.4a Trace element compositions of low-Ti lunar basalts.[†]

	Green Glass (15426)	Luna 24 VLT (24174)	Apollo 17 VLT (78526)	Apollo 12 olivine (12002)	(12009)	(15016)	Apollo 15 olivine (15545)	Apollo 12 pigeonite (12064)	Apollo 15 pigeonite (15597)	Apollo 12 ilmenite (12051)
I. Large Cations										
Cs	–	–	–	0.039	0.047	0.034	0.035	–	0.037	0.04
Rb	0.58	–	–	1.04	1.05	0.81	0.75	1.05	1.13	0.91
K	170	180	100	420	530	330	330	580	500	500
Ba	18	50	–	67	60	61	47	70	52	74
Sr	41	110	–	101	96	91	104	135	111	148
Pb	–	–	–	–	–	–	–	–	–	–
K/Rb	290	–	–	405	505	410	440	550	440	550
Rb/Sr	0.01	–	–	0.01	0.01	0.01	0.01	0.01	0.01	0.01
K/Ba	11	4.0	–	6.3	8.8	5.4	7.0	8.3	9.6	6.8
II. Rare Earth Elements										
La	1.20	2.87	1.2	6.02	6.1	5.58	4.93	6.76	4.86	6.53
Ce	–	8.6	–	17	16.8	15.6	13.9	17.5	13	19.2
Nd	–	7.0	1.0	12.3	16	11.4	9.82	16	9.3	15.4
Sm	0.83	2.10	1.0	4.24	4.53	4.05	3.29	5.51	3.09	5.68
Eu	0.24	0.83	0.30	0.84	0.94	0.97	0.90	1.16	0.84	1.23
Gd	–	–	–	5.65	5.2	5.4	4.48	7.2	–	7.89
Tb	0.21	–	–	–	1.11	0.9	–	1.27	0.69	–
Dy	–	2.9	2.0	6.34	7.13	5.74	4.68	9.03	4.51	9.05
Ho	–	–	–	–	1.4	1.1	–	–	0.86	–
Er	–	–	–	3.89	3.6	3.1	2.67	6.0	–	5.57
Yb	0.97	2.00	1.4	3.78	3.74	2.62	2.16	4.59	2.13	5.46
Lu	0.14	0.31	0.23	–	0.55	0.32	0.31	0.67	0.30	–
Y	7.2	–	–	39	34	21	33	41	–	48
La/Yb	1.24	1.44	0.86	1.59	1.63	2.13	2.28	1.47	2.28	1.20
Eu/Eu*	0.76	1.10	0.76	0.52	0.59	0.63	0.72	0.56	0.75	0.56

III. Large High Valence Cations

U	0.02	—	0.22	0.24	0.12	0.13	0.22	0.14	0.26
Th	0.08	0.2	0.75	0.88	0.50	0.43	0.84	0.53	1.0
Zr	22	50	105	105	86	—	115	—	130
Hf	0.57	—	—	4.0	2.6	2.2	3.9	—	—
Nb	1.5	—	8.5	6	10	—	7	—	7
Th/U	4.0	—	3.41	3.63	4.2	3.3	3.51	3.79	3.85
Zr/Hf	38	—	—	27	33	—	29	—	—
Zr/Nb	15	—	12.5	17.8	8.6	—	16.3	—	18.3

IV. Ferromagnesian Elements

Cr	—	—	5620	—	6400	—	—	—	—
V	165	—	175	155	200	170	119	—	100
Sc	38	—	38	46	39	42	63	—	58
Ni	153	—	64	52	—	51	7	—	6
Co	75	—	—	49	—	—	—	—	—
Cu	—	—	4.6	—	11	—	7	—	6
Fe%	16.4	14.5	16.9	16.4	17.7	17.3	15.5	15.8	15.8
Mn	2000	1850	2150	2150	2450	2310	2080	2310	2150
Zn	—	—	1.5	—	4	—	2	1.2	2
Mg%	10.0	7.32	8.93	6.93	6.78	6.71	3.89	5.24	4.21
V/Ni	1.08	—	2.7	2.9	—	4.0	17	—	17
Cr/V	—	—	32	—	32	—	17	—	—

†Values in ppm, except those indicated.

Table 6.4b Trace element compositions of high-Ti and high-Al lunar basalts.†

	High-Ti							High-Al		
	Apollo 11 Low-K (10003)	Apollo 17 A (75055)	Apollo 17 B (70215)	Apollo 17 C (74275)	U (70017)	Orange Glass (74220)	Apollo 11 High-K (10049)	Apollo 12 (12038)	Luna 16 (B1)	Apollo 14 (14053)
I. Large Cations										
Cs	0.022	0.019	—	—	—	—	—	—	0.054	0.09
Rb	0.62	0.58	0.36	1.2	0.28	1.1	6.2	0.60	1.58	2.20
K	500	660	420	500	420	750	3000	630	1200	830
Ba	110	76	57	67	43	76	330	130	220	145
Sr	160	190	121	153	168	210	161	190	440	100
Pb	—	—	—	—	—	—	—	1.3	—	—
K/Rb	800	1140	1170	420	1500	680	484	1050	760	380
Rb/Sr	0.004	0.003	0.003	0.01	0.002	0.01	0.04	0.003	0.004	0.02
K/Ba	4.6	8.7	7.4	7.5	9.8	9.9	9.1	4.8	5.5	5.7
II. Rare Earth Elements										
La	15.2	6.27	5.22	6.33	—	6.25	28.8	11.8	—	13.0
Ce	53	21.5	16.5	21.4	10.7	19.0	83	31.6	—	34.5
Nd	40	23.9	16.7	22.8	12.1	17.8	63	24.6	—	21.9
Sm	14.8	10.1	6.69	9.19	5.13	6.53	22.3	7.6	—	6.56
Eu	1.85	2.09	1.37	1.80	1.62	1.80	2.29	1.97	—	1.21
Gd	19.5	15.7	10.4	14.8	—	8.52	29.3	10.1	—	8.59
Tb	3.3	—	—	—	—	—	—	—	—	—
Dy	12.9	18.1	12.2	16.3	10.2	9.40	33.4	9.7	—	10.5
Ho	—	—	—	—	—	—	—	—	—	—
Er	13.6	10.7	7.4	9.66	6.31	5.10	30.9	5.0	—	6.51
Yb	13.2	9.79	7.0	8.47	6.25	4.43	0.2	4.8	—	6.0
Lu	1.0	—	1.03	—	0.95	0.61	—	0.69	—	0.90
Y	112	112	—	82	71	—	—	80	—	55
La/Yb	1.17	0.64	0.74	0.75	—	1.41	1.43	2.46	—	2.2
Eu/Eu*	0.33	0.51	0.50	0.47	0.77	0.74	0.27	0.68	—	0.49

III. Large High Valence Cations

U	—	0.14	0.13	0.13	0.06	0.16	0.81	0.25	0.30	0.59
Th	1.8	0.45	0.34	0.46	0.20	—	3.1	0.90	—	2.1
Zr	310	270	190	250	220	185	—	200	—	215
Hf	11.4	7.2	6.3	8.6	6.4	—	17	6.5	—	9.8
Nb	—	25	21	22	18.5	—	—	8	—	15.7
Th/U	—	3.3	2.62	3.4	3.30	—	3.8	3.6	—	3.55
Zr/Hf	27	38	30	29	34	—	—	31	—	22
Zr/Nb	—	11	9.2	11.2	11.8	—	—	25	—	13.7

IV. Ferromagnesian Elements

Cr	1390	—	3030	—	3490	4650	—	500	—	2860
V	63	—	50	79	145	—	—	120	—	135
Sc	78	83	86	75	80	—	81	50	—	55
Ni	—	2	2	< 3	< 2	—	—	1	—	14
Co	14	—	—	—	—	—	—	29	—	25
Cu	—	—	6.4	3	—	—	—	5	—	—
Fe%	15.44	14.1	15.3	14.2	14.2	17.2	14.6	13.8	15.1	13.1
Mn	2300	2230	2080	2000	1930	—	1930	1930	1540	2000
Zn	7.4	7	—	—	—	—	—	—	—	3.4
Mg%	4.01	4.23	5.06	6.22	5.99	8.66	4.22	4.1	4.23	5.09
Ga	3	—	—	—	—	—	—	—	—	—
V/Ni	—	—	17	26	—	—	—	120	—	—
Cr/V	22	—	61	—	24	—	—	4.2	—	—

†Values in ppm, except those indicated.

element, while potassium is volatile and the high lunar Ba/K ratios compared to chondritic or terrestrial values reflect this fact.

Strontium (another refractory element) is also strongly enriched by about fifteen times over the chondritic abundances in mare basalts. This element is mainly contained in plagioclase feldspar, where it readily substitutes for calcium. Divalent europium is similar in ionic radius to strontium and, like it, is concentrated in plagioclase feldspar. Strontium is less strongly enriched in the lavas than barium. This is consistent with its geochemical behavior, since it enters calcium sites rather readily. Hence, it will occur in the main mineral phases in the interior of the Moon and will be less readily partitioned into the liquid phase than barium.

Sodium is present at rather uniform abundance levels in contrast to the heavier alkalis, K, Rb, and Cs. This is a consequence of its entry into plagioclase, which the larger alkali elements enter only with difficulty. It is depleted relative to terrestrial basalts by a factor of about five. Since sodium will be partitioned into the liquid phase during partial melting, the source rocks will be depleted even more in sodium than their terrestrial counterparts.

The lunar lavas were extruded at temperatures approaching $1200°$ C into a hard vacuum of at least 10^{-9} torr. Although such conditions might be thought to favor loss of volatile elements, the lithospheric pressure of the lava exceeds that of the vapor pressure of the elements at depths greater than 10^{-3} cm and loss could occur only from a thin skin [80].

Since the lavas are quite uniform in sodium content, a very efficient mechanism would be needed. The mineralogy of the samples shows no such effects. The zoned plagioclases with minor exceptions [81] become richer in Na toward the exterior, the reverse to that predicted by volatile loss. Also, there seem to be no Na-rich concentrations in the soils or other likely repositories for the proposed very large amount of volatilized sodium. Loss of volatiles is, of course, inhibited by rapid freezing of the surface of the extruded lavas as noted above. The consensus is that volatile loss from the mare basalts during extrusion is trivial, and the Na depletion in lunar lavas is an inherent feature, due to the overall low content of volatile elements in the Moon.

The Rb/Ba ratio is a good example of the coherence between the volatile element Rb, and the involatile element Ba. Such low volatile/involatile element ratios in mare basalts indicate that the Moon is depleted in volatile elements relative to the abundances in carbonaceous chondrites. Isotopic systematics confirm that this volatile depletion, which increases for the highly volatile elements, occurred before accretion of the Moon. Rubidium/barium ratios (Fig. 6.31) provide a good example of the constancy of volatile/involatile element ratios. A striking characteristic is that the lunar highland samples and the mare basalts show similar volatile/involatile element ratios. This rules out models which called for heterogeneous accretion of the highland crust from a source enriched in refractory elements, relative to the lunar interior. If

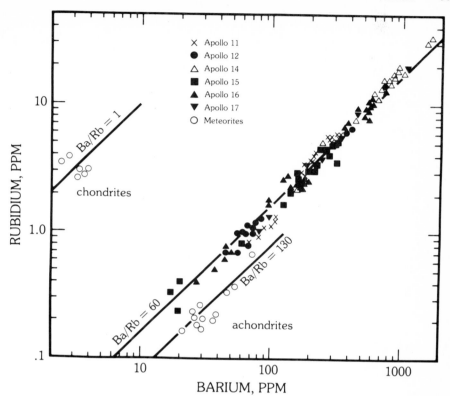

6.31 The correlation between the volatile element Rb and the refractory element Ba in mare basalts, highland samples and meteorites. This figure illustrates that the sources of the highlands and mare samples possess similar volatile/refractory element ratios. Thus these portions of the Moon (to a depth of several hundred km) have accreted from homogeneous material, or were homogenized following accretion. Note that the meteorite samples both differentiated (achondrites) and primitive (chondrite) have distinctly different ratios. [From *Basaltic Volcanism* (1981) Fig. 1.2.9.16.]

a heterogeneous accretion scenario is valid for lunar accretion, then homogenization must have occurred before the formation of the highland crust.

Rubidium and barium are an example of a pair of elements whose close association is predictable on geochemical grounds. In the mare basalts, many other close element correlations (e.g., K/U) arise from their presence together in residual phases or glassy mesostasis during crystallization of the basaltic melts. The many close element correlations observed among diverse groups of elements in the highland rocks, in contrast, appear to result from mechanical mixing of a few components.

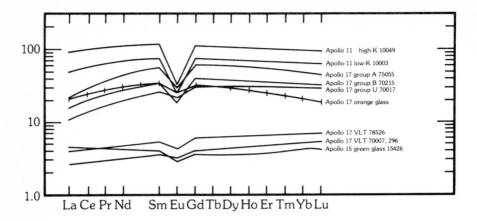

6.32 REE patterns for high-Ti basalts, with those for Apollo 15 green glass and Apollo 17 VLT basalts shown for comparison. Note the strong Eu depletion in the high-Ti suite, but also that the green glass, a very primitive basalt, also has an Eu depletion, indicating derivation from an already fractionated source. (Data from Table 6.4.)

6.3.3   Rare-Earth Elements and the Europium Anomaly

The rare-earth elements (REE) (Table 6.4) provide important insights into the nature of the source regions of the mare basalts. Typical REE patterns are shown in Figs. 6.32 and 6.33, in which nearly all classes of basalts are represented. With minor exceptions (Luna 24 and some VLT basalts), all mare basalts have negative europium anomalies. This includes the Apollo 15 green glass [73]. Although many hypotheses were advanced to account for this depletion [82], it is now recognized that it represents an intrinsic signature of the source regions, indicating prior removal of plagioclase that extracted Eu^{2+} (and Sr^{2+}) from the source regions of the mare basalts. These elements now reside principally in the highland crust.

Isotopic systematics date this event prior to 4.4 aeons, during the crystallization of the magma ocean. This primary evidence for the derivation of mare basalts from source regions which have already been differentiated is confirmed by a wide variety of other trace element evidence. The REE patterns commonly show light REE depletion. The heavy REE are commonly parallel to chondrites, but may be relatively depleted (high-Ti basalts) or enriched (Apollo 17 VLT basalts). The latter patterns provide evidence of derivation from a source dominated by olivine and orthopyroxene for this particular class of basalt. Among other constraints, the REE evidence indicates an extreme diversity of sources for the mare basalts. The question whether the

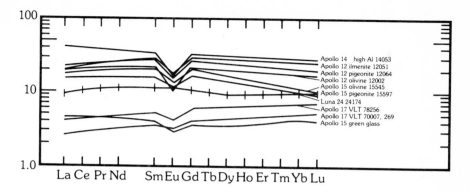

6.33 REE patterns for selected low-Ti and high-Al basalts, normalized to the abundances in chondritic meteorites. Note LREE depletion in the low-Ti basalts and the HREE enrichment in the VLT basalts. The Luna 24 high-Al basalt has no Eu anomaly, while the Apollo 15 high-Al basalt has LREE enrichment. (Data from Table 6.4).

samples with heavy REE depletion are reflecting an initial bulk lunar depletion in the heavy REE (Gd-Lu) has been frequently addressed [83]. However, it may merely reflect prior crystallization of orthopyroxene, which selectively concentrates the heavy REE, before the crystallization of the source regions of the mare basalts (see Section 6.5.2).

A significant feature of all lunar basalt REE patterns is that there is no evidence of any garnet signature. Garnet is strongly enriched in heavy REE. Even the most primitive proposed lavas (e.g., Emerald Green Glass, 15426) apparently derived from the greatest depths in the Moon (400–500 km), do not show the steep heavy-REE-depleted patterns typical of lavas formed from a garnet-bearing source region, but have a pattern slightly enriched in heavy REE, with a pronounced Eu depletion, consistent with derivation from a source region from which some olivine and plagioclase had been removed.

6.3.4 High Valency Cations

The large high valency cations (Table 6.4) are enriched in lunar basalts relative to terrestrial basalts. The enrichment of these refractory elements in the Moon, relative to the Earth, appears to be well established. These elements are all "incompatible" as well as refractory. During crystallization of the magma ocean, they will be excluded from the major mineral phases (olivine, pyroxenes, plagioclase). Increasing amounts will be trapped in interstitial liquids or in accessory minerals in the later stages of crystallization. Such concentrations provide heat sources for partial melting which produces the mare basalts, and the cumulate model concept provides a mechanism for so concentrating the heat-producing elements. There are adequate amounts of K, U and Th to provide for this secondary melting, which amounts to only about 0.1% of lunar volume. Accordingly, the source regions of the the high-Ti basalts formed late in the magma ocean crystallization contain greater concentrations of the incompatible elements than do the low-Ti basalts. Extreme concentrations occur in KREEP samples from the highland crust, derived most probably from the invasion of the highlands by the residual melts resulting from the crystallization of the magma ocean (Section 5.11). Partial melting within the source regions to generate the mare basalts will concentrate these incompatible elements in the melt, so that at least two stages of enrichment, over the whole-moon abundances, occur during the formation of these basalts. Differentiation, caused by fractional crystallization, may result in a further concentration of these elements.

Thorium and uranium are of interest both because of their use in radioactive dating and because they are important sources of heat production. The high U and Th contents of the lunar surface rocks would produce enough heat to melt the entire Moon. Since the shape of the Moon indicates that it has been rigid for probably the past 3 aeons, U and Th must have been concentrated near the surface.

Potassium/uranium ratios for many terrestrial rocks are constant at about 10,000. This is not the consequence of any profound geochemical coherence, but rather that both elements are strongly concentrated in residual systems by fractionation processes. The K/U ratios for chondrites are 50,000–80,000. The mare basalts, in contrast, average about 2500 (Fig. 8.4) [84].

The K/U ratios provide a useful method of distinguishing the geochemistry of the Moon from that of other units of the solar system. In contrast to the K/U ratios, the Th/U ratios for the Moon, meteorites, and the Earth are closely similar (3.5–4.0). Some exceptions occur on the Moon, the Apollo 17 basalts and some highland feldspathic rocks having lower Th/U ratios of about three.

The Zr/Nb ratio [85–86] is surprisingly constant at about 13–15 in mare basalts. This constancy holds as well for many highland rocks, emphasizing the close relationship for many element ratios between the highlands and mare chemistry, a fact of much significance for theories of lunar origin.

The high-Ti basalts from Apollo 11 and 17 show some exceptions to the constancy of the inter-element ratios (K/Zr, Zr/Nb, etc.) observed rather widely in lunar samples. The variations are in the direction of increased amounts of elements (e.g., Zr^{4+}) associated with Ti^{4+}. Thus Zr/K ratios increase with increasing Ti content. The evidence is consistent with the derivation of the high-Ti basalts from a source already enriched in Ti-Zr-Nb phases. This observation is consistent with at least a two-stage origin for the high-Ti basalts [85].

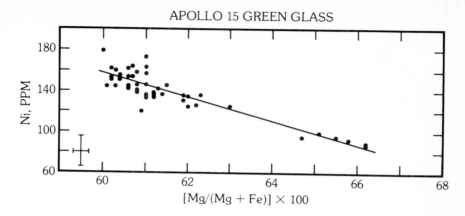

6.34 The nickel content of the green glass (15425-27) shows an inverse relation with Mg/(Mg + Fe) contrary to that predicted from a simple fractional crystallization model. The most probable explanation for this trend is that it is due to mixing of magmas generated by partial melting in a heterogeneous mantle [70]. (Courtesy J. W. Delano.)

6.3.5 The Ferromagnesian Elements

The ferromagnesian elements in lunar basalts show both extensive depletions and enrichments compared with chondritic meteorites or with terrestrial basalts (Table 6.4). Thus, Ni, Cu, Ga, and V are strongly depleted, and Cr, Sc, and Ti, enriched. The depletion in Cu and Ni may be ascribed in part to entry into metallic phases under strong reducing conditions during the partial melting and crystallization of the lunar lavas. In addition, Ni and the siderophile elements were probably depleted before accretion.

Nickel is strikingly depleted in the aluminous mare basalts and in the high-Ti basalts, where values of only 1–2 ppm are common. Iron/nickel ratios become extreme, reaching values on the order of 15,000–20,000, compared with 100–500 in terrestrial basalts and 20 in chondritic meteorites. The quartz-normative basalts, derived from source regions which crystallized before those of the high-Ti basalts, contain 10–20 ppm nickel (still very low by terrestrial standards). The olivine-normative basalts from still more primitive source regions have 40–50 ppm Ni, while the Emerald Green Glass (15426) has the highest amount (Fig. 6.34). These values must all be distinguished from the high (100–300 ppm) Ni concentration in the soils, due to addition of a meteoritic component. The sole example of a lunar dunite (74215) has an Fe/Ni ratio of 500, a factor of about 20 higher than that for typical terrestrial ultrabasic rocks. The available lunar samples accordingly show both lower Ni contents and higher Fe/Ni ratios than do their terrestrial counterparts. Does this reflect an overall depletion in Ni in the Moon? If the green glass or primitive basalts such as 12002 or 12009 are derived from a primitive unfractionated interior, then these source regions are highly depleted in nickel relative to the accessible portions of the terrestrial mantle. However, if the mare basalts are derived from differentiated source regions, as proposed in the cumulate model, then the Ni content in the mare basalt sources will already be depleted in nickel. Nickel enters early crystallizing olivine and orthopyroxene lattice sites and accordingly will be stored in the early cumulate phases in a crystallizing magma ocean, particularly if much adcumulus growth occurs at that stage. The Taylor-Jakeš model [86], for example, predicts the existence of large volumes of olivine and orthopyroxene at depth below the source regions of the mare basalts. Whether the Ni content of the bulk Moon is inherently lower than that of the terrestrial mantle thus turns on the amount of Ni trapped in the deep and inaccessible olivine and orthopyroxene cumulates.

The hazards of extrapolating from terrestrial conditions are well illustrated by the geochemical behaviour of chromium on the Moon. In contrast to Ni, Cr is strongly enriched in mare basalts, relative to terrestrial basalts, by about an order of magnitude. The low abundance of nickel can be readily explained as due to a combination of an inherently low lunar abundance, as a siderophile element, and to removal of Ni^{2+} in early crystallizing olivine in the magma ocean. On the Earth, trivalent chromium would also preferentially enter early crystallizing phases, and be retained in residual phases during later partial melting. The presence of chromium concentrations of 2000–4000 ppm in lunar basalts compared to abundances of a few hundred ppm in terrestrial lavas was thus remarkable. There is no cosmochemical process available to enrich the bulk moon in chromium, relative to other ferromagnesian elements. The resolution of this paradox is due to the highly reducing conditions in the Moon. A large proportion of Cr in the lunar interior is present as Cr^{2+} [62, 87, 88]. Somewhat contrary to expectation, but probably on account of its larger size, this ion strongly concentrates in the melt relative to olivine, so that during crystallization of the magma ocean, chromium will be selectively enriched in later pyroxenes [62, 87]. High concentrations can thus be built up in the source regions of the mare basalts. During crystallization of the mare basalts, in contrast, chromium behaves mainly as the trivalent ion, entering phases such as spinel and chromite. The redox conditions on the Moon thus span the transition from divalent to trivalent chromium and provide a solution to the apparently enigmatic behaviour of this element. Vanadium shows a similar distribution to chromium, while Sc parallels the titanium content. The ferromagnesian element distributions are entirely consistent with derivation of the mare basalts from source regions which have undergone prior crystallization of olivine, pyroxene and plagioclase.

6.3.6   Sulfur and the Chalcophile Elements

The cosmochemical behaviour of sulfur and the chalcophile elements (Table 6.5), which in the terrestrial geochemical environment enter sulfide phases, is strongly influenced by their volatile character. The chalcophile elements are strikingly depleted in ordinary chondrites relative to the carbonaceous chondrites. In the lunar basalts they are generally low in abundance and are grossly depleted relative to the cosmic abundances and to the terrestrial abundances. The basic questions are whether these elements were ever accreted, were lost subsequent to accretion, were retained in the interior during partial melting, or entered an Fe-FeS core [89].

Surface loss of volatiles during extrusion or meteorite impact has not found much favor. During surface extrusion of the lavas, freezing of the surface will rapidly occur, and loss of volatiles will not occur. Lead is not more strongly depleted than other much more volatile elements, such as mercury, which would be more readily lost under such conditions. The case for some movement of lead has been discussed earlier (see Sections 4.5.2 and 5.5). The two most likely mechanisms to account for the depletion of the chalcophile elements are loss before accretion or retention in an Fe-FeS-type core. Since the chalcophile elements are volatile, it is more probable that they were never accreted in the Moon. Definitive answers to these questions are difficult, since no refractory chalcophile elements, which could provide a key, are known.

The total lead abundances are low in comparison with terrestrial average crustal lead concentrations of 15 ppm and with the carbonaceous chondrite values of 3–4 ppm. The low abundance of lead parallels that of the other volatile elements and indicates loss of such elements at or before accretion of the Moon. The comparatively high uranium and thorium contents mean that a major portion of the lunar lead is radiogenic in origin, so that the amount of common lead is very low. Contamination by terrestrial lead can accordingly be a serious problem in analytical work on returned lunar samples. The ratio $^{238}U/^{204}Pb$ (μ), is distinct for terrestrial and lunar rocks. It is commonly in the range 5–10 for terrestrial rocks and rarely exceeds 30. The mean value is about 9. In contrast, the values in the lunar rocks are, with minor exceptions, very much higher. For the mare basalts, the usual range is from 100 to 1000.

The extreme depletion of the chalcophile elements in the mare basalts allows their use as indexes of the contribution by carbonaceous chondrites. The abundances in the lunar rocks are about a factor of 10^3–10^4 less than observed in the Type I carbonaceous chondrites (Sections 4.5.3, 5.6 and 8.2). These elements are abundant in the regolith, where they have been supplied during meteorite impact. All sulfur appears to be present in the mare basalts as troilite (FeS). The abundance of the element varies with basalt type, being lower in the low-Ti Apollo 12 and 15 basalts. These typically contain 500–700

Table 6.5 Siderophile, volatile and chalcophile elements in mare basalts.†

| | | Low-Ti | | | | | High-Ti | High-Al | |
| | | Apollo 12 | | Apollo 15 | | | Apollo 17 | Apollo 12 | Apollo 14 |
	(ave.)	Olivine (12002)	Olivine (12009)	Olivine (15016)	Olivine (15545)	Ilmenite (12051)	A (75055)	(12038)	(14053)
Os	≤0.026	—	—	—	—	0.09	—	—	—
Ir	0.01	—	0.018	0.018	0.015	—	0.035	0.04	0.017
Re	0.0025	—	0.0017	0.0033	0.0009	—	0.0031	—	0.007
Ni (ppm)	41	64	52	—	51	6	2	2	14
Au	0.027	0.024	0.02	0.025	0.005	0.0075	0.007	—	0.11
Sb	0.76	—	2.2	3.8	1.3	—	0.99	—	0.64
Ge	4.6	—	3.3	4.4	3.8	—	2.5	—	—
Se	120	140	110	114	117	200	120	120	140
Te	2.9	10	4.1	2.4	2.8	—	—	—	—
Ag	0.94	0.81	1.2	0.84	—	0.81	0.76	—	0.60
Zn (ppm)	1.0	0.70	0.90	1.1	0.99	0.53	1.5	—	2.1
Sn	45	—	87	—	—	—	—	—	—
In	0.60	1.9	—	0.34	1.5	1.2	0.57	—	—
Cd	2.0	1.4	3.6	2.1	1.8	—	1.9	—	—
Bi	0.23	—	0.27	0.36	0.21	0.53	—	0.76	0.29
Tl	0.34	0.25	0.30	0.32	0.31	0.37	0.37	—	1.4
Br	14	10	—	—	—	—	—	10	—

† Elements arranged in order of increasing volatility. Data in ppb, except Ni and Zn.
Sources: Anders, E., et al. (1971) *PLC 2*: 1022.
Morgan, J.W., et al. (1972) *PLC 3*: 1379.
Wolf, R., and Anders, E. (1980) *GCA* 44: 2113.
Wolf, R., et al. (1980) *PLC 11*: 2110.

ppm, and have δ^{34}S values of –0.1 to –0.3 per mil. In Apollo 11 and 17 high-Ti basalts, the sulfur values are about three times higher, averaging 2200 ppm with δ^{34}S values of about +1.2 per mil [90]. There is an inverse correlation with μ values (^{238}U/^{204}Pb), consistent with the presence of lead in a sulfide phase in the lunar interior. Much of the lead in mare basalts is derived from such sources, where it has been isolated from U and Th since the initial lunar differentiation.

The inverse correlations of metallic iron with sulfur suggest that some of the metallic Fe in lunar basalts comes from reduction of FeS. The δ^{34}S values overlap those in meteoritic troilite and in terrestrial basalts. The values for the Apollo 12 basalts resemble those in terrestrial olivine alkali basalts [91]. Such compositional differences are consistent with different source regions for the various types of mare basalts.

The high sulfur abundances in lunar basalts contrast strongly with those in terrestrial basaltic rocks, which are typically in the range 100–200 ppm. What is the cause of this apparent paradox? Sulfur is volatile and is depleted in the moon mantle relative to that of the Earth: either it was not accreted, or it is buried as FeS in a small lunar core. Crystallization of a magma ocean, however, will concentrate the chalcophile elements into sulfide layers, as is observed in terrestrial layered intrusions. These will be concentrated in the later stages of crystallization, so that high-Ti basalt source regions will be richer in sulfides than the low-Ti basalt source regions. The sulfide phases will be preferentially concentrated into the melt during partial melting, thus enriching the lavas in sulfur.

6.3.7   The Siderophile Elements

The trace siderophile element abundances (Table 6.5) are very low in lunar basalts. The basalt data is critical for our understanding of the content of siderophile elements in the Moon, since the lavas are derived from deep source regions which have escaped the meteoritic bombardment of the lunar surface. This has added a variable siderophile element signature to the highland rocks and makes suspect the use of such samples to establish indigenous lunar abundances for these elements (see Section 5.6). The basalt rock samples do not suffer from this problem, but nevertheless are derived from source regions that have undergone a previous fractionation history. Since the partial melting process that produces the basalts and the formation of a metal phase during cooling will also fractionate the siderophile elements, many factors have to be considered in evaluating the abundance data.

Gold and iridium contents range from 0.1 to 0.001 ppb. Rhenium is even more depleted, being a factor of about 10 lower. Nickel abundances range from less than 2 to 64 ppm. Silver abundances are typically about 1 to 0.1 ppm. Germanium values are typically about 1–10 ppb [92, 93].

It is predicted that Ni would enter early-crystallizing olivines during magma ocean solidification, and so it would be depleted in the source regions of most mare basalts. However, no such mechanism is available to explain siderophile trace element abundances. There is no indication in mare basalt chemistry that the source regions contained metallic phases. The Fe metal, which is observed in lunar basalts, crystallized generally late in the paragenetic sequence, so its presence is a near-surface phenomenon, and indicates that metal was not present in the source regions.

The relation of the siderophile element content of the Moon to that of the Earth is of critical importance for theories of lunar origin. The extreme depletion in the mare basalt source regions is probably due to two causes: (a) loss of siderophile elements before accretion, as is shown by the overall low iron content of the Moon, and (b) possible depletion of the interior by formation of an iron or FeS core.

6.3.8   Oxygen and Carbon

The $\delta^{18}O$ variation in mare basalts is very small, ranging from $+5.4$ to 6.8. The $\delta^{18}O$ values for the minerals increase in the sequence ilmenite, olivine, clinopyroxene, plagioclase and silica minerals. The range is from 3.8 to 7.2. The $\delta^{18}O$ values for the minerals from different rocks are very similar, and there is no evidence of oxygen isotope exchange or equilibration during cooling following crystallization at temperatures estimated about 1120°C. This effect is attributed to the effective absence of water in the lunar rocks. Although the rocks crystallized rapidly, and the minerals are often inhomogeneous, there is no sign of isotopic zoning in the minerals [94, 95].

This lack of diversity in oxygen isotope chemistry in the Moon contrasts sharply with the variations seen in the Earth or the metamorphosed meteorites. The oxygen isotopic compositions in the lunar basalts are determined by igneous processes and are locked in, due to the absence of alteration processes [96]. Since the variations are very small, it is a reasonable conclusion that the $^{18}O/^{16}O$ ratio is rather constant in the Moon. It is the "best-known chemical parameter for the moon" [96], and it is accordingly a constraint that theories of lunar origin must meet.

The carbon content in lunar basalts is low. The Apollo 11 samples average about 70 ppm, the Apollo 12 basalts from 16 to 45 ppm, the Apollo 15 lavas from 10 to 30 ppm, and the Apollo 17 basalts from 40 to 80 ppm [97] (see Section 6.2.2). These values are all much lower than those found in the soils, on which most interest has been focused. The carbon is apparently present in the lunar rocks as carbides, although mineralogical evidence of their presence is lacking. Their existence is inferred from the release of methane and other gases during acid hydrolysis. The $\delta^{13}C$ values are in the range -20 to -30, closely resembling terrestrial basalts. The soils, in contrast, have values of $+10$ to $+20$ [94, 98].

6.4   Ages and Isotopic Systematics of the Mare Basalts

6.4.1   The Commencement of Mare Volcanism

The oldest recognizable mare basalts on the Moon post-date the Imbrium and Orientale collisions since no debris from these impacts occurs on the basaltic surfaces. This stratigraphic evidence does not preclude the existence of earlier basalts [99], particularly since sample 10003 from Apollo 11 gives an age of 3.85 aeons, a little older than the age accepted for the Imbrium collision (3.82 aeons). The first evidence for Pre-Imbrian basaltic volcanism was the identification of basalts in the Apollo 14 highland breccias. Samples 14053 (251 gm) and 14072 (45 gm) were recognized as basalts, as were several fragments from breccias [100]. The Apollo 17 highland breccia 73255 contains 3 basalt clasts with ages of about 3.9 aeons [101]. Other examples are known from Station 2, Boulder 1, and the Station 6 boulders at Apollo 17. The ages of all these are in the 3.9–4.0 aeons range. All these basalts are high-Al_2O_3 types. They differ from younger basalts of similar chemistry by having been derived by partial melting from more evolved sources having higher $^{87}Sr/^{86}Sr$ ratios, compatible with their higher Rb contents. Younger examples include 12038 and the Luna 16 and 24 samples. The question of whether plagioclase is present in the sources is not yet resolved, but appears probable for 12038, Luna 24 and 16 basalts [102]. Clearly high-Al basalts in different areas are derived from differing sources so that the amount of plagioclase in the source regions may vary [103]. The ancient high-Al basalts have 500 ppb Ge as do the Luna 16 basalts, but normal mare basalts have less than 40 ppb Ge. This high abundance probably is due to their derivation from a very fractionated source.

Contamination with a KREEP type component appears inadequate to account for these basalts since the negative Eu anomalies are either small or non-existent. It should be noted that the high-Al basalts, such as 14053, have high U, Rb, Cs, Tl, Br, Bi and especially In and Cd. These indicate a late stage cumulate source [104]. Siderophile abundances are low, as in most mare basalts. Accordingly, they are considered to represent true mare basalts derived from a fractionated source region by partial melting.

Impact ejecta blankets from more recent craters (e.g., Copernicus) have covered mare basalt surfaces. Smaller impact craters penetrating through these ejecta blankets and excavating mare basalts from beneath, appear as "dark-halo" craters. Dark haloed craters exist on the ejecta blanket of Mare Orientale, suggesting that some mare volcanism predates that collision. Dark-halo craters occur on some light-plains units, south of Mare Humorum and in the eastern hemisphere of the Moon. The orbital geochemical data indicate a high Mg content. Accordingly, some of the smooth highland plains may represent mare basalts covered by ejecta from the last basin-forming impacts (P. Schultz, pers. comm., 1981).

The extent of such early mare basalt volcanism is discussed in Section 5.2.6. It is considered to have been of minor extent, perhaps adding 20% to the observed post-basin basalts since clasts are rare in highland breccias, and there is no distinctive chemical signature in the highland breccias which could confidently be ascribed to the presence of a substantial component of mare basalt type. The view is taken here that mare volcanism developed slowly (on account of the time required for heat sources in the interior to generate partial melting) and is not a feature of initial planetary differentiation. It should be noted that the closure of the isotopic sources of the mare basalts at about 4.4 aeons implies solidification of the mineral phases involved.

6.4.2   Radiometric Ages for the Mare Basalts

A summary of ages is given in Table 6.6, employing the new age constants [105]. The Apollo 11 low-K basalts are old, but not all were formed simultaneously. As noted above, 10003 has an age of 3.85–3.86 aeons, while other ages for the low-K suite range from 3.55 to 3.83 aeons. Very old Rb-Sr and Sm-Nd ages of 3.92 and 3.80 aeons for 10062 are not matched by the ^{40}Ar–^{39}Ar ages of 3.78 aeons. The evidence is nevertheless clear that the Apollo 11 low-K suite is close to and possibly overlaps in part with the Imbrium collision. The low-K basalts have initial Sr ratios of 0.69906, but the high-K rocks have initial ^{87}Sr/^{86}Sr ratios of 0.69930. The high-K rocks have ^{40}Ar–^{39}Ar ages of 3.57 aeons, distinctly younger than those of the low-K suite. The filling of the later stages of Mare Tranquillitatis thus occupied at least 300 million years. Exposure ages correlate with the petrology and the age sequence [106].

The other group of high-Ti basalts at Apollo 17 have ^{40}Ar–^{39}Ar ages ranging from 3.66–3.79 aeons consistent with Rb-Sr dating. The mean ^{87}Sr/^{86}Sr is 0.69912 and, accordingly, all Apollo 17 basalts appear to have been derived from the same source at about 3.72 aeons (Fig. 6.35). An Sm-Nd isochron for 75075 yields a similar age of 3.70 aeons.

The low-Ti basalts from Apollo 15 have ^{40}Ar–^{39}Ar ages of 3.21–3.37 aeons, consistent with the Rb-Sr ages. The initial ^{87}Sr/^{86}Sr is 0.69899, a little above BABI. The Apollo 12 low-Ti basalts show a slightly younger set of ages (3.08–3.29 aeons) with a range in initial ^{87}Sr/^{86}Sr ratios from 0.6992 to 0.6996.

The lunar basalts thus range from about 3.9 to 3.1 aeons—800 million years—a respectable period of time even by terrestrial standards, but indicating that the production of lava or its eruption at the surface ceased on the Moon during Archean or Early Proterozoic time on the Earth. Flooding of the mare basins and irregular mare areas occupied some hundred of millions of years. There is no real evidence of large lava lakes, and no apparent connection between lunar magmatic activity and the excavation of the basins.

It is clear from the isotopic systematics that the sources for the mare basalts are heterogeneous, and that this heterogeneity was established at

Table 6.6 Radiometric ages of lunar basalts, listed in approximate order of age.

Mission	Location	Basalt Type	Sample Number	Age (aeons)	±	Method
Apollo 14	Cone	High-Al	14072	3.96	0.05	Ar–Ar
(Fra Mauro			14072	3.91	0.09	Rb–Sr
	Cone	High-Al	14321	3.87	—	Ar–Ar
			14321	3.87	0.04	Rb–Sr
	Cone	High-Al	14053	3.85	0.05	Ar–Ar
Apollo 11	LM	Low-K	10003	3.86	0.07	Ar–Ar
(Mare			10003	3.76	0.08	Rb–Sr
Tranquillitatis)	LM	Low-K	10020	3.72	—	Ar–Ar
	LM	Low-K	10047	3.69	0.03	Ar–Ar
	LM	Low-K	10062	3.79	0.04	Ar–Ar
			10062	3.92	0.11	Rb–Sr
			10062	3.88	0.06	Sm–Nd
Apollo 17	Camelot	High-Ti (A)	75055	3.73	0.04	Ar–Ar
(Taurus-			75055	3.69	0.06	Rb–Sr
Littrow)			75055	3.70	0.07	Sm–Nd
	Camelot	High-Ti (A)	75075	3.69	0.02	Ar–Ar
			75075	3.70	0.07	Sm–Nd
	LM	(B)	70215	3.79	0.04	Ar–Ar
	LM	(U)	70017	3.59	0.08	Rb–Sr
	LM	(U)	70035	3.69	0.07	Ar–Ar
			70035	3.67	0.09	Rb–Sr
			70035	3.77	0.06	Sm–Nd
Apollo 11	LM	High-K	10072	3.57	0.05	Ar–Ar
(Mare			10072	3.56	0.05	Rb–Sr
Tranquillitatis)			10072	3.57	0.03	Sm–Nd
Luna 16	—	High-Al	L26-B1	3.41	0.04	Ar–Ar
(Mare Fecunditatis)			L26-B1	3.35	0.18	Rb–Sr
Apollo 15	Rhysling	Olivine	15016	3.34	0.08	Ar–Ar
(Hadley-Apennines)			15016	3.33	0.04	Rb–Sr
	Hadley Rille	Olivine	15555	3.27	0.03	Ar–Ar
			15555	3.25	0.04	Rb–Sr
	Dune	pigeonite	15499	3.30	0.08	Ar–Ar
	Hadley Rille	pigeonite	15597	3.37	0.07	Rb–Sr
Luna 24	—	VLT	24170	3.26	0.04	Ar–Ar
(Mare Crisium)			24170	3.30	0.05	Sm–Nd
Apollo 12	N of Head	Olivine	12002	3.21	0.05	Ar–Ar
(Oceanus			12002	3.29	0.10	Rb–Sr
Procellarum)	Bench	Olivine	12040	3.23	0.04	Rb–Sr
	NE of Head	pigeonite	12021	3.26	0.06	Rb–Sr
	Head	pigeonite	12052	3.15	0.09	Rb–Sr
	Bench	pigeonite	12056	3.20	0.14	Sm–Nd
	Surveyor	pigeonite	12064	3.15	0.01	Ar–Ar
			12064	3.11	0.04	Rb–Sr
	N of Head	Ilmenite	12022	3.08	0.06	Ar–Ar
	Surveyor	Ilmenite	12051	3.23	0.05	Ar–Ar
				3.09	0.09	Rb–Sr
				3.13	0.06	Ar–Ar
				3.19	0.10	Rb–Sr
	Bench	High-Al	12038	3.08	0.05	Ar–Ar

Sources: All data from *Basaltic Volcanism* (1981) p. 950, Table 7.3.1, except Apollo 14 values from Ryder, G. and Spudis, P. (1979) *Lunar Highlands Crust*, p. 360, Table 3.

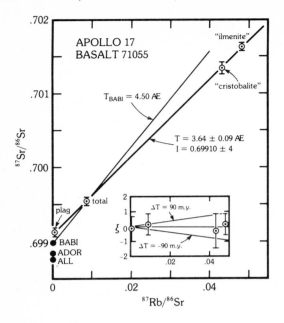

6.35 Rb-Sr evolution diagram for Apollo 17 high-Ti basalt 71055 which was erupted on to the lunar surface 3.64 aeons ago. The initial ratios for basaltic achondrites (BABI), Angros dos Reis (ADOR), and Allende (ALL) show that the Allende ratio is significantly lower than lunar, a stumbling block for models making the Moon from Allende-type material. "One picture is worth 1,000 words" [110]. (Courtesy G. J. Wasserburg.)

about 4.4 aeons, with only mild degrees of elemental fractionation accompanying partial melting and eruption. Although the source regions from which the lavas were derived are very ancient, being established at about 4.4 aeons, they are not primitive, but had already undergone a prior fractionation. The significance of these events is discussed in the next section.

What age are the youngest mare basalts on the Moon? It seems to be indicated from crater density data that basalts as young as perhaps 2.0 aeons exist in the western maria [107]; however, doubt as to the accuracy of this estimate will persist until radiometric ages are obtained. Other data from albedo studies and spectral reflectivity show that many distinctive mare units, distinguished by differing parameters, were not sampled by the Apollo missions [107] (Fig. 6.17).

6.4.3   Isotopic Indexes of Mantle Heterogeneity and Basalt Source Ages

The $^{87}Sr/^{86}Sr$ data for the Apollo 11 low- and high-K rocks indicate that they are derived from distinct source regions, and are not related by fractional crystallization [108]. The Apollo 17 high-Ti basalts all appear to be derived from sources with similar Rb/Sr ages. The initial ratios for the Apollo 12 low-Ti basalts could be interpreted to indicate the existence of a least four distinct mantle sources but this conclusion does not appear probable on petrological or chemical evidence. The Apollo 15 data indicate a similar situation, except that there is definite evidence of enrichment in Rb relative to Sr at the time of melting. This yields model ages of about 4.1 aeons as a consequence. The enrichment factor is 1.4, in accordance with reasonable estimates from distribution coefficients.

The U-Pb data are in agreement with the Sr isotopic systems, and with a simple two-stage history in which there was only minor U-Pb fractionation during magma formation. The measured μ values (^{238}U/^{204}Pb) range from 300 to 1000, consistent with values in the source regions of 60–500. This range of μ in the source regions indicates that the mantle was differentiated and was not similar, for example, to primary carbonaceous chondrites [108].

The Sm-Nd initial ratios likewise indicate that the mantle source of the basalts was evolved relative to primitive chondritic values [109] (Fig. 6.36). The Rb-Sr data (except for the Apollo 11 high-K basalts) yield model ages (T_{BABI}) of about 4.5 aeons. As noted earlier, this result indicates that only minor fractionation of Rb from Sr accompanied the partial melting which formed the mare basalts. The combination of low Rb, initial ^{87}Sr/^{86}Sr ratios only a little above BABI, and a heterogeneous mantle means that the Rb-Sr system provides only an approximate value for the time of differentiation of the mantle (Fig. 5.37).

The U-Pb system provides a more sensitive index. Internal U-Pb isochrons are known for three basalts with ages ranging from 3.3–3.8 aeons [110]. The isochrons intersect "concordia" at 4.47 aeons for these three distinct basalts, which suggest that this is the time of U-Pb enrichment in the source, and marks the time of mantle differentiation.

These data agree with measurements on highland samples [111], which define an array with an age of 4.42 aeons. The 4.42 aeon age is interpreted here as dating the closure of the istopic systems following planetary-wide differentiation, and so it dates the formation of the lunar highland crust and the mantle sources of the mare basalts.

6.5   Origin of the Mare Basalts

The source of the mare basalts which flood many multi-ring basins and the irregular maria have historically been the subject of much debate. An early theory proposed that the lavas were produced by impact melting of preexisting surface rocks or of underlying mantle during the large impact events that formed the ringed basins; vestiges of this theory occasionally reappear [112]. The fact that the density of the basalts is close to that of the whole Moon has always lent a facile attraction to this hypothesis. However, the mare basalts cannot be equivalent to bulk Moon compositions for a wide variety of reasons, a principal one being that such compositions would transform to the

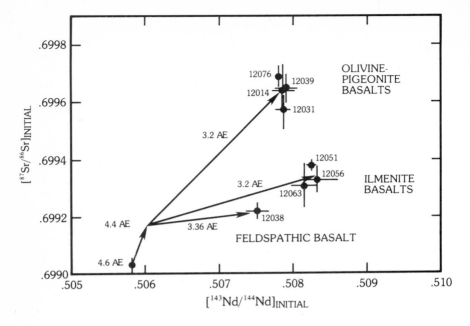

6.36   Model evolution paths for Sr and Nd isotopic composition in (i) the bulk moon (lunar magma ocean) between 4.6 and 4.4 aeons ago, and (ii) the mantle source regions of Apollo 12 mare basalts. The isotopic data clearly require the lunar mantle to be heterogeneous. The figure implies the existence of lunar reservoirs whose products will have Sr and Nd isotopic compositions complementary to the mare basalts, i.e., for which initial Sr and Nd isotopic ratios lie above and to the left of an extension of the bulk moon evolution line. Lunar KREEP basalts satisfy this latter criterion [102]. (Courtesy L. E. Nyquist.)

dense rock type eclogite and produce a greater overall density for the Moon than observed [113].

The major objection, however, is that the composition of the mare basalts is now known to be totally distinct from that of the highland crust. Such an anorthositic composition cannot produce the mare basalts by partial or complete melting, nor by fractionation processes following melting. It has been suggested that the rocks could have been near their melting point, in order to facilitate the production of large amounts of lava [114]. The evidence of a cool surface sufficient to support the Apennine Mountains and the mascons provides many difficulties for such ideas. Melting of the mafic substratum by large impacts also encounters various difficulties [115]. The principal objection is that the stratigraphic evidence, now confirmed by the age dating of the lavas, indicates that the eruption of the observable mare

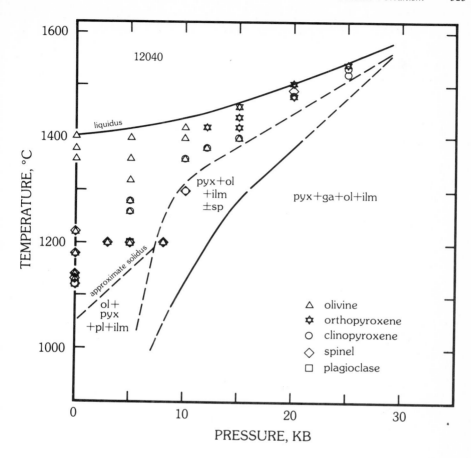

6.37 Experimental crystallization of Apollo 12 olivine basalt 12040 under subsolidus conditions at pressures up to 25 kbar [118].

basalts occurred in general much later than the creation of the multi-ring basins (see Section 2.1). Thus, the hypothesis of impact-related generation of mare basalts can be considered to be disproved [116]. Some localized melting occurs within impact craters (Section 3.9), and melt rocks are common in the highlands (Section 5.3). No such impact melts younger than highland material were sampled by the Apollo missions, but such sheets of impact melt might underlie the mare basalt fill in the large ringed basins. This impact-derived material should not be confused with mare basalts. A consensus has been reached from geochemical and experimental petrological data that mare basaltic lavas originate by partial melting deep within the interior of the Moon (Figs. 6.37 and 6.38). The precise depth of origin remains a matter of some

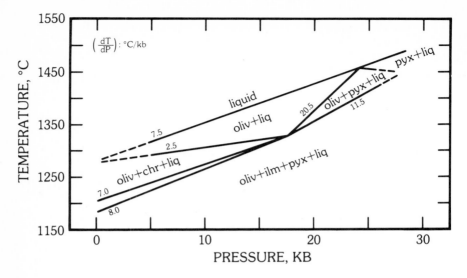

6.38   The liquidus phase relationships for the Apollo 15 red glass. The numbers near each boundary are the slope in units of °C/kbar. Olivine is the liquidus phase from 0 to 24 kbar. A multiple saturation point occurs at 24 kbar where olivine and orthopyroxene coexist on the liquidus implying a depth of melting for these high-Ti (13.8% TiO_2) glasses of 450 km [129]. (Courtesy J. W. Delano.)

dispute, as indeed does the precise mineralogical composition of the mantle and the depths of the seismic discontinuities. However, these debates are over second order questions.

6.5.1   Petrogenetic Schemes for Basalt Origin

Various degrees of partial melting of material deep within the Moon can explain the chemistry, mineralogy and isotopic characteristics of the lavas. Partial melting produces liquids with densities of about 3 g/cm³ at depths of up to 400 km. These accordingly rise to the surface, where they crystallize to form basalts with densities in the range 3.3–3.4 g/cm³. On the lunar far side, the additional thickness of the highland crust mostly prevents the lavas reaching the surface.

Initial results from high-pressure experimental studies [117–122] showed that material of high-Ti basalt composition transforms to the dense rock type, eclogite, at pressures above 12 kbars at 1110°C. At higher temperatures, the pressure needed to effect the transformation rises slowly, increasing at the rate of about 20 bars per degree centigrade.

The transition from basalt to eclogite involves changing from the mineral assemblage ilmenite-clinopyroxene-plagioclase typical of the Apollo 11 and

17 lavas to one composed of garnet-rutile-clinopyroxene. Thus, the transition is not sharp, but is smeared out over a pressure and temperature interval as the different mineral phases react. However, in comparison with terrestrial rocks, the interval is surprisingly narrow (2 kbars at 1100° C). Thus, if the Moon were composed of high-Ti lunar basalt, it would possess a central core of dense eclogite. These experimental studies, therefore, comprise good evidence along with the evidence from the REE patterns for an origin of lunar basalts by *partial* melting from material less dense than eclogite.

The degree of partial melting at lunar depths depends on models of the lunar interior. Thus, if a primary unfractionated composition is inferred at depth, differing amounts of melting are required to account for the variations in the basalt types discussed earlier. The wide variety of basaltic compositions already sampled on the lunar surface provides one of the many difficulties which this model faces. An alternative view is that the lunar interior is zoned, and that the different species of mare basalts come from differing depths, which differ significantly in composition. A number of constraints limit the depth of partial melting. The presence of mascons indicates that the basalts must have been emplaced on a crust strong enough, and hence cool enough, to support them. The mascons are due both to the excess mass resulting from the emplacement of the mare basalts and to probable uplift of a high density plug beneath the center of the multi-ring basins. Isostatic adjustment has not occurred since their formation. The mascon constraint sets an upper limit of about 100 km on partial melting. Melting at depths greater than 400 km is unlikely, since the experimental data indicate that at that depth a melt of the composition of the mare basalts would be in equilibrium with clinopyroxene and garnet (Figs. 6.37 and 6.38). This mineral assemblage would have a density greater than 3.6 g/cm^3, a density inconsistent with our knowledge of the lunar interior. It should be noted here that the total volume of mare basalt is about 0.1% of lunar volume. Partial melting episodes spread over several hundred million years are thus likely to involve very small volumes and so not affect the strength of the lunar interior.

Integration of major and trace element data, isotopic systematics, experimental phase equilbria data both on primitive mare basalt liquids and whole Moon compositions, and geophysical constraints are necessary to construct meaningful models of mare basalt petrogenesis.

Three basic types of models for mare basalt source regions have been presented: primitive source, assimilation, and cumulate source models. They are discussed below.

Primitive Source Models

The primitive source model proposes that mare source regions were composed of primitive fractionated material of lunar bulk composition at the time of magma genesis [117, 118]. The range of basalt chemistries is attributed to different degrees of partial melting of this homogeneous source. Some problems confronting this model are:

(a) It cannot explain the geochemical characteristics and the wide variety of high- and low-Ti basalts; e.g., their REE patterns and TiO_2 contents.

(b) The nearly ubiquitous presence of a negative Eu anomaly in mare basalts indicates prior separation of plagioclase from the source regions.

(c) It cannot account for the early differentiation event which is recorded in mare basalt Sm-Nd [109], U-Pb [110], and Rb-Sr isotopic systematics [123].

(d) The high-Ti basalts contain in general no nickel (< 2 ppm), but the low-Ti basalts contain up to 60 ppm, with the Apollo 15 green glass containing 180 ppm. Thus, the high-Ti basalts carry no memory of or contact with a primitive lunar nickel content.

Assimilation Models

A second class of models involves hybridization or assimilation. One suggestion is that magmas from the primitive interior selectively assimilated parts of the 4.6–4.4 billion year late-stage residuum, thereby giving rise to mare basalt hybrid liquids [124, 125]. Such a process should produce basalt liquids with extremely high Fe/Mg ratios; however, these ratios are not observed. The inverse correlation observed between trace element abundances in KREEP and those in mare basalts [126] rules out any involvement of late-stage residual KREEP compositions, derived from crystallization of the magma ocean, in the genesis of mare basalts.

A "dynamic assimilation" model [125] proposed that the outer few hundred kilometers of the Moon were involved in the 4.6–4.4 billion year lunar differentiation event. The Fe, Ti and incompatible element-rich late-stage liquids thus formed were segregated in "pods" (of dimensions 5–20 km) beneath the highland crust and above early cumulates and refractory residua. This gravitationally unstable situation resulted in the sinking of ilmenite-pyroxenite pods through the underlying cumulate and residual sequence. Meanwhile, temperatures in the primitive undifferentiated interior were beginning to rise due to radiogenic heating, and the sinking pods start to melt and react to give hybrid patches in the lunar mantle which *later* experienced partial melting to give high-Ti basalts. The Fe and Ti-rich liquids thus produced reacted and reequilibrated with a large surrounding area giving rise to high-Ti mare basalts but still having the isotopic imprint of the ilmenite pyroxenites. Continued radioactive heating resulted in partial melting of a zone near the base of the differentiated sequence. Smaller and less-efficiently differentiated ilmenite-pyroxenite pods sank much more slowly through the differentiated sequence. The infall of these relatively cool pods into the partially molten zone allegedly produced low-Ti hybrid magmas. These low-

Ti hybrid liquids would also carry the imprint of the 4.6–4.4 billion years differentiation and the ilmenite pyroxenite cumulates. In this model [125], high-Al_2O_3 mare basalts were produced at the crust-mantle interface when parental magmas from the deep interior assimilated limited amounts of plagioclase-rich crustal material. The high-Al_2O_3 basalts, however, have very low nickel contents (< 2 ppm), and other geochemical characteristics (e.g., small Eu anomalies) in conflict with the predictions.

Other factors argue against such models. Two particularly decisive arguments are: (a) the nickel abundances of high-Ti basalts are effectively zero. No amount of contamination or assimilation could produce this effect. (b) There is good inverse correlation between the abundances of incompatible elements in mare basalts and highland rocks (including KREEP) [126]. Those elements most depleted in mare basalts are most enriched in the potential contaminating material. This effect is the reverse of that predicted by assimilation models.

The Cumulate Model

The cumulate model was proposed by Smith, Philpotts and Schnetzler [115] and has been developed by Taylor and Jakes [86]. This model explains the complementary geochemical nature of the lunar highlands and mare basalt source regions [127]. In the Taylor-Jakeš model, which invokes extensive early melting and differentiation of the Moon, the source regions of low-TiO_2 mare basalts (200–400 km depths) are composed of olivine-orthopyroxene cumulates. Late-stage differentiates contain clinopyroxene and ilmenite and are the proposed source for high-TiO_2 basalts. Prior extraction of plagioclase to form the feldspar-rich lunar highlands crust depletes the source regions in Eu, Sr and Al, accounting for the low abundances of these elements in mare basalts.

The hypothetical crystallization sequence is illustrated in Fig. 6.39 which shows a layered sequence ranging from olivine and orthopyroxene at the base, to clinopyroxene and ilmenite cumulates at higher levels. Adcumulus growth (exclusion of trapped liquid) will predominate in the Ol-Opx layers, while orthocumulus growth (trapping of pore liquids) will occur in the later crystallizing portions, by analogy with terrestrial layered intrusions [128]. However, many factors will combine to upset this idealized situation. Sinking of dense cumulates, convective overturn and the effects of large planetesimal infall will combine to produce a heterogeneous interior.

New experimental work on depth of melting of mare basalts [129] suggests that they may possibly be derived from a complex zone of cumulates at about 400 km depth, consistent with sinking and convective overturn of the cumulate layers during crystallization of the magma ocean although earlier work provided a variety of depth estimates from less than 200 km for the high-Ti suite to 1000–500 km for the low-Ti suite [118] (Fig. 6.38).

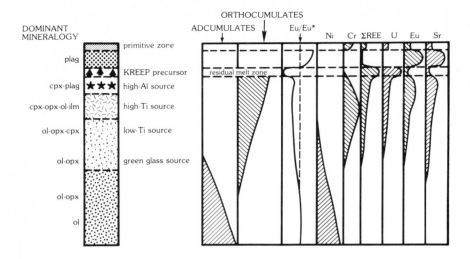

6.39   A schematic diagram of the crystallization of the lunar magma ocean at 4.4 aeons. This idealized section will be greatly modified by sinking of dense cumulates, convective overturn, and by planetesimal impacts, leading to a heterogeneous lunar interior source region for mare basalts. Idealized element distributions of Ni, Cr, ΣREE, Eu depletion or enrichment (Eu/Eu* where Eu* is the abundance for no Eu anomaly), U and Sr are shown.

The isotopic data indicate that mantle heterogeneities were frozen in the Moon at 4.4 aeons. Subsequent partial melting produced the mare basalts. The depths at which this partial melting occurred, as deduced from experimental petrology, although still somewhat uncertain, may be as great as 400 km [129].

This result has implications for the initial depth of lunar melting. Crystallization of large amounts of olivine and orthopyroxene (and plagioclase, which migrates upward to form the crust) occur before the formation of the mare basalt source regions. Accordingly, an uncertain but probable major thickness of olivine and orthopyroxene underlie the source regions of the mare basalts. This implies deep initial or even whole-moon melting. In the cumulate model, the variety of lunar lavas erupted can be related to the various mineral zones resulting from crystallization of the initial magma ocean. In the model, a thick zone of olivine and orthopyroxene, which contains very little Th, U, or K, is first formed. This zone will incorporate nickel, which makes it somewhat more difficult to answer the question whether the Moon as a whole is depleted in nickel relative to the Earth. However, all lunar basalts have very high Fe/Ni ratios compared to terrestrial basalts, so that an overall lunar depletion in nickel is reasonable. Although the

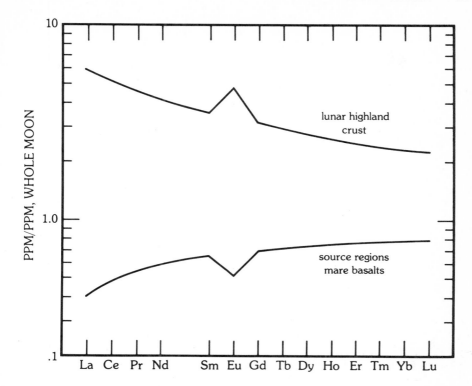

6.40 The reciprocal relationship between REE abundances in the highland crust and the source region of the mare basalts. The mare basalt source region pattern is obtained by subtracting the highland REE abundances (Table 5.5) from the bulk moon composition (Tables 8.1, 8.2). Patterns are normalized to the whole moon abundances and show that the extraction of the highland crust produces on average light REE depletion and a negative Eu anomaly in the lunar mantle.

relative constancy of $Mg/(Mg + Fe)$ values in high- and low-Ti basalts is sometimes raised as an objection to cumulate models [125], such major element features probably result from buffering during crystallization of such a large system. Reasonable models of such processes have been constructed (e.g., [121]). Cr^{3+} is excluded from early-crystallizing olivine and orthopyroxene. Cr^{2+} is present in the highly reducing conditions in the Moon, and is likewise excluded from these early crystallizing minerals. Thus, the effect of crystallization of the magma ocean is to concentrate Cr in the source regions of the mare basalts. As a result, a highly differentiated lunar interior, with respect to trace elements, will be formed. Europium and strontium will be principally concentrated in the feldspar-rich highland crust (Figs. 6.39 and 6.40).

Many incompatible elements are concentrated in the residual phases which form the precursor to KREEP during crystallization of the magma ocean. These elements include the REE, Zr, Hf, Nb, Rb, Ba, Cs, Th, and U, which concentrate in the residual melt that finally invades and pervades the highlands crust and forms the KREEP component. Portions of these elements are trapped in interstitial phases during crystallization of the source regions of the mare basalts and the K, U and Th concentrations supply the heat for later partial melting. The abundances of these incompatible elements in mare basalts, in which they are concentrated to a further degree by partial melting processes, effectively refutes the claim that the cumulate model produces only barren cumulates [125].

The several new types of mare basalts which have been identified (see Section 6.2) provide additional support for the cumulate model. For example, the very primitive Apollo 17 VLT basalts possess, in addition to the ubiquitous Eu depletion, an enrichment in heavy REE consistent with derivation from an olivine-orthopyroxene source region which has already see plagioclase removal [48]. Such patterns cannot be reconciled with either melting of primitive lunar mantle, or contamination or assimilation of late stage lunar differentiates. Even such "primitive" lunar lavas as the Apollo 15 Emerald Green Glass (15426) contain a negative Eu anomaly, indicating prior removal of plagioclase from their source regions. The high-Al basalts appear to be derived by 15–30% partial melting of a clinopyroxene-plagioclase source different from that which produced the Apollo 11 and Apollo 17 high-Ti basalts. These source minerals appear to have crystallized from a heavy REE depleted source [102]. This result is consistent with derivation from a cumulate source region, probably at shallow depths. The heavy REE depletion is the most marked of any lunar basalt, and contrasts with the heavy REE enriched patterns of the Apollo 17 VLT basalts. The absolute REE content is high, Fe/Mg is high, K, Rb and Ba are enriched and the Cr content is low, consistent with the crystallization of the source region late in the history of the magma ocean. Thus, the Luna 16 basalts appear to be derived by 15–30% melting of a clinopyroxene-plagioclase source which was depleted in heavy REE [79] and Ni; therefore, the high-Al basalts may be derived from shallow levels which contain some plagioclase. They are not derived from the highland crust, but are true mare basalts [102]. The Apollo 17 VLT basalts come from a 90% olivine-10% orthopyroxene source with flat REE patterns [48], while the Apollo 17 high-Ti basalts come from a source dominated by clinopyroxene [130].

Considering the various models for the source of mare basalts, derivation from a cumulate source region developed during the 4.4 billion year differentiation event appears the most likely explanation. No mare basalts have been identified as coming from primitive unfractionated source regions. All bear the isotopic and trace element imprint of the early lunar differentiation event.

As our understanding of the details of trace element behavior in the Moon increases, so the fit of the basalt chemistry to the cumulate model improves. (For example, the predominately divalent state of chromium on the Moon in contrast with the trivalent state in the Earth accounts for the curious behaviour of this element, relative to nickel.) It is probable that most mare basalts have experienced some olivine removal en route to the surface (even for such samples as 12009). It is likewise possible that many of the twenty types of mare basalts may be derived from depths of about 400 km [128] although the assignment of depths of melting from the experimental petrological data is subject to considerable uncertainties, with depths from 150 to 400 km being suggested (e.g., [118]). Accordingly, the simple layered sequence depicted in Fig. 6.39 will be much disrupted by sinking, convective overturn [131], and possibly by the effects of giant planetesimal impacts into an early magma ocean.

6.6   Cooling and Crystallization of the Lavas

6.6.1   Primary Magmas

What changes in the chemical composition of the lavas occur between the production of the melt and its crystallization as mare basalt? During the cooling and crystallization of the lava, separation of early formed mineral phases may change the composition. This may occur during the ascent of the lava to the surface and during the spreading of the lava across the maria, as well as during the final crystallization.

The extremely fluid nature of the lavas (viscosity is in the range of 5–10 poises, comparable to that of heavy engine oil at room temperature and about an order of magnitude lower than terrestrial lavas [132]) enables them to spread out widely in thin sheets. They cool rapidly and crystallize quickly. Nevertheless, crystal settling, for example, of olivine, may occur [133]. Whenever a deep lava lake occurs, the conditions are present for slow cooling and crystal fractionation, leading to compositional changes. Such topographical conditions occur only occasionally on the Moon, perhaps most frequently where the lavas are confined within large craters (e.g., in the flooded craters such as Archimedes or Plato). The Taurus-Littrow valley and the Apollo 15 sites are other favorable structural situations for ponding lavas but there is little geological evidence (Section 6.1.3) to support extrusion as other than thin flows.

Similar conditions might apply in the earliest flooding of the mare basins, when extensive ponding could occur. As the basins are filled, the extreme fluidity of the lavas allows them to flow freely over the lunar surface. Even vast outpourings will not form deep lakes, but instead they will flow widely across

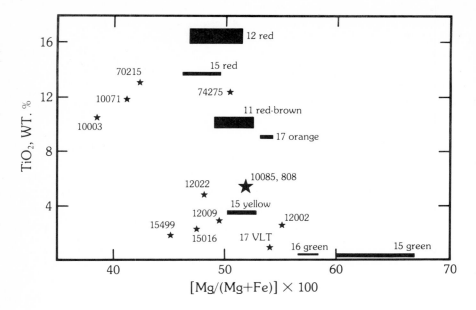

6.41   Volcanic glasses have higher Mg/(Mg + Fe) ratios for any given abundance of TiO$_2$ than most mare basalts (black stars) (15 = Apollo 15). The glasses are thus probably better candidates for primary igneous samples than the majority of mare basalts [70, 129]. (Courtesy J. W. Delano.)

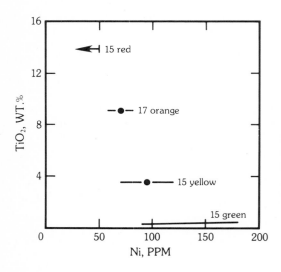

6.42   The abundance of Ni in mare volcanic glasses, regarded as primary magmas, varies strongly as a function of TiO$_2$. This data is an important test of two competing models of mare petrogenesis. The data support cumulate-remelting models. Hypotheses invoking hybridization or assimilation are not consistent with these data [70, 129]. (Courtesy J. W. Delano.)

the lunar surface forming the irregular seas, spreading out from the circular basins. Oceanus Procellarum is the type example. These irregular maria are of shallow depth compared with the circular maria, and they do not have mascons. Thus, conditions for the preservation of primary magmas are probably more favorable on the Moon than on the Earth [133].

Several workers have concluded that rocks such as 12022 and 12009 represent primary magmas, derived by partial melting from the lunar interior, which have arrived at the surface still totally liquid. Other candidates for primary magmas are 15555 (olivine normative), 15597 (quartz normative), 15016, 70215, 70017, 71055 and 74275, although this last sample may not be a good choice on account of contamination. Delano, from a study of glass compositions [134], concludes that 74275, 12002 and 10085, 808 are the best candidates, and that 12009, 12002, 15016, Apollo 17 VLT, and Apollo 11 Group D high-Ti lavas are possible candidates (Figs. 6.41 and 6.42). This variety of compositions reflects the diversity and heterogeneity of the lunar interior.

Liquid Immiscibility

A characteristic feature of lunar magmas is that at very late stages of crystallization, when about 95% of the rock has solidified, the residual liquid splits into two immiscible liquids [135]. Typical compositions are (a) high in SiO_2, K_2O, and other selected elements, similar in many ways to a potassic granite; and (b) low in SiO_2 and K_2O, but very high in FeO and other elements characteristic of pyroxenite. In Table 6.7 the average chemical analyses of two immiscible phases, determined from analysis of globules, are presented and clearly show the extreme contrast in composition of the two phases. Such a process produces the long sought "lunar granites." Possibly some of the "granitic glass" compositions in the soils result from segregation of this material by impact. The process is quantitatively not important, since the amount of "granite" produced in this fashion is minute (see Section 4.4.3).

6.6.2   Fractional Crystallization

The petrological and mineralogical studies indicate that internal element fractionation of great diversity occurs during the crystallization of the basaltic

Table 6.7 Chemical analysis (normalized to 100%) of high and low-silica glasses resulting from silicate liquid immiscibility [135].[†]

Type	SiO_2	Al_2O_3	FeO	MgO	CaO	Na_2O	K_2O	TiO_2	Total
High-silica	76.2	11.6	2.6	0.3	1.8	0.4	6.6	0.5	100.0
Low-silica	46.4	3.1	32.3	2.2	11.3	0.1	0.3	4.3	100.0

[†]Data given in wt. %.

lavas. How much has this affected the bulk compositions? The observed trends resemble those familiar on the Earth during the crystallization of basaltic magmas, but the overall effect is less. The changes in bulk composition are relatively small and do not produce the extremely fractionated rocks typical of terrestrial basalts. Thus, there are no lunar equivalents of the basalt differentiation sequences of Hawaii or the Skaergaard intrusion. It is difficult to find a lunar rock with more than 48% SiO_2. In many of the rocks (e.g., Apollo 12 and 15 samples) there is good chemical and petrological evidence for removal of olivine and some clinopyroxene [46, 133, 136]. A detailed study of the Apollo 12 olivine basalt suite indicates that settling of olivine crystals in a cooling unit perhaps 30 m thick can account for the petrographic and chemical variations observed [133].

Some of the lunar lavas (e.g., Apollo 12 and 15 pyroxene-porphyritic quartz-normative basalts) have textures that have been interpreted as indicating a two-stage crystallization [137]. For example, the chemical trends in the lunar pyroxenes could be correlated with the sequence of crystallization, the bulk composition of the rock, and the oxygen fugacity. In basalts with high-Ti contents (Apollo 11 and 17), the pyroxenes show continuous chemical variation from augite to pyroxferroite. In the quartz-normative Apollo 12 and 15 basalts, with lower Ti/Al ratios, a break occurs in the pyroxene compositions apparently due to the crystallization of plagioclase.

Two models were presented to account for the break in clinopyroxene compositions and the crystallization of plagioclase. The first, involving initial crystallization at depth followed by surface extrusion, was narrowly favored over in situ crystallization of all phases with delayed nucleation of plagioclase. More recent experimental studies on cooling histories have resolved the problem [137]. The porphyritic textures characteristic of the Apollo 12 and 15 quartz-normative basalts were reproduced during single-stage monotomic cooling at rates on the order of 10°C per hour.

It has been suggested that the high titanium content of the Apollo 11 and 17 rocks might be a consequence of accumulation of crystals of ilmenite [119]. However, there is no evidence from the rock fabric to indicate "cumulate" textures resulting from selective accumulation of crystals during cooling of the high-Ti basalts. There is much evidence of intense melt-mineral fractionation at the microscopic level. The overall uniformity in composition of the rocks is in marked contrast, since separation of the complex minerals would produce a heterogeneous rock suite. Many of the minerals present (ilmenite, sulfides, and metallic iron, for example) are much denser than the melt, and would readily separate, thereby depleting the melt (in titanium, for example). The failure of the dense phases to separate indicates either very rapid crystallization or late appearance of the phases. The textural and experimental evidence indicate that for much of the period of crystallization, the three main rock-forming minerals—ilmenite, pyroxene, and plagioclase—were precipitating together.

Finally, in most lunar basalts, plagioclase is not a liquidus phase; therefore, near-surface crystallization and removal of the phase is not a possible explanation for the ubiquitous depletion in Eu, which is inherited from the source regions. After considering all the petrological and chemical evidence outlined in this chapter, the consensus is that the mare basalts are derived from the lunar interior by partial melting, with local near-surface fractionation. The source region itself has undergone at least one previous stage of element fractionation, following accretion of the Moon.

6.7 The Record Elsewhere

6.7.1 Mercury—Are Basalts Present?

The geological evidence was assessed in Section 2.5 where it is concluded that there was no definitive evidence for basaltic rocks. The spectral reflectance data obtained by remote sensing indicate an overall crustal composition closely similar to that of the lunar highlands [138]. The low spatial resolution of this technique does not permit the observation of small amounts of mare-type basalts, but the albedo similarities among the various topographic units on Mercury does not encourage this speculation. The Mercurian surface is older than 4.0 aeons. Whether basaltic lavas could be generated within the Mercurian mantle before that time depends on unknowns such as the content and distribution of radioactive heat sources following the major Mercurian fractionation. Mercury differs from the Moon in bulk density, and in possessing a large core. Whether or not basalts, if present on Mercury, would resemble lunar mare basalts sufficiently to be recognized with the present observational techniques is debatable.

6.7.2 Mars

In contrast to Mercury, the evidence for volcanism on Mars, probably of basaltic type, is unequivocal. Descriptions of the volcanic landforms are given in Section 2.6.2. In addition to this morphological evidence (for example, the profile of Olympus Mons is close to that of Mauna Loa), the chemical evidence from the Viking Lander (Table 6.8) is consistent with weathering from basaltic parent materials [139]. The spectral reflectance data is in general agreement with this interpretation [140], and indicates the presence of pyroxene and olivine in the darker (unweathered?) regions of Mars. There is no chemical, spectral or morphological evidence to suggest the presence of other

Table 6.8 Analytical data for the Martian surface.

a. Gamma-ray orbital data (ppm)

	K(2)	U(1)	Th(1)	Th/U	K/U
Young volcanic terrain	2500	1.1 ± 0.8	5.0 ± 2.5	4.5	2200
Ancient crust	—	0.2 ± 0.14	0.7 ± 0.35	3.5	—

b. Viking Lander XRF data (3)

wt. %	Chryse Planitia	Utopia Planitia	Igneous Component from Chryse Planitia	Instrument Precision	Matrix Limitation
SiO_2	43	43	53.3	± 4	$-6 + 4$
TiO_2	0.63	0.54	0.78	± 0.1	± 0.25
Al_2O_3	7.2	—	8.9	± 1	$-3 + 4$
Fe_2O_3	18.0	17.5	2.23	± 0.5	$-5 + 5$
MgO	6	—	7.4	± 1	$-3 + 5$
CaO	5.7	5.6	7.1	± 0.2	± 2
SO_3	7.6	8.0	—	± 0.7	$-2 + 6$
Cl	0.75	0.45	—	± 0.25	$-.5 + 1.5$

ppm		
Rb	< 30	< 30
Sr	60 ± 30	100 ± 40
Y	70 ± 30	50 ± 30
Zr	< 30	30 ± 20

(1) Surkov, Yu. A. (1980) *PLC 11*: 669.
(2) Surkov, Yu. A. (pers. comm, 1980).
(3) Clark, B. C., et al. (1982) *JGR*. In press.

than basaltic volcanism: no sign appears of acidic, silicic or granitic compositions (C. A. Wood, pers. comm., 1981).

6.7.3 Venus

The possibility that basaltic rocks occur on Venus is reinforced by two observations: (a) large structures resembling shield volcanoes (e.g., Beta Regio) occur (Section 2.7), and (b) the Russian gamma-ray data (Table 6.9) provides K, U and Th values for Venera 9 and 10 consistent with those expected for basaltic rocks. The Venera 9 and 10 data were acquired on the flanks of Beta Regio. These data encourage the speculation that this feature is indeed a large shield volcano, analogous to Mauna Loa. Possibly the smooth rolling plains of Venus are floored with basaltic rocks, analogous to the terrestrial ocean floor [141].

Table 6.9 Analytical data for the surface of Venus.

	Venera 8	Venera 9	Venera 10
K, %	4.0 ± 1.2	0.47 ± 0.08	0.30 ± 0.16
U, ppm	2.2 ± 0.7	0.60 ± 0.16	0.46 ± 0.26
Th, ppm	6.5 ± 0.2	3.65 ± 0.42	0.70 ± 0.34
K/U	20,000	8,000	7,000
Th/U	3.0	6.1	1.5

Gamma-ray data from Surkov, Yu. A. (1977) *PLC 8*: 2665.

6.7.4 Io

The spectacular eruptions observed on Io appear to result from explosive volcanism, powered by SO_2 [142, 143]. Many of the surface flows resemble those observed at terrestrial volcanoes (Fig. 6.43). However, all of them display colors indicative of a high sulfur content. Whether volcanism on Io

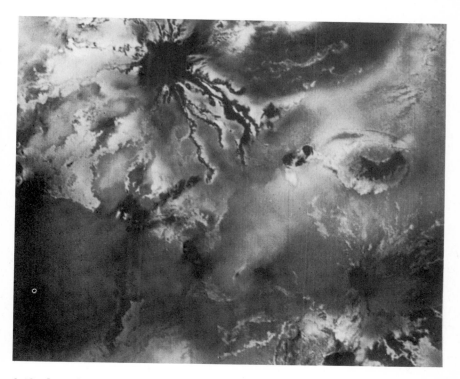

6.43 Lava flows from a volcanic caldera on Io. Width of photo 225 km (NASA JPL P.21277).

has the familiar silicate composition or is more exotic remains as a fascinating subject for investigation.

6.7.5 Vesta

The reflectance spectra of Vesta, the second largest asteroid (269 km in radius), are dominated by the pyroxene, pigeonite [144, 145]. There is no indication of the presence of olivine. The spectra match that of basaltic achondrites (e.g., Nuevo Laredo) and this has led to the concept that the surface of Vesta consists of basalt. This has in turn led to the view that Vesta is the parent body for the eucritic meteorites [146, 147], although serious dynamic difficulties remain in deriving meteorites from Vesta. Possibly other smaller asteroids have basaltic surfaces, but the identification is difficult. Excellent reviews of asteroidal reflectance spectra, and of the difficulties of matching these with the meteorite data, are given by Chapman and Gaffey [148] and Gaffey and McCord [149]. The latter reference contains a tabulation of possible meteorite analogues of many asteroids. Whether or not the eucrites come from Vesta, the spectral evidence indicates that basaltic volcanism is possible on asteroidal sized objects. This raises questions about appropriate thermal regimes and the possibility of early heating by ^{26}Al.

References and Notes

1. E.g., Baldwin, R. B. (1949) *The Face of the Moon*, Chicago; Kuiper, G. (1954) *U.S. Nat. Acad. Sci.* 40: 1096; Fielder, G. (1963) *Nature*. 198: 1256.
2. Wilson, A. T. (1962) *Nature*. 196: 11; Gold, T. (1955) *Mon. Not. Roy. Astr. Soc.* 115: 585; Gilvarry, J. J. (1968) *Nature*. 218: 336; A last-ditch defense of the sedimentary hypothesis was offered by J. J. Gilvarry 1970, *Radio Sci.* 5: 322 following the Apollo 11 sample return.
3. Latin mare = sea.
    I am indebted to Carroll Ann Hodges for instruction in Latin grammar.

	Singular	Plural
Nom.	mare	maria
Gen.	maris	marium
Dat.	mari	maribus
Acc.	mare	maria
Abl.	mari	maribus

4. *Basaltic Volcanism* (1981) Section 5.4.
5. Whitford-Stark, J. L. (1981) LPS XII: 1182.
6. De Hon, R. A. (1974) *PLC 5*: 53.
7. De Hon, R. A., and Waskom, I. D. (1976) *PLC 7*: 2729.
8. De Hon, R. A. (1979) *PLC 10*: 2935.
9. Hörz, F. (1978) *PLC 9*: 3331.
10. McCord, T. B., et al. (1972) *JGR*. 77: 1349.
11. Zisk, S. H., et al. (1978) LPS IX: 1298.
12. Head, J. W., et al. (1978) *Mare Crisium*, p. 43.

13. Peeples, W. J., et al. (1978) *JGR.* 83: 3459.
14. McCauley, J. F., et al. (1981) *Icarus.* In press.
15. *Basaltic Volcanism* (1981) Table 7.3.1.
16. Baldwin, R. B. (1974) *Icarus.* 23: 97.
17. Mutch, T. A. (1972) *Geology of the Moon*, Princeton, p. 201.
18. Schaber, G. G. (1973) *PLC 4*: 73; (1976) *PLC 7*: 2783.
19. *Basaltic Volcanism* (1981) Chap. 5; See also Wilson, L., and Head, J. W. (1981) *JGR.* 86: 2971.
20. Howard, K. A., et al. (1972) *PLC 3*: 1.
21. Schaber, G. G., et al. (1976) *PLC 7*: 2783.
22. Gifford, A. W., and El-Baz, F. (1981) *Moon and Planets.* 24: 391.
23. Greeley, R. (1976) *PLC 7*: 2747.
24. Greeley, R., and Spudis, P. D. (1978) *PLC 9*: 3333.
25. NASA SP 330 (1974).
26. Heiken, G., et al. (1974) *GCA.* 38: 1703.
27. Meyer, C., et al. (1975) *PLC 6*: 1690, Fig. 10.
28. El-Baz, F., and Evans, R. E. (1973) *PLC 4*: 139.
29. Beaty, D. W., and Albee, A. L. (1980) *PLC 11*: 23.
30. Hawke, B. R., et al. (1979) *PLC 10*: 2995.
31. Head, J. W., and Gifford, A. (1980) *Moon and Planets.* 22: 235.
32. *Basaltic Volcanism* (1981) Section 5.4.3.
33. Wood, C. A. (1979) *PLC 10*: 2815.
34. Bryan, W. B. (1973) *PLC 4*: 93.
35. Mattingly, T. K., and El-Baz, F. (1973) *PLC 4*: 55.
36. Solomon, S. C., and Head, J. W. (1979) *JGR.* 84: 1667.
37. Howard, K. A., et al. (1972) *PLC 3*: 1.
38. Schubert, G. G., et al. (1974) LS V: 675.
39. Gaddis, L. R., and Head, J. W. (1981) LPS XII: 321.
40. Nakamura, Y., et al. (1979) *PLC 10*: 2299.
41. Luchitta, B. K., and Watkins, J. A. (1978) *PLC 9*: 3459.
42. Haskin, L. A., et al. (1977) *PLC 8*: 1731.
43. *Basaltic Volcanism* (1981) Section 1.2.9.
44. Petrographic and chemical studies on the Apollo 11 basalt collection have been carried out by the following workers since 1974: Beaty, D. W., and Albee, A. L. (1978) *PLC 9*: 359; Gamble, R. P., et al., ibid., 495; Guggisberg, S., et al. (1979) *PLC 10*: 1; Beaty, D. W., et al., ibid., 41; Rhodes, J. M., and Blanchard, D. P. (1980) *PLC 11*: 49; Ma, M. -S., et al., ibid., 37.
45. Similar studies on Apollo 17 basalts have been carried out by: Brown, G. M., et al. (1975) *PLC 6*: 1; Dymek, R. F., et al., ibid., 49; Rhodes, J. M., et al. (1976) *PLC 7*: 1467; Warner, R. D., et al. (1979) *PLC 10*: 225.
46. Revised studies of the Apollo 12 basalt suite have been carried out by: Dungan, M. A., and Brown, R. W. (1977) *PLC 8*: 1339; Nyquist, L. E., et al., ibid., 1383; Baldridge, W. S., et al. (1979) *PLC 10*: 141; Rhodes, J. M., et al. (1977) *PLC 8*: 1305; Walker, D., et al. (1976) *PLC 7*: 1365.
47. The Apollo 15 basalt suite has been studied by Lofgren, G. E., et al. (1975) *PLC 6*: 79; Ma, M. -S., et al. (1978) *PLC 9*: 523.
48. VLT basalts have been extensively studied by Vaniman, D. T., and Papike, J. J. (1977) *PLC 8*: 1443; Norman, M. D., and Ryder, G. (1980) *Moon and Planets.* 23: 271; Wentworth, S., et al. (1979) *PLC 10*: 207 and references in the Mare Crisium volume (1978).
49. References to high-Al basalt suites include: Ridley, W. I. (1975) *PLC 6*: 131; Kurat, G., et al. (1976) *PLC 7*: 1301; Beaty, D. W., et al. (1979) *PLC 10*: 115; Warner,

R. D., et al. (1980) *PLC 11*: 87; and the initial Luna 16 report in (1972) *EPSL*. 13: 223.

50. *Mare Crisium* (1978).
51. Frondel, Judith W. (1975) *Lunar Mineralogy*, Wiley, N.Y.
52. Smith, J. V. (1974) *Amer. Mineral.* 59: 231.
53. Smith, J. V., and Steele, I. M. (1976) *Amer. Mineral.* 61: 1059.
54. *Basaltic Volcanism* (1981) Section 1.3.2.
55. Papike, J. J., et al. (1976) *Rev. Geophys. Space Phys.* 14: 475.
56. Haggerty, S. in *Balsaltic Volcanism* (1981) Section 1.3.4.4.
57. Sato, M. (1976) *PLC 7*: 1323; See *Basaltic Volcanism* (1981) Section 1.3.4.4.
58. Sato, M. (1977) *EOS.* 58: 425.
59. Housley, R. M. (1978) *PLC 9*: 1473.
60. Brett, R. (1976) *GCA.* 60: 997.
61. Sato, M. (1973) *PLC 4*: 1061.
62. Schreiber, H. D. (1977) *PLC 8*: 1785.
63. Huneke, J. C., et al. (1973) LS IV: 404.
64. Huneke, J. C., et al. (1976) LS V: 377.
65. Butler, P. (1978) *PLC 9*: 1459.
66. Wasson, J. T., et al. (1976) *PLC 7*: 1583.
67. Gibson, E. K., and Moore, C. B. (1973) *EPSL.* 20: 404.
68. Schaeffer, O. A., and Husain, L. (1973) *EOS.* 54: 613.
69. Tera, F., and Wasserburg, G. J. (1976) LS VII: 858.
70. Delano, J. W. (1970) *PLC 10*: 275; (1980) *PLC 11*: 251; (1981) *EPSL*. In press.
71. Morgan, J. W., and Wandless, G. A. (1979) *PLC 10*: 327, following the earlier suggestion by Roedder, E., and Weiblen, P. (1973) *Nature.* 244: 210.
72. Reid, A. M., et al. (1973) *EOS.* 54: 606.
73. Grove, T. L. (1981) LPS XII: 374; *PLC 12*: 935; Ma, M. -S., et al. (1981) LPS XII: 637; *PLC 12*: 915.
74. Housley, R. M. (1978) *PLC 9*: 1473.
75. Rhodes, J. M., and Blanchard, D. P. (1980) *PLC 11*: 49 (see also [45]).
76. Beaty, D. W., and Albee, A. L. (1978) *PLC 9*: 359 (see [45]).
77. Papanastassiou, D. A., et al. (1977) *PLC 8*: 1639.
78. Warner, R. D., et al. (1980) *PLC 11*: 87 (see [49]).
79. Ma, M. -S., et al. (1979) *GRL.* 6: 910.
80. Gibson, E. K., and Hubbard, N. J. (1972) *PLC 3*: 2003.
81. Brown, G. M., and Peckett, A. (1971) *Nature.* 234: 262.
82. A review of the many interesting and ingenious hypotheses was given in *Lunar Science: A Post-Apollo View* (1975) pp. 154-159 and need not be repeated here.
83. E.g., Nyquist, L. E., et al. (1977) *PLC 8*: 1401.
84. Eldridge, J. S. (1973) *PLC 4*: 2115.
85. Duncan, A. R., et al. (1973) *PLC 4*: 1097; (1974) *PLC 5*: 1147; (1976) *PLC 7*: 1659.
86. Taylor, S. R., and Jakes, P. (1974) *PLC 5*: 1287.
87. Schreiber, H. D., and Haskin, L. A. (1976) *PLC 7*: 1221.
88. Huebner, J. S., et al. (1976) *PLC 7*: 1195.
89. Brett, P. R. (1973) *GCA.* 37: 165.
90. Kaplan, I. R., and Petrowski, C. (1971) *PLC 2*: 1402. The $^{34}S/^{32}S$ values are expressed as deviations from a standard sample, as follows:

$$\delta^{34}S = \left(\frac{^{34}S/^{32}S \text{ sample} - ^{34}S/^{32}S \text{ std.}}{^{34}S/^{32}S \text{ std.}} \right) 1{,}000.$$

Std. = FeS (troilite from Canyon Diablo meteorite).

91. *Basaltic Volcanism* (1981) Section 1.3.5.
92. Wolf, R., et al. (1979) *PLC 10*: 2107. A summary of low-Ti lunar basalt data is given in Table 1 of this paper.
93. Anders, E., et al. (1971) *PLC 2*: 1021. Gives data for Apollo 11 and 12 samples.
94. Taylor, H. P., and Epstein, S. (1973) *PLC 6*: 1657.
95. The $^{18}O/^{16}O$ data are customarily expressed as deviations from a standard sample, in parts per 1,000:

$$\delta^{18}O = \left(\frac{(^{18}O/^{16}O) \text{ sample}}{^{18}O/^{16}O \text{ std. ocean water}} - 1 \right) 1,000.$$

$^{17}O/^{16}O$ values are similarly treated.
96. Clayton, R. N., et al. (1973) *PLC 4*: 1535.
97. Gibson, E. K. (1977) PCE 10: 57.
98. The variation in $^{13}C/^{12}C$ are given as:

$$\delta^{13}C = \left(\frac{^{13}C/^{12}C \text{ sample} - ^{13}C/^{12}C \text{ std. limestone}}{^{13}C/^{12}C \text{ std. limestone}} \right) 1,000.$$

99. This was clearly recognized by Taylor, S. R. (1975) *Lunar Science* (pp. 16, 182, 309) despite occasional assertions to the contrary.
100. E.g., clasts from 14321, 14063, 14082, 14312 and 14318; Ryder, G., and Spudis, P. (1980) *Lunar Highlands Crust*, p. 353.
101. James, O. B., and McGee, J. J. (1980) *PLC 11*: 67.
102. Nyquist, L. E., et al. (1981) *EPSL*. 55: 335.
103. Schultz, P., and Spudis, P. D. (1979) *PLC 10*: 2899.
104. Wolf, R., et al. (1979) *Lunar Highlands Crust*, p. 201.
105. All ages quoted are based on the following decay constants:
$\lambda(^{87}Rb) = 1.42 \times 10^{-11}/yr.$
$\lambda(^{147}Sm) = 6.54 \times 10^{-12}/yr.$
$\lambda(^{40}K) = 4.96 \times 10^{-10}/yr.$
$\lambda_e(^{40}K) = 0.581 \times 10^{-10}/yr.$
$\lambda(^{238}U) = 1.55 \times 10^{-10}/yr.$
$\lambda(^{235}U) = 9.84 \times 10^{-10}/yr.$
$\lambda(^{232}Th) = 4.95 \times 10^{-11}/yr.$
(Steiger, R. H., and Jäger, E. (1977) *EPSL*. 36: 359.
106. Beaty, D. W., and Albee, A. L. (1980) *PLC 11*: 23; Geiss, J., et al. (1977) *Phil. Trans. Roy. Soc.* A285: 151.
107. *Basaltic Volcanism* (1981) Section 8.6; see also Boyce, J. M. (1975) *Origins of Mare Basalts* (LPI Contribution No. 234), p. 11.
108. *Basaltic Volcanism* (1981) Section 7.3.4.
109. Lugmair, G. W., et al. (1975) *PLC 6*: 1419; (1978) *EPSL*. 39: 349.
110. Tera, F., and Wasserburg, G. J. (1975) LS VI: 807.
111. Oberli, F., et al. (1979) LPS X: 940.
112. Birck, J. L., et al. (1974) LS V: 64.
113. Wetherill, G. W. (1968) *Science*. 160: 1256.
114. Opik, E. J. (1969) *Ann. Rev. Astron. Astrophys.* 7: 473.
115. Smith, J. V. (1970) *PLC 1*: 897; Philpotts, J. A., and Schnetzler, C. C. (1970) *PLC 1*: 1471.
116. Such hypotheses occasionally appear in popular scientific articles. A more detailed refutation is given in *Lunar Science*, pp. 188–191.
117. Ringwood, A. E., and Essene, E. (1970) *PLC 1*: 796.

118. Green, D. H., et al. (1971) *PLC 2*: 601; (1975) *PLC 6*: 871.

119. O'Hara, M. J., et al. (1970) *PLC 1*: 695.

120. Biggar, G. M., et al. (1971) *PLC 2*: 617.

121. Longhi, J., et al. (1974) *PLC 5*: 447; (1980) *PLC 11*: 289.

122. Walker, D., et al. (1975) *GCA*. 39: 1219; (1975) *PLC 6*: 1103; (1976) *EPSL*. 30: 27.

123. Nyquist, L. E., et al. (1976) *PLC 7*: 1507.

124. Hubbard, N. J., and Minear, J. W. (1975) LS VI: 405.

125. Ringwood, A. E., and Kesson, S. E. (1976) *PLC 7*: 1697.

126. Taylor, S. R., and Bence, A. E. (1975) Mare Basalt Conf., 159.

127. Among the many papers which have addressed chemical petrological and isotopic aspects of the cumulate model, the following may be cited.
Brown, G. M. (1977) *Phil. Trans. Roy. Soc.* A285: 169; Walker, D., et al. (1975) *GCA*. 39: 1219; Shih, C. -Y., et al. (1975) *PLC 6*: 1255; Shih, C. -Y., and Shonfeld, E. (1976) *PLC 7*: 1757; Drake, M. J., and Consolmagno, G. J. (1976) *PLC 7*: 1633; Nyquist, L. E., et al. (1977) *PLC 8*: 1383; (1978) *Mare Crisium*, 631; (1979) *PLC 10*: 77; (1981) *EPSL*. 55: 335; Lugmair, G. W. (1975) *Mare Basalts*, 107; Lugmair, G. W., et al. (1975) *PLC 6*: 1419; see also [85].

128. Wager, L. R., and Brown, G. M. (1968) *Layered Basic Intrusions*, Oliver and Boyd; Brown, G. M. (1977) *Phil. Trans. Roy. Soc.* A285: 169.

129. Delano, J. W. (1980) *PLC 11*: 251.

130. Warner, R. D., et al. (1979) *PLC 10*: 225.

131. E.g., Rice, A. (1981) *JGR*. 86: 405.

132. Murase, T., and McBirney, A. R. (1970) *Science*. 167: 1491.

133. Walker, D., et al. (1976) *PLC 7*: 1365.

134. Delano, J. W. (1981) *EPSL*. In press.

135. Roedder, E., and Weiblen, P. W. (1973) *PLC 4*: 681.

136. LSPET (1970) *Science*. 167: 1325.

137. Lofgren, G. E. (1980) in *Physics of Magmatic Processes* (ed., R. B. Hargraves), Princeton Univ. Press, Chap. 11 provides an extensive review of this topic; see also Nathan, H. D., and Van Kirk, C. J. (1978) *J. Petrol*. 19: 66.

138. *Balsaltic Volcanism* (1981) Section 2.2.3.

139. Toulmin, P., et al. (1977) *JGR*. 82: 4625.

140. *Basaltic Volcanism* (1981) Section 2.2.4.

141. Phillips, R. J., et al. (1981) *Science*. 212: 879.

142. Smith, B. A., et al. (1979) *Science*. 204: 951; ibid., 206: 927; *Nature*. 280: 738.

143. Carr, M. H., et al. (1979) *Nature*. 280: 729.

144. McCord, T. B., et al. (1970) *Science*. 168: 1445.

145. McFadden, L., et al. (1977) *Icarus*. 31: 439.

146. Consolmagno, G. J., and Drake, M. J. (1977) *GCA*. 41: 1271.

147. Gaffey, M. J., and McCord, T. B. (1977) *PLC 8*: 113.

148. Chapman, C. R., and Gaffey, M. J. (1979) in *Asteroids* (ed., T. Gehrels), Univ. Ariz. Press, p. 655.

149. Gaffey, M. J., and McCord, T. B. (1979) in *Asteroids* (ed., T. Gehrels), Univ. Ariz. Press, p. 688.

Chapter 7

PLANETARY INTERIORS

7.1   The Advantages of Geophysics

The present understanding of the structure and evolution of the Moon depends heavily on the program of geophysical measurements carried out during the Apollo missions. Geophysical measurements have the advantage that many of the parameters can be established by instruments in orbit as well as from those landed on planetary surfaces. Sample return is a useful but secondary part of the geophysics program, enabling the direct measurements of magnetic properties, thermal conductivities, bulk density and so on. Most of our present knowledge of planetary bodies other than the Moon, and Mars to a very minor extent, is based on the assessment of the geophysical data.

Gravity and magnetic measurements acquired from orbiting spacecraft can provide information not otherwise obtainable about planetary interiors. Magnetic measurements may provide evidence of the past history of a planet. Precise measurements of the gravitational field provide constraints on such fundamental properties as the moment of inertia, a basic parameter whose measurement was "eagerly awaited by aficionados of the moon for some 200 years" [1]. However, one must also recall the lament of Lambeck [2] that "it is rather curious and unfortunate that some of the most readily observed geophysical quantities are also the most tantalizing, but ambiguous ones."

7.2   Radii, Densities and Moments of Inertia

Current values for these parameters are given in Table 7.1. The precise determination of radii has allowed accurate mapping of the elevation of planetary surfaces by spacecraft. The circular maria on the Moon are filled to varying depths with lavas, but generally lie at lower elevations than the irregular maria, which do not occupy deep basins. The far-side highlands are exceedingly rough, but the mare surfaces are smooth, with differences in elevation of only ±150 m over hundreds of kilometers [3]. In general, there is little evidence of sagging of the lavas in the centers of the circular maria. The lava surfaces are remarkably flat, although local subsidence by normal faulting occurs (e.g., in Crisium). Information from the laser altimeters led to the mapping of ancient ringed basins on the lunar far side originally photographed by Russian spacecraft [4, 5]. Detailed surface elevations on both Venus and Mars have also been obtained, allowing, for Venus, our first detailed understanding of the surface topography of that planet [6] (see Section 2.7).

The coefficient of moment of intertia is critical for understanding the density distribution in the lunar interior. The current value of 0.391 ± 0.002 for the Moon indicates that, in addition to a low density crust, it is probably necessary for an increase in density to occur in the deep interior. This may occur in three ways: (a) a primitive iron-rich interior with only the upper half of the Moon differentiated, (b) phase changes to denser minerals such as garnet at depth, and (c) whole-moon melting and formation of an Fe or FeS core. The first two alternatives encounter various difficulties. Melting the outer half of the Moon, leaving a primitive unmelted deep interior, is probably not sufficient to provide a thick enough feldspathic crust (Sections 5.11.3 and 8.4.2). No evidence of garnet signatures is observed in the REE patterns of mare basalts, which may be derived from depths of 400–500 km accordingly (Section 6.5.2). Initial melting and fractionation (removing garnet constituents Ca and Al into the plagioclase-rich crust) probably extended to greater depths and the possibility of a garnet-rich interior is thus diminished. Accordingly, the third alternative of a core is favored [7].

Tolerances in lunar core densities corresponding to an uncertainty of ±0.0025 in the coefficient of moment of inertia are equivalent to about 4.2 g/cm^3 for lunar cores 300 km in radius; hence core compositions of either Fe or FeS are allowable. The alternative of density changes with depth due to initial heterogeneities appears unlikely in view of the evidence for large scale melting of the Moon. In both the Earth and Mars, the values of the coefficient of moment of inertia clearly indicate the existence of a core. The values for Venus and Mercury are not known, but other evidence suggests differentiation and core formation [8, 9].

7.3   Lunar Center of Mass/Center of Figure Offset

The center of mass of the Moon (CM) is displaced from its center of figure toward the Earth (CF) by about 2 km [10]. Three explanations have been forthcoming, and are presented below in order of increasing credibility:

(a) Mare basalt flows are more dense than the crust. They are common on the near side, but less so on the far side. Their volume is insufficient, however, by about an order of magnitude, to account for the offset.

(b) A lunar core might be displaced from the center of mass [11]; however, this model is unrealistic since the displacement would create shear stresses which could not be supported by the hot interior [12]. A variation of this model calls for mantle asymmetry developed as a consequence of crystallization of the magma ocean [13]. In this model, more rapid crystallization of the magma ocean on the far side results in a greater thickness of Mg-rich (low density) cumulates in the far-side mantle. It seems unlikely that such irregularities could persist during magma ocean crystallization or be maintained during cooling if whole-moon melting occurred. All such models are subject to stress relaxation in the high temperature interior unless maintained by convection which could then account for the offset. In addition, the geochemical evidence indicates that any initial heterogeneities due to accretion which might account for the offset were smoothed out.

(c) The conventional explanation is that the far-side highland crust is thicker [10]. It is massive enough, comprising about 10% of the volume of the Moon, and sufficiently irregular in density and thickness to account for the observed center of mass/center of figure offset [3, 14].

Various other suggestions, such as movement of basalt magma from the lunar interior or removal of near-side crust relative to the far side by asymmetric bombardment [15], have not been generally accepted. An important effect arises from this displacement of the center of figure from the center of mass (Fig. 7.1). A particular equipotential surface will be closer to the lunar surface on the Earth-facing hemisphere. Multi-ring basins and craters on this face will flood with lava much more readily than those on the farside for a given depth of melting. The diversity of lunar basalt compositions indicates many different source regions, probably at differing depths, but in general, it is more difficult for lavas to reach the surface on the farside.

7.4   Gravity

The gravity field of the Earth and planets, and the problems of and interpretations possible from the data, are discussed in an extensive review by Phillips and Lambeck [16].

7.4.1   Lunar Gravity

Estimating the lunar gravity field is not a straightforward process for a number of reasons, not the least of which is the paucity of far-side spacecraft

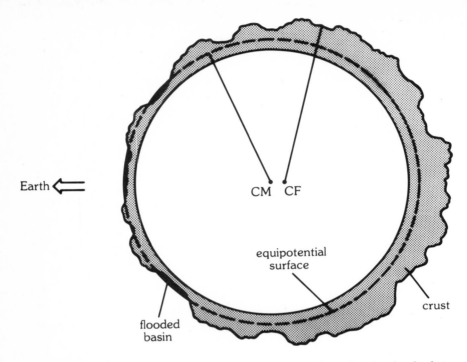

7.1 Cross-section through the Moon in the equatorial plane showing the displacement of the center of mass towards the Earth relative to the center of figure. This is to illustrate that a given equipotential surface will be closer to the lunar surface on the near side than on the far side. Magmas originating at equal depths below that surface will have greater difficulty in appearing as surface lavas on the far side. (Courtesy Susan Pullan.)

tracking data. The lunar gravity field possesses many interesting features, the most notable of which are the mascons, whose discovery in 1968 [17] caused much interest in the state of stress of the lunar crust and the length of time for which it could support such loads.

The early crust appears to be generally compensated, and no effects are discernible from the pre-4.2 billion year bombardment. Formation of the highland crust by flotation in a magma ocean is consistent with this gravity picture. By about 4 aeons the crust had become thick enough and the interior cool enough to support substantial loads for the next 4 billion years. Thus, contrary to earlier interpretations, the Apennine Mountains, which rise 7 km above the present surface of Mare Imbrium, are uncompensated [18]. This provides additional evidence for a relatively cool lunar interior, and that crystallization of the magma ocean was complete by 4.4 aeons. An extensive

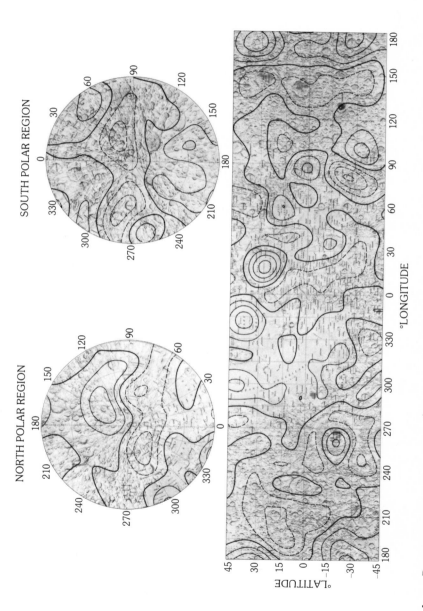

7.2 Lunar Bouguer gravity anomalies. The heavy solid curve is the zero contour. Light solid curves show positive anomalies. Dashed curves indicate negative anomalies. The contour interval is 100 mgal [19]. (Reproduced from *JGR*. 85: 1022, Fig. 7; Courtesy A. J. Ferrari and B. G. Bills).

period of KREEP volcanism in the highlands, with shallow sources implying high near surface temperatures, at about 4.0 aeons must be judged unlikely. (The view is taken throughout this book that the source of KREEP is related to the final stages of crystallization of the magma ocean at about 4.4 aeons. Younger "KREEP basalts" are interpreted as impact melts.) The subsequent partial melting in the deep interior to produce mare basalts between about 4.0 and 3.0 aeons had no effect on the strength of the crust and upper lithosphere, since the observed volume of basalt produced is about only 0.1% of total lunar volume.

The most spectacular Bouguer gravity anomalies are displayed by the Orientale Basin (Fig. 7.2), which has a "bull's eye" appearance, with a central positive region (+200 mgal) surrounded by a ring of negative anomaly (−100 mgal) with an outer positive apron (+30 to +50 mgal). These conclusions, although based on limited data, are firm. Orientale had only restricted coverage during the missions, and for this reason, the topography is not well constrained. With the exception of Orientale and the other mascons, the lunar gravity field is mild with variations typically of ±30 mgal. The far-side topography shows a higher mean elevation, 3–6 km above the mean radius. The basins on the near side, flooded with basalt, are 1–4 km below the mean lunar radius. This lends support to the explanation that the near-side–far-side visual asymmetry for mare lavas is due to the height which the lavas can reach on the basis of density differences. The far-side gravity is less well-known but shows rather broad scale positive features with local negative free-air anomalies over the major ringed basins [18, 19] such as the Mendeleev, Moscoviense, Korolev and Apollo basins which all appear to have strong negative anomalies (about −100 mgal). Basins such as Ingenii have relative gravity lows. The Mendeleev basin, 4 km deep and 300 km in diameter, has a negative anomaly of −116 mgal. Corrections for the topography indicate a positive Bouguer anomaly of +100 mgal. Thus, some isostatic readjustment took place following the basin impact. Mendeleev is Nectarian in age (Table 3.1). This positive gravity anomaly must be due to mantle rebound during basin formation. This seems to be consistent with the Orientale basin evidence. Since there is only a small amount of mare basalt fill in the center of the Orientale ringed basin, most of the positive gravity anomaly must also be due to mantle rebound. The positive anomaly of Grimaldi (+60 mgals) is too large for a 150 km crater, even if filled with lava; other craters of this size show negative anomalies. The paradox is resolved by the recognition that Grimaldi is, in fact, a ringed basin with a diameter of 420 km [20] and a significant part of the positive anomaly is due to mantle rebound. This basin, for which good gravity and topographic data exist, thus provides a key to understanding much of the mascon problem (Fig. 7.3) (see Table 3.1).

A summary of the gravity data indicates that:
(a) Young ray craters have negative Bouguer anomalies, a consequence

both of the mass defect due to excavation and of low density breccia and fall-back within the crater.

(b) Craters less than about 200 km in diameter, filled with lava, also have negative anomalies. Sinus Iridum, the Bay of Rainbows on the coast of the Sea of Rains (Mare Imbrium), has a negative anomaly of –90 mgal.

(c) The older lunar highlands are in isostatic equilbrium and the Imbrian and Pre-Imbrian craters which saturate the highland crustal surface have no Bouguer anomalies indicating probably that there was no mantle rebound.

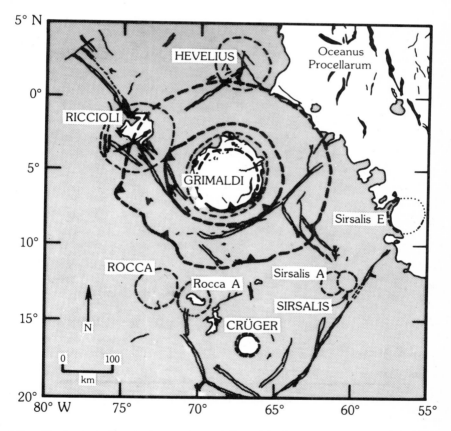

7.3a Tectonic and geologic structure of Grimaldi. Concentric ring structures are shown for Grimaldi as heavy lines, dashed where inferred or extrapolated, with scarps indicated by arrowheads. The innermost ring shown is probably a normal fault resulting from the volcanic load in the central basin. Ridges are shown in the mare-covered central basin of Grimaldi and in Oceanus Procellarum to the northeast. Other structures depicted are rilles [20].

GRIMALDI

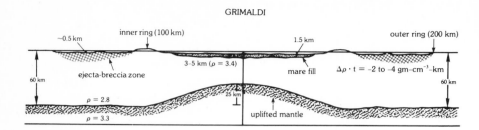

7.3b Schematic cross-section of the Grimaldi basin and inferred subcrustal structure, determined from the gravity data. Horizontal and vertical scales are equal. The ejecta-breccia zone is expressed as a density-thickness product [20].(Courtesy R. J. Phillips.)

    (d) The large mountain ranges formed around the young multi-ring basins (e.g., Apennines) are uncompensated and have large positive anomalies, i.e., they are supported by stresses in the rigid lithosphere.

    (e) The circular maria on the near side have large positive gravity anomalies. The largest are those of Mare Imbrium and Mare Serenitatis (+220 mgal), with smaller positive anomalies under the circular mare, Crisium, Nectaris, Humorum, Humboltianum, Orientale, Smythii; and under Sinus Aestuum and Grimaldi.

    (f) The Marius Hills, widely interpreted as volcanic domes, superimposed on the basaltic maria, show a positive anomaly of +65 mgal, indicating that virtually no late isostatic adjustment has occurred in the mare basalt flooding stage and that these volcanic features are supported by the crust without yielding.

7.4.2   The Mascons

The excess masses represented by the positive gravity anomalies over the circular maria have excited much interest, and yield useful information on the thermal state of the Moon. A key piece of information is the fact that the Apennine ridge is uncompensated [18], showing that the crust was cold enough to support the basin topography at 3.8 aeons. Earlier interpretations indicated that some compensation had occurred, so that the mascons were interpreted to result from later basaltic flooding of the multi-ring basins. However, the thickness of lava fill which would be necessary to account for the mascons is typically 15–20 kilometers. The greatest thickness of mare basalt in multi-ring basins is unlikely to exceed 5 km at most and may be much thinner (Section 6.1.1). Accordingly, although the mare basalt fill does contribute significantly to the excess mass observed, it is inadequate to account for it all.

This rules out ideas that the mascons are due to the accumulation of dense iron-titanium oxides on the bottom of large lava lakes during cooling and crystallization of the lunar basalts. This suggestion encounters the difficulty that there is only limited evidence of near-surface fractionation or sinking of such phases (see Section 6.6.2). Possibly such a mechanism might operate where the lavas are ponded, as in the flooded craters. Such ponding might also occur during the first stages of filling of the impact basins, but there is little evidence of such processes even in such apparently favorable sites as the Taurus-Littrow Valley (Apollo 17) or at the Apollo 15 site (see Section 6.1.3).

The possibility that the mascons are due mainly to the transformation of the lunar basalts to the denser rock, eclogite, at the base of the circular maria is also very unlikely. This transformation, at shallow depths and low temperatures depends critically on the kinetics of the reactions, but the depth of mare basalt fill (< 5 km) is nowhere sufficient to induce this transformation.

One mechanism for the production of the mascons proposes that the ejecta of the ringed basins blankets the surrounding highlands with a low thermal conductivity blanket [21]. This induces melting, and the lavas so generated move laterally to fill the fractured ringed basin. The small amount of fill in Mare Orientale is thought to be due to the excavation of the basin "late in the history of the Moon" [21]. Three objections may be noted: (a) old unflooded basins have been discerned on the far side, (b) the thicker crust on the far side should have caused higher rises in temperature and extensive outpouring of lavas, and (c) the lavas are not thick enough anyway.

The early notion that the mascons are the remnants of the colliding objects (large meteorites, asteroids, or primitive earth satellites), which produced the great mare basins by impact, has been rendered unlikely by our greater understanding of impact mechanics. The projectile does not survive [22].

Slow isostatic mantle response, involving upwelling of the dense lunar mantle, has been shown to be unlikely [23], since many very large Imbrian-age craters (e.g., Neper and Humboldt) should also show positive, not zero, Bouguer anomalies. The case of Grimaldi, whose central crater is the same size as Humboldt, has been noted earlier, but it is now known to represent the central portion of a 430 km diameter basin (Fig. 7.3).

The detailed gravity structures over mascons appear to include a central positive anomaly, flanked by negative anomalies, as are apparently observed at Orientale. Clear examples exist for Grimaldi and Mare Serenitatis, which were directly overflown [20, 24]. The gravity anomalies appear to be modelled best by a rather narrow high density pipe. Such central uplift features are common in complex terrestrial craters (Section 3.5). The negative anomaly surrounding the central positive anomaly may be due to locally thickened crust (S. Pullan, pers. comm., 1981), or a residual from the initial depression filled with breccia. This latter possibility seems more reasonable. Accordingly, the evidence suggests that the mascons are principally due to rapid mantle uplift (formed immediately as a consequence of the impact and not due to a long-term isostatic response), followed by an additional component due to mare basalt filling (Fig. 7.3).

The loading of the mare basalts on the crustal surface induces the rather mild form of adjustment observed, in the form of wrinkle ridges inside the mare basin due to compressive stress and the formation of curved grabens on the outer edges of the basins due to tensional stresses [24]. Most of the deformation of the lunar crust which is not attributable to impact processes is due to this basaltic loading on the surface.

From these observations, calculations of the thickness of the elastic lithosphere have been made. Some estimates call for a relatively thin (50 ± 25 km) layer at 3.6–3.8 aeons, thickening to about 100 km after 3.6 aeons [25]. Such small thicknesses would impose important constraints on lunar thermal models. Graben type rilles are observed only on mare surfaces older than 3.6 aeons, but wrinkle ridges persist to younger periods [26]. This evidence is often interpreted as the consequence of a global stress system, evolving from a relatively tensional to a compressive system. The very flat floors of the mare lavas in the basins (e.g., Serenitatis) may indicate that this very mild tectonic regime is local rather than of moon-wide extent. Theoretical calculations employing sophisticated models indicate that the low values of 100 km or so for the thickness of the elastic lithosphere may be serious underestimates and that the stress-bearing layer may be much thicker [27]. Accordingly, the use of these small lunar surface tectonic features to deduce lunar thermal history does not provide tight constraints [27]. A final comment on mascons relates to the possibility of bringing mantle material within reach of sampling. Successive impacts may bring deep material near to the surface. Possibly the observation that the central peak of Copernicus (Chapter 3) may be composed of olivine can be explained by such processes. Perhaps also some of the Mg-suite rocks owe their origin to such uplifts.

7.4.3 Mars

Martian gravity is dominated by the Tharsis plateau [2, 16] (Fig. 7.4a). Most of the other Martian topography appears to be in a state of isostatic compensation. A basic geophysical question is how the excess mass represented by the Tharsis Plateau is supported. The age of the plateau from crater counting is possibly 3 aeons. If this load is supported statically, a compensation depth (i.e., thickness of the lithosphere) is about 300 km. An alternative suggestion is that the plateau is supported dynamically, possibly by convection. The Tharsis Plateau has the most anomalous gravity features of all the terrestrial planets.

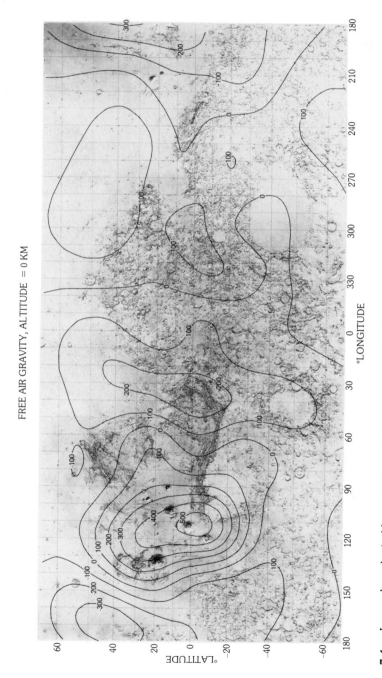

FREE AIR GRAVITY, ALTITUDE = 0 KM

7.4a Low-order spherical harmonic free-air gravity field for Mars showing the dominance of the Tharsis anomaly. Contour values in mgals. (Courtesy R. J. Phillips.)

The large volcano, Olympus Mons, may be uncompensated [28], but this is uncertain. The other large shield volcanoes may be locally compensated [29]. Olympus Mons should exert tensional stresses of the order of one kilobar, for a lithospheric thickness of 200 km. Possibly there is a local thick lithosphere under Olympus Mons [30, 31] and possibly there is some subsidence, the evidence for which is concealed beneath the younger flows which drape the scarp around the foot of the mountain. Clearly many more data for Mars are required, particularly on the elevation of the Tharsis region. Recent radar data support the concept of a trough-like subsidence around Olympus Mons (P. Mouginis-Mark, pers. comm., 1981) and indicate that the Tharsis Plateau relief is some 2–5 km lower than previously thought.

7.4.4   Venus

The Venus gravity field shows a strong positive correlation with topography (Fig. 7.4b) in contrast to the Earth. A very thick lithosphere (i.e., a low temperature gradient) would be required to support these loads statically if the surface is ancient. If the circular features on the plains are impact craters, then the mean surface age is of the order of 2 billion years [31]. If the radioactive element concentrations in Venus approximate those of the Earth, then it appears unlikely that such topography can be supported against creep for such periods. Accordingly, dynamic support of the topography, in particular the high standing masses of Ishtar Terra and Aphrodite Terra, is called for.

7.5   Seismology

7.5.1   The Lunar Record

The lunar seismic signals have very low attenuation [32] and a large degree of wave scattering [33–34]. The scattering occurs mostly in the upper few hundred meters, which is highly heterogeneous, but significant scattering extends to depths of 10–20 km and is due to the extensively fractured nature of the crust. The lunar crust appears to be heterogeneous at all size ranges up to several kilometers. The absence of water and volatile constituents allows the seismic signals to propagate with little or no attenuation [34, 35].

The observed moonquakes are all less than about three on the Richter scale and so compare with only the smallest of terrestrial earthquakes [33]. The number recorded per year is less than 3000. The total energy release per year by moonquakes is 2×10^{13} ergs, compared to 10^{24}–10^{25} ergs by earthquakes. The total lunar energy release is less than that of about 500 g of TNT. About 10% of them are repetitive, and many sets of nearly identical wave trains have been observed [35]. This important observation means that the

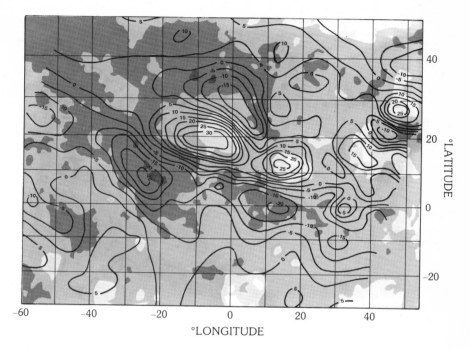

7.4b Line-of-sight gravity measurements in mgals for the central portion of Venus (cf. Fig. 2.25) where the line-of-sight direction for the Pioneer Venus spacecraft was essentially vertical to the surface. Gravity data are superimposed on topography shaded at about 1 km intervals. Lighter shading indicates higher elevation. Note the very clear correlation between gravity and topography for Venus. (Adapted from [16].)

locations and mechanisms are fixed. In addition, it is noted that the deep-focus moonquake activity occurs usually at a fixed phase of the lunar tidal cycle.

Thirty-seven active moonquakes foci have been located. These lie in two belts along great circles. The eastern belt intercepts the western one at about 80° in a T-junction. The western belt lies along one of the rays of Tycho, which also follows a great circle. Most occur at depths of around 1000 km, centered on the lithosphere-asthenosphere boundary. A few occur at depths of less than about 100 km. These so-called high frequency teleseismic events are probably associated with thermal and tidal stresses acting on the deep fractures surrounding the multi-ring basins [36]. As such, they are the only present form of lunar tectonism. These events are the most energetic of the moonquakes.

The Moon is thus almost inert compared with the Earth. Tidal energy, not heat energy as in the Earth, is the driving force for most of the weak lunar seismic events. The thick lunar lithosphere is almost stable, and moonquake activity is concentrated at the transition region between lithosphere and asthenosphere. There is thus a great contrast between the active, unstable thin lithosphere of the Earth, which is driven by an active thermal engine and the stable lithosphere of the Moon, which is an order of magnitude thicker.

The paucity of shallow seismic activity means that the Moon is neither rapidly expanding nor contracting at present. No seismic activity is observed along the lunar rilles within the limits of detection of the lunar seismometers.

7.5.2   The Martian Record

The seismometer on the Viking Lander 2 in Utopia Planitia is mounted on the lander itself (for weight reasons), and so is considerably affected by wind noise. Nevertheless, it has demonstrated that Mars is extremely quiet seismically. The background microseism level is below the detection threshold, and wind noise dominated the data. The instrument operated for 19 months but only 2100 hours (16%) of low noise level data was obtained. One probable natural seismic event of intensity three on the Richter scale, at a distance of 110 km, was detected on SOL 80 in November 1976, three months after the landing. This is probably a local seismic event or meteorite impact, although a wind noise origin cannot be completely excluded [37, 38].

7.6   Internal Structures

7.6.1   The Moon

The structure of the lunar regolith, as revealed by seismic data, has been discussed in Section 4.3. The lunar crust (Section 5.1), which is less dense and chemically distinct from the interior, varies from an average thickness of 64 km on the near side to 86 km on the far side, with an average thickness of 74 km. Recent refinements of the seismic data may indicate a near-side crustal thickness of 50 km (Y. Nakamura, pers. comm., 1981). This would indicate an overall thickness of about 60 km, 10% of lunar volume. The mare basalts, although prominent visually, constitute less than 1% of the crustal volume (0.1% of lunar volume). A 20-km discontinuity within the crust does not reflect the base of the mare basalts, but rather the depth of fracturing or of closure of cracks produced by the large collisions, thus representing the change from fractured to competent rock (Section 5.3). The subsurface structure in the mare basins has been revealed by the Lunar Sounder experiment [39] (see Section 6.1.1).

For depths down to 150 km, information comes mainly from data recorded on the seismic network from the Saturn IV B booster and LM ascent

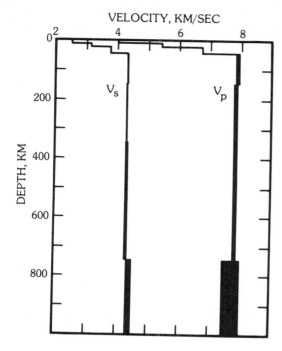

7.5 The crustal and mantle velocity structure of the Moon. The crustal model is derived from artificial impact data, and the mantle is mainly from natural event data. Solid regions represent the uncertainty of the estimated velocities [36]. The most recent interpretation of the seismic data does not allow resolution of the small changes in V_p and V_S in the mantle. The lunar crust may be about 10 km thinner than previous estimates (Y. Nakamura, pers. comm., 1981).

stage impacts. Source to station distances range from 8.7 to 1750 km. Seismic signals from distant meteorite impacts and from moonquakes provide data for regions below 150 km [33–36].

The lithosphere extends to about 1000 km depth [40] (Fig. 7.5). The average P-wave velocity (V_p) is 7.7 km/sec and the average S-wave velocity (V_S) 4.45 km/sec, with negative velocity gradients. There is some inconclusive evidence for lateral heterogeneity [41].

On the basis of the lunar seismic data, Goins and coworkers [40] proposed a model for the structure of the lunar interior. Their preferred model calls for a minor discontinuity, possibly located at a depth of 400–480 km, with a change to lower seismic wave velocities ($V_p = 7.6$ km/sec; $V_S = 4.2$ km/sec) down to depths of about 1000 km. Whether this discontinuity really exists, or whether it is located at a different depth, does not appear to be resolvable from the present seismic data. The model is based mainly on the differences in S-wave velocities and the P-wave data do not show such a marked change. If there is a real change in the V_S values above and below 400–480 km, then many models of the lunar interior structure would satisfy the data. The geochemical and petrological consequences of such a discontinuity, if firmly established, are considerable. It could mark the base of the outer melting zone, and thus might represent a fundamental break between upper fractionated material and lower primitive compositions. Since the lower regions would contain their original complement of radioactive elements, some partial melting might be expected, as well as the presence of denser phases such as garnet. Since the preferred model of Goins and coworkers [40] of a transition zone between 400 and 480 km depth does not appear to be demanded by the seismic data (see discussion by Cleary [42]), it is not adopted here as a constraint. It is clear that the problem will not be resolved like so many others, until seismometers are once again operating on the lunar surface. The geochemical evidence can be interpreted to indicate that melting of the Moon was deeper than 400–480 km (Sections 5.11 and 6.5). If a core exists, then whole-moon differentiation probably was involved and the lunar mantle structure might be complex in detail, as is suggested by the variety of lunar basalts which have been collected. If solid state convection existed until 3 aeons, then the structures revealed by seismic studies might represent only the final condition, and provide no evidence on the earlier structures, as is assumed here.

The mantle P-wave and S-wave velocities continue to about 1100 km without much change [40]. Deep within this zone (800–1000 km depth) occur most of the moonquakes. Below 1100 km, the structure becomes less certain. P and S waves become attenuated ($V_S = 2.5$ km/sec), consistent with the presence of some melt. This marks a major discontinuity, with a very sharp decrease in the S-wave velocities. The evidence came from a single event: impact of a large meteorite (weighing about 1 ton) on the far side of the Moon in July, 1972. P waves were transmitted, with slightly lower velocities through the Moon, but S (shear) waves, which are not transmitted through liquids, were missing, indicating a central zone of a few hundred kilometers radius, below a depth of about 1000 km, with differing properties. Either the shear waves did not propagate at all or were so highly attenuated as not to be recorded.

7.6.2   A Lunar Core?

The possible existence of a lunar core has been much debated. Conclusive evidence for its existence has not been obtained from seismic, magnetic field, or moment-of-inertia data. The value for coefficient of moment of inertia (0.391 ± 0.002) (Table 7.1) is low enough to require a density increase in the deep interior, in addition to a low density crust. It would be satisfied by a core, the presence of dense phases such as garnet, or a more primitive iron-rich silicate interior. A recent reassessment of the deep lunar electrical conductivity profile [43] places an upper limit of 360 km on a metallic core, but the error limits are large. An assessment of the magnetic data from the Apollo subsatellite magnetometer strongly suggests the presence of a central conducting

Table 7.1 Radii, densities and moments of inertia for some solar system planets and satellites.

	R (km)	±	Density g/cm³	±	I/MR²	±
Mercury	2439	2.0	5.435	0.008	—	—
Venus	6051	0.1	5.245	0.005	—	—
Earth	6731	0.005	5.514	0.005	0.3315	0
Moon	1738	0.1	3.344	0.003	0.391	0.002
Mars	3390	1.0	3.934	0.004	0.365	0.005
Phobos	27 × 21 × 19	—	1.9	0.6	—	—
Deimos	15 × 12 × 11	—	—	—	—	—
Io	1816	5	3.55	0.03	—	—
Europa	1536	10	2.99	0.06	—	—
Ganymede	2638	10	1.93	0.02	—	—
Callisto	2410	10	1.83	0.02	—	—
Mimas	195	5	1.2	0.1	—	—
Enceladus	250	10	1.1	0.6	—	—
Tethys	525	10	1.0	0.1	—	—
Dione	560	10	1.4	0.1	—	—
Rhea	765	10	1.3	0.1	—	—
Titan	2572	26	1.9	0.06	—	—
Hyperion	145	20	—	—	—	—
Iapetus	720	20	1.2	0.5	—	—
Vesta	269	—	2.9	0.5	—	—

region of 400–500 km radius [44]. Limits on core sizes vary somewhat from different workers, with estimates of up to 300–400 km radius for a metallic core and 500–600 km for one of FeS [45]. In addition, zoned cores, comprising an inner solid iron core surrounded by a molten FeS core, have been proposed [46]. (For further discussions of lunar cores, see [47, 48].)

In summary, the geophysical evidence for the existence of a core has become slightly stronger, but it is still not demanded by the data. Only further lunar missions will ultimately solve this problem. The view is adopted here that the presence of a core is consistent with the overall evaluation of the geochemical data and is not inconsistent with the geophysical evidence. Several possible models of the lunar interior are shown in Fig. 7.6.

7.6.3 Mars, Venus and Mercury

The internal structures of these three planets are heavily model dependent. An extended discussion is given in Chapter 4 of *Basaltic Volcanism on the Terrestrial Planets* (see Table 8.5 for mantle and core compositions of these planets) and only an outline indicating the various limits is given here.

It is generally agreed that Mercury must possess a metallic core. Estimates vary from 65 to 68 wt.% of the planet, with an overlying silicate mantle.

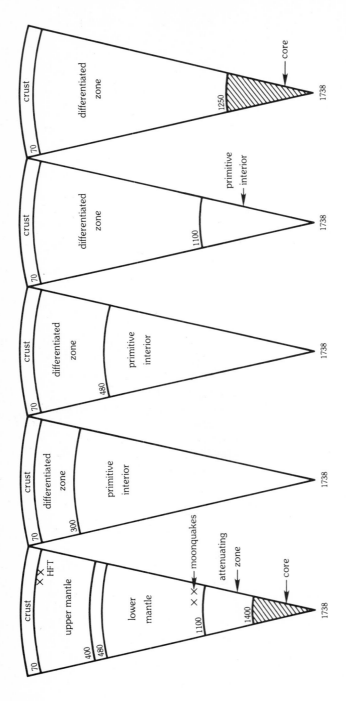

7.6 Five possible internal structural models for the Moon. The model on the left shows an interpretation of the seismic data due to Goins et al. [40]. The division into upper and lower mantle is non-unique [42]. The second model shows a lunar structure due to a shallow depth of melting, leaving a primitive undifferentiated interior [25]. In the third model the depth of initial melting is 480 km, consistent with the Goins et al. [40] interpretation, leaving a primitive undifferentiated interior. The fourth model illustrates an initial depth of melting to the base of the attenuating zone. In the fifth model on the right, the entire Moon has differentiated. A small core has formed and the mantle is comprised of cumulates (olivine, orthopyroxene and clinopyroxene), overlain by the feldspathic crust. This model is preferred as being most consistent with the geochemical data.

The low density and high value for the coefficient of the moment of inertia of Mars compared to the Earth indicate a differing bulk composition (Chapter 8) and internal structure. The geophysical constraints limit mantle densities to between 3.43 and 3.54 g/cm^3. Using reasonable estimates for bulk compositions, core densities are constrained between 5.6 and 7.9 g/cm^3, and the weight percent of cores to about 20%.

The bulk density of Venus is about 2% lower than if it had the same bulk composition as the Earth. Core densities in various models range from 6.7 to 7.9 g/cm^3 and the wt.% of the core from 28 to 32%. Phase transitions in the mantle, as on Earth, are predicted at depths of 450–750 km. Densities and radii of the Galilean and Saturnian satellites are given in Table 7.1.

7.7 Heat Flow and Internal Temperatures

7.7.1 Heat-Flow Data

The lunar heat-flow measurements were less fortunate than any other aspect of the Apollo missions. A heat-flow probe was on board the aborted Apollo 13 mission. It was not carried on Apollo 14. The first measurements were made at the Hadley Apennines site (Apollo 15) on the mare surface. The long awaited result from the highlands site at Descartes was lost due to a broken cable. Some variation with depth was encountered in one probe at the Apollo 17 site, but the heat flow at the bottom of the holes (1.3 and 2.3 m in depth) was similar to that at the Apollo 15 site.

The revised heat-flow values at the Apollo 15 Hadley Rille and Apollo 17 Taurus-Littrow sites are 2.1 and 1.6 $\mu w/cm^2$ [49, 50].

7.7.2 Electrical Conductivity

Lunar electrical conductivity measurements can be made by recording the electromagnetic response of the Moon to the magnetic field variations in the solar wind. It is also possible to invert the electrical conductivity profiles to temperatures at depths within the Moon. The basic information is obtained from the surface magnetometers and orbiting spacecraft. Conversion to a temperature profile is beset with difficulty, since the composition of the interior must be inferred from other evidence.

The first results showed a conductivity "spike" at a depth of 240 km [51], but other work showed that a two-layer model with a boundary at 160 km depth was consistent with the same data, and the spike is not a unique or necessary feature [52]. The early temperature profiles [51] are much too low, due to the fact that olivine containing ferric iron was used for calibration purposes.

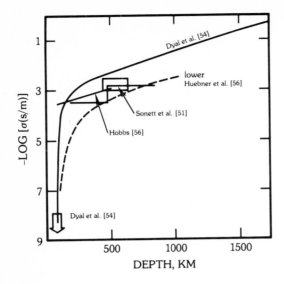

7.7a The variation of lunar electrical conductivity with depth based on electromagnetic sounding [51, 54, 56].

Later values for the conductivity of the lunar interior were provided by Dyal [53, 54], Vanyan [55], Huebner et al. [56], Herbert et al. and Hood et al. [57]. The latest estimate of conductivity with depth is given in Fig. 7.7a, based on these various estimates.

7.7.3   The Lunar Temperature Profile

Various methods are available to estimate the temperature in the lunar interior. These include the following:
(a) Calculations from the heat-flow data, assuming a steady-state thermal model for the upper mantle [50].
(b) Thermal model calculations [58].
(c) Utilizing the stress state calculated from the lunar gravity and topography data [59, 60].
(d) Obtaining electrical conductivity profiles from the magnetic data, and using these to obtain temperatures [54, 56, 61].

Various constraints are available from the mascon data (Section 7.4.2) which imply significant lithospheric strength at 3.8 aeons. Most of the lunar gravity anomalies are near surface. If the internal temperatures are above a certain value, the presently observed topography and gravity anomalies should have vanished in the past 3.8 aeons. Pullan and Lambeck [12] conclude that the "temperature at 300–400 km depth must be 200–300° C less than that proposed by Duba et al." [60]. It should be recalled that all models of load support are dependent on the laboratory data for strain rates in dry olivine. They also conclude that the Moon has undergone little thermal evolution in the past 3 aeons (e.g., [58]), in contrast with other thermal evolution models, which had thin (100 km) lithosphere at 4 aeons. The thickness of the lithosphere at the time of mascon formation was probably about 400 km in order to support the stress differences of 70 bars [61]. It should be recalled that the mare basalts comprise only 0.1% of the volume of the Moon and do not require that the entire interior, or even more than about 1%, approach melting temperatures. Partial melting induced by localized concentrations of the heat-producing elements is adequate to account for the volcanic activity. The large surface area/volume ratio of the Moon indicates that much of the Moon's internal heat has been lost, which would argue for a near steady-state heat flow, and thus for uranium values of up to 45 ppb (Section 8.6). The value of uranium adopted here (Table 8.1) for the bulk moon is 33 ppb, which is a highly conservative estimate. From the various isotopic constraints, it is clear that substantial proportions of the lunar heat producing elements were concentrated near the surface at 4.4 aeons, while the remainder were mainly concentrated in the source regions of the mare basalts. If the Moon was totally differentiated, and possesses a small core, then little K, U and Th is left in the very deep interior. This would also be in accord with low present-day internal temperatures. Thus, Lambeck and Pullan [59] argue for temperatures as low as 800° C at depths of 300 km, compared with values from 900 to 1100° C estimated from the electrical conductivity models. These low temperatures are in accord with the observed low state of lunar activity

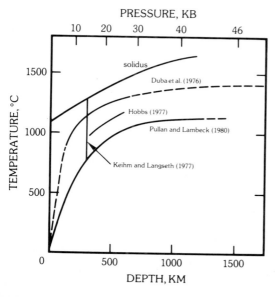

7.7b Lunar temperature profiles with depth, calculated from the conductivity data [56, 60] and by considering the internal stress state [12]. The limits suggested by Keihm and Langseth [50] from the heat flow data are shown.

for the past 3 aeons, but new data will be required to resolve this question. The ranges in internal lunar temperatures from these constraints are given in Fig. 7.7b.

7.8 Magnetic Properties

7.8.1 Lunar Magnetism

The present global lunar magnetic field, as measured by the early Explorer satellites, is vanishingly small, but among the surprises contained in the returned Apollo 11 samples was evidence of a stable natural remanent magnetism (NRM) in the rocks, suggestive of an ancient field, now vanished [62–65].

The surface magnetometer experiment on Apollo 12 measured a remanent magnetic field of 38 gamma [63]. Only 6 gamma was recorded at the Hadley-Apennine site, again on mare basalt. The highland sites, in contrast to the maria, showed evidence of higher fields. At the Apollo 14 Fra Mauro location, values of 43 and 103 gamma were obtained at two separate locations, and 121 to 313 gamma at the Apollo 16 Descartes site. The last value was measured at North Ray crater rim and is the highest recorded on the Moon. These relatively strong surface fields in the highlands affect both the implantation of outgassed ions and the accretion of solar wind particles [64, 65].

A third set of observations of lunar magnetism were made by orbiting magnetometers and charged-particle detectors on the "subsatellites" launched from the Apollo 15 and 16 command modules [66]. The Apollo 15 system operated for 6 months. Local magnetic anomalies on the lunar surface were observed for the first time. Most of the measurements of lunar fields were made during the five days around full-moon when the Moon is in the geomagnetic tail of the Earth; this provides a relatively low noise environment for mapping the lunar surface magnetic fields.

7.8.2 Remanent Magnetism

Magnetic data provide many potential constraints on planetary evolution and history. For example, the presence or absence of a conducting core may be inferred. Hence, the report of a stable remanent magnetism (NRM) in the Apollo samples excited much interest.

Many components of magnetism are present in the lunar rocks. One is a "soft" component, most of which is acquired through exposure to magnetic fields during sample return from the Moon. This was demonstrated by returning a demagnetized Apollo 12 basalt (12002, with a very low "hard" component) to the Moon during the Apollo 16 mission. The sample reacquired its "soft" component, indicating that this was at least in part an artifact of the trip [67].

Metallic iron has been identified as the dominant magnetic material in lunar samples [68]. Most Curie temperatures are around 750–770°C, close to that for pure iron. This is in contrast to the Earth, where the major magnetic materials are the Fe oxides within the system FeO-TiO_2-Fe_2O_3. Most of the lunar iron is very fine and is attributed, because of its low nickel content, to autoreduction in glass spherules formed during impact (see Section 4.5.1). No magnetic effects attributable to troilite have been noted, indicating that troilite is stoichiometric FeS.

The identification of the magnetic phase in lunar fines as metallic iron appears to have been confirmed by electron spin resonance (ESR) techniques [69, 70] (see Section 4.5.1), refuting earlier suggestions that hematite, magnetite, or ferric oxides with magnetite-type lattices were present. The absence of Fe^{3+} in ilmenite and ulvöspinel is indicated by their lack of ferromagnetic properties. Magnetite appears to be absent, except for one possible occurrence in the Apollo 17 orange soil. The magnetic data accordingly support the petrological interpretations of extreme reducing conditions (Section 6.2.2).

A hard component of magnetization in the samples is of lunar origin. The basalts have stable remnant magnetizations with intensities of 1–2×10^{-6} emu/g. Some breccias have much larger intensities, on the order of 1×10^{-4} emu/g, accounting for the observed surface anomalies. This increase is correlated with the content of total metallic iron in the breccias and to changes in grain size. The samples collected on the present surface had random magnetic orientation. Accordingly, they were magnetized before being excavated by meteorite impact and thrown onto the lunar surface.

The magnetic properties measured in the highlands breccias have the following characteristics.

(a) The unwelded breccias resemble the soils. Most of the iron particles are less than 150 Å in diameter, and they have low thermoremanance. The magnetic component in the soils is in the finest-grain size [71].

(b) The intermediate grades of breccias are dominated by single domain iron (150–300 Å grain size) and possess a strong thermoremanance.

(c) The highly crystallized breccias contain multi-domain iron (300 Å grain size) and do not carry a strong remanent magnetism. In this they resemble the igneous rocks from the maria, except that they tend to be magnetically inhomogeneous and contain a high content of metallic iron.

Using these criteria, the anorthositic gabbro 68415 for example is not igneous but a strongly recrystallized breccia, a view in accord with the chemical and age characteristics of the sample. Experiments have shown that annealing at temperatures below 900°C produces the fine-grained iron (100

Å) typical of soils and soil breccias [71]. At higher temperatures, coarser grains up to micron size are produced by autoreduction processes.

This correlation of iron grain size and content with metamorphic grade indicates that the breccias have both higher metallic iron contents than the mare basalts and higher thermoremanant magnetism. These effects occur in recrystallized samples that have been heated in impact events indicating that such processes cause enhancement of the sample magnetization.

7.8.3   Lunar Magnetic Field Paleointensities

Considerable efforts have been made to determine the paleointensity of the vanished field. A basic problem in making lunar paleomagnetic measurements is to avoid inducing chemical changes in the samples during heating experiments. The lunar magnetic memory is carried by metallic iron, much of it in a size range of a few hundred angstroms. Many of the problems encountered in the early stages of magnetic studies on the lunar samples were probably due to alteration of this fine-grained iron in the strongly oxidizing and moist terrestrial environment. Recent sample preparation techniques have been devised so that oxidation or reduction can be avoided during thermal treatment [72].

A review of magnetic paleointensity methods [73] which compares four methods of determining paleointensity concludes that only the Thellier method applied using the new sample preparation method [73] appears to provide reliable estimates. Of the five published results considered to be reliable (Table 7.2), none show any clear evidence of a strong correlation with

Table 7.2 Reliable estimates of lunar magnetic paleointensity.

Sample	Paleointensity (Oersteds)	Method	Age (aeons)	Reference
70019	0.025	Thellier	0.003	1
15498	0.021–0.05	Thellier	<3.3	2
70215	0.02–0.075	Thellier	3.84	3
	0.06	ARM†		3
62235	1.2	Thellier	3.9	4
	1.4	ARM†		3
60255	0.005	Thellier	3.9	5

†ARM = Anhysteretic remanent magnetization.

1. Sugiura, N., et al. (1979) *PLC 10*: 2189.
2. Gose, W. A., et al. (1973) *Moon 5*: 196.
3. Stephenson, A., et al. (1974) *PLC 5*: 2859.
4. Collinson, D. W., et al (1973) *PLC 4*: 2977.
5. Sugiura, N. and Strangway, D. W. (1980) *PLC 11*: 1801. (Table adapted from this source.)

time for an ancient lunar magnetic field. More data are clearly required (Fig. 7.8).

A field intensity of 2500 nT or gamma (= 0.025 Oersteds) was measured for a very young impact glass lining a 3-m crater at the Apollo 17 landing site [74]. The exposure age of the glass is less than 3 million years [75]. The source of the magnetizing field can be neither the extinct dynamo, nor an internal primordial memory, since both sources are no longer extant. Two explanations are possible: (a) there is today a local permanent field of 2500 gamma at the Apollo 17 site, which magnetized the impact glass as it cooled through its Curie point. Unfortunately, no measurements of the magnetic field were carried out [76]. (b) Plasma processes in the impact which produced the glass may have enhanced a much weaker local field [77], but probably a larger impact is needed for this mechanism to operate. Some combination of shock and rapid thermal quenching might account for the observed magnetization. The association of strong magnetic anomalies with features such as the Reiner Gamma swirl [79, 80] which are in some way connected with impact events [81] lends some credence to this view. The surprising result is that fields of the order of a few tenths of an Oersted were present or enhanced during the formation of some of the soil breccias, which often carry multiple weak components of magnetization. These remagnetizations have widely scattered directions, suggesting that multiple events such as major impacts are responsible for their features.

7.8.4 Crustal Magnetic Anomalies

Observations both by surface-based magnetometers and by those aboard the subsatellites revealed the moon-wide nature of crustal magnetic anomalies. Some correlations with surface geological features have been observed. Many of the anomalies appear to be correlated with ejecta blankets from the

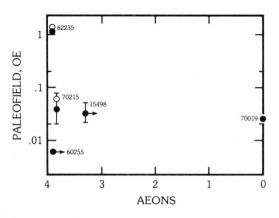

7.8 Ancient lunar magnetic field intensities, plotted as a function of time. Sample 70019 is a young impact glass formed about 3 m.y. ago [73, 74].

multi-ring basins [76]. The youngest identifiable area with an anomaly is the swirl-like feature Reiner Gamma, dated stratigraphically at 100 million years [81] but it may be as old as 2900 million years [79].

A similar correlation is reported with parts of the Fra Mauro Formation, which appears to be magnetized very heterogeneously, in contrast to the rather uniform magnetization observed at Reiner Gamma. Other correlations of surface magnetic anomalies associated with swirls have been reported [80]. Whether these are associated with impact (the swirls show a tendency to be antipodal to large multi-ring basins) remains conjectural [81] (see Section 3.15). This surface field variability seems consistent with impact-induced enhancement in a pre-existing weak ambient field. Temperatures above the Curie temperature of iron (750°C) must have been reached. A possible scenario is that autoreduction of iron oxides takes place during impact glass formation. The pre-existing field is enhanced momentarily during the impact event. Cooling of the glass and the tiny iron spherules occurs rapidly enough for this field to be recorded.

7.8.5   Origin of the Lunar Magnetic Field

A strong moon-wide field, with intensities of up to 1 Oersted, appears to have been in existence at about 4 aeons, but is no longer present.

A wide variety of hypotheses have been offered to account for this vanished phenomena. Just as it is difficult to infer the form of the Cheshire Cat from its smile, so it is equally difficult to produce a credible image of the nature of the magnetic field. The only consensus appears to be that the lunar rocks became magnetized both by weak (< 20 kbar) shock processes and as they cooled through the Curie temperature for metallic iron (780°C) in the presence of a magnetic field.

The resolution of the magnetic debate is of interest to all lunar scientists. If a core is shown to be present, it has considerable geochemical as well as geophysical consequences, implying probable total melting of the Moon, and providing a possible sink for the siderophile elements.

The origins proposed for the lunar magnetic field encompass a large number of processes [82] that fall within the following categories: (a) random fields due to localized near-surface origin; (b) external uniform fields; (c) internal dipole field. Some comments on various proposed models follow.

Impact-generated fields.
Many suggestions have been made that the magnetic effects are in some manner connected with impact processes. A review of these has been given by Fuller [64]. The large scale and number of impact events lends some credence to these theories, but experimental evidence for strong fields in impacts is lacking [83].

Raisin model.

In this variation, small pockets of molten iron collect during differentiation, and act as small dynamos [84]. A variation of this would be the interaction of molten magma with the solar wind [85]. Again, one faces the problem of maintaining such fields over long periods. They do not explain the magnetization of the breccias. Most of the mare basalts were extruded as thin flows so that accumulation of molten iron pockets or the formation of deep lava lakes (except very locally) is petrologically very unlikely.

Earth's field.

The present magnetic field of the Earth at the Roché limit (2.4 earth radii) is 4000 gamma. If the Moon had been just outside the Roché limit, the appropriate field would be available. A major difficulty lies in keeping the Moon at such a close distance for more than 1 aeon. Very large tidal effects would have been present during the extrusion of the mare basalts. There seems to be no evidence of this or other features indicating such a close approach.

Enhanced solar wind field.

It is possible that the solar wind field, at present about 10 gamma, was 100 times more intense [86]. While such strong fields may exist in early stages of stellar evolution, it would be necessary to maintain this high value for more than 1 aeon, with a steady component along the axis of rotation of the Moon. This seems unlikely. Such activity is unlikely to occur from a main sequence star and persist for this period of time, when all geological evidence points toward a sun similar to the present-day one at least during the period 3.5–3.0 aeons (Chapter 9). A major difficulty with this and with all external field theories is that it is very difficult to provide a steady field in a single direction over expected cooling times.

Permanent magnet model.

For more than 300 years, the only explanation of the Earth's magnetic field was that it was due to permanent magnetism. This theory eventually failed to account for rapidly varying terrestrial field changes and many other features, and was discarded and effectively forgotten until recently resurrected for the Moon [87]. The model depends on accretion of the Moon from a sphere of dust and gas in which the solids settled. As contraction occurs, the temperature rises, and eventually the outer solid layers melt. High magnetic fields will occur in the contracting gas sphere, perhaps due to an enhanced solar field; the remnant magnetization of some meteorites may be evidence of this. A variation of this hypothesis supposes that the Moon became magnetized by an early field of perhaps 20 gauss, either through close approach to the Earth or because of enhanced solar wind reaction. The cold inner portion of the Moon (below 780°C) is magnetized while the exterior is hot. This would most likely be at 4.6 aeons or shortly thereafter. Subsequently, the crust cools and becomes magnetized. Next the interior heats above 780°C and erases the internal field, so that no field is now observed [88].

This is an ingenious suggestion, although it is linked to a specific model of lunar formation, not easily reconciled with the evidence in favor of the planetesimal hypothesis, discussed in Chapter 9.

Fluid core model.

This model deserves close scrutiny since the origin of the Earth's magnetic field is due to the dynamo action in the Ni-Fe conducting core [63–65]. The geophysical data (Section 7.6.2) permit this process, but restrict core radii to less than 400–600 km. Whether such a small core (1–4% of lunar volume) could produce a dynamo has also been debated.

The imprint of a magnetic dipole field on the lunar crust as it cooled through the Curie point would imply the existence of a very early field, but such a distribution is not seen [89–90]. Even if only a weak field were present, it could be pumped up locally by impact-induced magnetism, which can operate provided an ambient field is present [63–65]. Such a mechanism could account for the random and highly variable fields observed at present on the surface (Section 7.8.1). In retrospect, the study of lunar magnetism has provided much stimulus to scientific thinking, as is attested by the volume of literature on the topic. Great contrasts exist with the terrestrial magnetic signatures, carried by oxide minerals in contrast to the very fine-grained iron in the lunar samples. The association of the magnetic anomalies with breccia sheets on the Moon is instructive. Fine-grained iron formed by reduction in impact glasses is preserving a record of perhaps very transient fields. The random orientations observed on kilometer scales are presumably due both to the ejection process and to subsequent smaller cratering events.

7.8.6 Planetary Magnetic Fields

The information obtained from observations of magnetic fields on other planets can help shed light on the problem of the origin of the lunar field.

Mercury has a small but measurable field, with a field strength about one percent of that of the Earth [91], but strong enough to deflect the solar wind and to form a detached bow shock wave. The magnetic dipole moment is 6×10^{-4} that of the Earth. As is the case for the Moon, the origin of this field is uncertain. Active dynamos and remanent fields have both been suggested. The field is clearly of internal origin. It is too large to be caused by a present-day induction process. If it is due to a permanently magnetized crust resulting from an ancient external field, then the crust must have cooled through the Curie point in the presence of this field. However, the lengthy time taken for this cooling rules out unidirectional external fields. If a former

Table 7.3 Planetary magnetic dipole moments.[†]

Planet	Core Radius, km	Spin Rate, day^{-1}	Dipole Moment G/cm^3
Mercury	~1800	0.017	~4×10^{22}
Venus	~3000	0.0041	<1×10^{22}
Earth	3486	1.003	8×10^{25}
Mars	~1700	0.975	≤ 2.5×10^{22}
Jupiter	~52,000	2.44	1.55×10^{30}
Saturn	~28,000	2.39	5×10^{28}

[†]Adapted from Russell, C. T. (1980) *Rev. Geophys. Space Phys.* 18: 101.

internal dipole field *were* responsible, no external field should be present, since a dipolar-magnetized shell in theory does not produce an external magnetic field [89, 90]. Of course, craters and multi-ring basins produce anomalies in this ideal situation, but the strength of the field appears high enough to make the operation of a presently active dynamo, and hence a present fluid core, a distinct possibility [64].

The probable radius of the Mercurian core is about 1800 km. Mars is predicted to have a core of about the same size. The rotation period of Mars is close to that of the Earth, whereas the rotation period of Mercury is 59 days. Accordingly, conditions for the existence of a currently active dynamo on Mars should be more favorable than on Mercury but there is considerable controversy over the size of the magnetic moment of Mars, or whether it is, in fact, detectable at present [64, 92]. The solar wind apparently impinges on the Martian ionosphere.

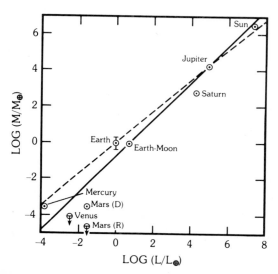

7.9 The magnetic moment of the planets and of the sun plotted against angular momentum. Both quantities are normalized to Earth. This is a magnetic version of Bode's Law [63]. Use of both the Earth and the Earth-Moon angular momentum results in two possible relationships.

Venus, which is a little smaller than the Earth, is thought to possess a core of about the same size. The rotation is retrograde with a period of 243 days, about four times that of Mercury. The large size of the core should produce a weak, but measurable field. No field has been detected, providing yet another instance of the difference between Venus and the Earth. Possibly the motions in the core, if it is liquid, are not sufficient to drive a dynamo. The absence of the magnetic field allows the solar wind to interact with the ionosphere, as is the situation with respect to Mars. It would be of interest to discover how long the surfaces of these two planets have been exposed to the solar wind. Figure 7.9 shows the correlation between angular momentum and magnetic moments of the sun and the planets.

Early magnetic fields recorded in the meteorites [93–97] are not discussed in this review. It should be noted that attempts to relate the lunar field to the presumed interplanetary field, responsible for the fossil magnetism observed in meteorites, may be misleading. Among other caveats, the fields responsible for the magnetic effects in meteorites were in existence about a billion years earlier than those responsible for the lunar magnetism [96]. Whether strong early magnetic fields were present is still not settled from the meteoritic evidence.

References and Notes

1. Coleman, P. J. (1979) *Science.* 205: 1082.
2. Lambeck, K. (1979) *JGR.* 84: 6241.
3. Bills, B. G., and Ferrari, A. J. (1977) *Icarus.* 31: 244.
4. El-Baz, F. (1973) *Science.* 180: 1173.
5. Mohan, S. N. (1979) *Icarus.* 38: 317.
6. Phillips, R. J., et al. (1981) *Science.* 212: 879.
7. Ananda, M. P., et al. (1977) *Moon.* 17: 101.
8. Kaula, W. M. (1979) *GRL.* 6: 191.
9. *Basaltic Volcanism* (1981) Section 4.2.1.
10. Kaula, W. M., et al. (1972) *PLC 3:* 2189; (1974) *PLC 5:* 3049.
11. Core offset was suggested by Ransford, G., and Sjogren, W. L. (1972) *Nature.* 238: 260. Kobrick, M. (1976) *Moon.* 15: 83, suggested random density anomalies would explain the effect.
12. Pullan, S., and Lambeck, K. (1980) *PLC 11:* 2031.
13. Wasson, J. T., and Warren, P. H. (1980) LPS XI: 1220.
14. Haines, E. L., and Metzger, A. E. (1980) *PLC 11:* 689.
15. Wood, J. A. (1973) *Moon.* 8: 73.
16. Phillips, R. J., and Lambeck, K. (1980) *Rev. Geophys. Space Phys.* 18: 27.
17. Muller, P. M., and Sjogren, W. L. (1968) *Science.* 161: 680.
18. Ferrari, A. J., et al. (1978) *JGR.* 83: 2863; compare Sjogren, W. L., et al. (1974) *Moon.* 11: 41.
19. Bills, B. G., and Ferrari, A. J. (1980) *JGR.* 85: 1013; Ferrari, A. J. (1975) *Science.* 188: 1297.
20. Phillips, R. J., and Dvorak, J. (1981) *Multi-ring Basins,* p. 91.

21. Arkani-Hamed, J. (1973) *PLC 4*: 2673.
22. See Chapter 3.
23. Dvorak, J., and Phillips, R. J. (1978) *PLC 9*: 3651.
24. Muller, P. M., et al. (1974) *Moon*. 10: 195; Scott, D. H. (1974) *PLC 5*: 3025.
25. Melosh, J. (1978) *PLC 9*: 3513; Solomon, S. C., and Head, J. W. (1979) *JGR*. 84: 1667; (1980) *Rev. Geophys. Space Phys*. 18: 107.
26. Lucchitta, B. K., and Watkins, J. A. (1978) *PLC 9*: 3459.
27. Pullan, S., and Lambeck, K. (1981) *PLC 12*: 853; Golombek, M. P. (1979) *JGR* 84: 4651.
28. Blasius, K. R., and Cutts, J. A. (1981) *Icarus*. 45: 87.
29. Smith, J. E., et al. (1980) NASA TM 81776, 79.
30. Comer, R. P., and Solomon, S. C. (1981) LPS XII: 166.
31. *Basaltic Volcanism* (1981) Section 4.5.5.
32. This corresponds to high values of the seismic parameter Q, which has typical values in the range 10–300 for the Earth, but is an order of magnitude higher (3000–5000) for the Moon. Values up to 5000 are given for the upper 5 km. See Tittmann, B. R., et al. (1979) *PLC 10*: 2131.
33. Latham, G., et al. (1973) *PLC 4*: 2518.
34. Kovach, R. L., and Watkins, J. S. (1973) *PLC 4*: 2550; Watkins, J. S., and Kovach, R. L. (1973) *PLC 4*: 2561.
35. Dainty, A. M., et al. (1974)*PLC 5*: 3091; Toksöz, M. N., et al. (1973) *PLC 4*: 2529; (1977) *Science*. 196: 979.
36. Nakamura, Y., et al. (1974) *PLC 5*: 2883; (1978) *PLC 9*: 3589; (1979) *PLC 10*: 2299.
37. Lazarewicz, A. R., et al. (1981) NASA Contractor Report 3408; Goins, N. E., and Lazarewicz, A. R. (1979) *GRL*. 6: 368.
38. Anderson, D. L., et al. (1977) *JGR*. 82: 4524.
39. Peeples, W. J., et al. (1978) *JGR*. 83: 3459.
40. Goins, N. R., et al. (1979) *PLC 10*: 2421; (1981) *JGR*. 86: 5061.
41. Nakamura, Y., et al. (1977) *PLC 8*: 487; *JGR*. 81: 4818.
42. Cleary, J. R. (1982) *JGR*. In press.
43. Hood, L. L., et al. (1981) LPS XII: 457.
44. Russell, C. J., et al. (1981) LPS XII: 914.
45. Solomon, S. C. (1979) *PEPI*. 19: 168.
46. Stevenson, D. J., and Yoder, C. F. (1981) LPS XII: 1043.
47. Brett, R. (1973) *GCA*. 37: 165; Goldstein, B. E., and Phillips, R. J. (1976) *PLC 7*: 3321.
48. Levin, B. J. (1979) *PLC 10*: 2321.
49. Langseth, M. G., et al. (1976) *PLC 7*: 3143.
50. Keihm, S. J., and Langseth, M. G. (1977) *PLC 8*: 499.
51. Sonett, C. P., et al. (1971) *Nature*. 230: 359; (1972) *PLC 3*: 2309.
52. Dyal, P., and Parkin, C. W. (1973) *PEPI*. 7: 251.
53. Dyal, P., et al. (1974) *PLC 5*: 3059.
54. Dyal, P., et. al. (1976) *PLC 7*: 3077; (1977) *PLC 8*: 767.
55. Vanyan, P. (1980) *Geophys. Surveys*. 4: 173.
56. Huebner, J. S., et al. (1979)*JGR. 84*: 4652; Hobbs, B. A. (1977) *Geophys. J. Roy Astr. Soc*. 51: 727.
57. Herbert, F., et al. (1981) LPS XII: 436; Hood, L. L., et al. (1981) LPS XII: 457.
58. Toksöz, M. N., et al. (1978) *Moon and Planets*. 18: 281.
59. Lambeck, K., and Pullan, S. (1980) *PEPI*. 22: 12.
60. Duba, A., et al. (1976) *PLC 7*: 3173.
61. Arkani-Hamed, J. (1974) *PLC 5*: 3127.

62. A report of a conference on planetary magnetism will be found in *PEPI*. 20: 75–384 (1979).
63. An excellent review of planetary magnetism is by Russell, C. T. (1980) *Rev. Geophys. Space Phys*. 18: 77.
64. Reviews of lunar magnetic measurements are given by Fuller, M. (1974) *Rev. Geophys. Space Phys*. 12: 23; Dyal, P., et al. (1974) *Rev. Geophys. Space Phys*. 12: 568.
65. Dyal, P. (1972) NASA SP 315, 11-l.
66. Russell, C. T., et al. (1973) *PLC 4*: 2833.
67. Pearce, G. W., and Strangway, D. W. (1972) NASA SP 315, 7–55.
68. Pearce, G. W., et al. (1973) *PLC 4*: 3045.
69. Tsay, F., et al. (1973) *GCA*. 37: 1201.
70. Morris, R. V. (1980) *PLC 11*: 1697.
71. Usselman, T. M., and Pearce, G. W. (1974) *PLC 5*: 597.
72. Taylor, L. A. (1979) *PLC 10*: 2183.
73. Sugiura, N., and Strangway, D. W. (1980) *PLC 11*: 1801.
74. Sugiura, N., et al. (1979) *PLC 10*: 2189.
75. Yokoyama, U., et al. (1974) *PLC 5*: 2231.
76. Hood, L. L., et al. (1979) *Science*. 204: 53.
77. Srnka, L. J. (1977) *PLC 8*: 785.
78. Cisowski, S. M., et al. (1977) *PLC 8*: 725, Fig. 7.
79. Hood, L. L. (1980) *PLC 11*: 1879.
80. Hood, L. L. (1981) LPS XII: 454.
81. Schultz, P. H., and Srnka, L. J. (1980) *Nature*. 284: 22.
82. Daily, W. D., and Dyal, P. (1979) *PEPI*. 20: 255.
83. Srnka, L. J. (1977) *PLC 8*: 785.
84. Murthy, V. R., and Bannerjee, S. K. (1973) *Moon*. 7: 149.
85. Nagata, T., et al. (1971) *PLC 2*: 2461.
86. Nagata, T., et al. (1972) *PLC 3*: 2423.
87. Runcorn, S. K., and Urey, H. C. (1973) *Science*. 180: 633.
88. Strangway, D. W., et al. (1973) LS IV: 697.
89. Runcorn, S. K. (1975) *PEPI*. 10: 327.
90. Runcorn, S. K. (1979) *Science*. 199: 771.
91. Ness, N. F. (1979) *PEPI*. 20: 209.
92. Russell, C. T. (1979) *PEPI*. 20: 237.
93. Nagata, T. (1979) *PEPI*. 20: 324.
94. Sugiura, N., et al. (1979) *PEPI*. 20: 342.
95. Brecher, A., and Fuhrman, M. (1979) *PEPI*. 20: 350.
96. Brecher, A., and Leung, L. (1979) *PEPI*. 20: 361.
97. Strangway, D. W. (1978) *Moon and Planets*. 18: 273.

Chapter 8

THE CHEMICAL COMPOSITION
OF THE PLANETS

8.1 The New Solar System

Two books set the stage for the scientific exploration of the solar system. The first was *The Face of the Moon*, by R. B. Baldwin (Chicago, 1949) who correctly identified the lunar craters as impact produced, and the maria as filled with basaltic lava flows. The second, *The Planets*, by Harold C. Urey (Yale, 1952), introduced the new chemical and isotopic constraints, which provided the first really new synthesis of data and intellectual stimulus to thinking about the solar system since the appearance of "*Exposition du Systeme du Monde*" by Laplace in 1796.

In 1982, following two decades of intensive exploration of the solar system from Mercury to Saturn, we have acquired much knowledge of the chemical composition of the planets and satellites, for which the groundwork was laid by Urey. In previous chapters, the new knowledge of the geology and stratigraphy of the Moon, planets, and satellites, the nature of their surfaces, crusts, lavas and interiors has been probed in detail. In this chapter, the problem of their bulk composition is addressed. This depends on an assessment of all the preceding information, so that such planetary compositions are consistent with the details of the analyzed material and the inferences from geophysical experiments and remote-sensing data. Once this stage is reached, then the planetary chemistry can be used as a basis for discussions about conditions during formation of the individual planets, and to constrain ideas about the composition of the solar nebula, and the origin of the solar system (see Chapter 9). The close correspondence between the solar and C1 abundances provides a starting point to constrain solar nebula compositions (Fig. 8.1).

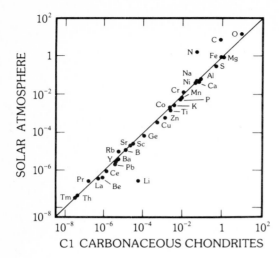

8.1 The abundances of elements in the sun compared with those in the Type I carbonaceous chondrites (C1). Values are in terms of atoms per Si atom. This figure shows the close correspondence in composition for a wide variety of elements. [After *Basaltic Volcanism* (1981) Fig. 4.3.1.]

As in the other portions of this book, the Moon serves as an invaluable reference point because of the wealth of data available to us, mainly from the Apollo missions. The Earth is dealt with a little more fully in this section, since the chemical composition of the Earth is a vital factor in interplanetary comparisons and for theories of origin of the solar system, although the bulk composition of the Earth has only recently become reasonably well understood.

Attempts to compare chemical compositions among the inner planets are complicated by many factors. The surfaces or outer regions from which samples or remote-sensing data are available generally have undergone a variety of element fractionating processes. These may include:

(a) Pre-accretion element fractionation. The widespread variations in Mg/Si ratios among meteorites and in K/U ratios among planetary bodies are typical examples.

(b) Post-accretion planet-wide melting. This could involve homogenization of a heterogeneously accreted planet, core formation, and early crust formation. The Moon serves as the best example of the latter process, where the highland crust, enriched in LIL elements, dates from this stage. The evolution of Mercury appears to form another example.

(c) Meteoritic bombardment, which was endemic throughout the solar system from Mercury to Callisto, and which lasted on the Moon until 3.8 aeons. This caused mixing and overturning of early crusts and the addition of meteoritic debris. The chondritic ratios for trace siderophile elements (Ir, Os, Pd) in the terrestrial upper mantle may be a relic of this event [1].

(d) Subsequent production of lavas from planetary interiors, which will transfer incompatible elements and low melting point components toward the planetary surface. This process, caused by the buildup of radiogenic heat in planetary interiors, and which produces partial melts, must be clearly distinguished from initial differentiation. The lunar crust, as an example, was formed by initial melting, but the mare basalts were produced by partial melting in the solid lunar interior as a consequence of secondary heating by K, U and Th. The terrestrial basalts provide another example of this secondary differentiation process. This has been carried a stage further on the Earth by generation of the continental crust through multi-stage melting processes. These have resulted in large near-surface enrichment in LIL elements, with concomitant large-scale depletion of the mantle in elements such as Cs, Rb, K, Ba, and the light REE.

(e) Other factors which cannot be addressed here may include interactions with atmospheres, hydrospheres and biospheres.

Various geophysical constraints on planetary compositions have been discussed in Chapter 7, and include many vital parameters. The basic information of planetary radii and densities is reinforced by knowledge of the moment of inertia, which provides crucial information about the distribution of density with depth. The seismological data enables models to be constructed for planetary interiors while heat flow data places some constraints on the abundance of the radioactive heat-producing elements within a planet. This information, freely used by geochemists up until three or four years ago to equate uranium content with heat flux, depends heavily upon whether the heat flow is in equilibrium with production. This is almost certainly not so on the Earth [2, 3], but the Moon may more nearly approach this equilibrium situation [4]. The heat production estimates provide limits on lunar uranium abundances of 30–50 ppb [5] or 33–46 ppb [6].

Several problems remain. The thickness of the lunar crust is a critical parameter in assessing bulk lunar compositions, since it constitutes over 10% of the lunar volume. The temperature profile in the interior of the Moon is another important property, which is still not completely agreed upon [7] (Section 7.7.3).

The question of planetary cores, central to the problem of magnetism and to abundances of the siderophile elements, is assessed for the Moon by a combination of seismic data, moment-of-inertia values, and magnetic measurements. The geophysical data, although suggestive, are not decisive in answering this question. This situation serves to remind us just how much information is needed to understand planetary bodies.

One of the principal geochemical constraints on models of planetary compositions comes from the oxygen isotopic data [8–11]. This was perhaps first realized from the analysis of oxygen isotopes on the Moon, where extreme uniformity was observed. No signs of isotopic fractionation or exchange below crystallization temperatures appear to have been observed, and it was concluded that "the $^{18}O/^{16}O$ ratio is the best known geochemical parameter on the moon" [12]. These ratios were distinct from those observed in high-temperature refractory Ca-Al phases, for example, in the Allende meteorite. This was an early piece of evidence which was not favorable to theories for lunar origin that proposed to accrete such refractory Ca-Al phases. Even the lunar dunite (72417) and the norite (78235) from Apollo 17 were not distinct. The $\delta^{18}O$ values were similar to those of the other lunar samples (average about +6) so that moon-wide uniformity for the oxygen isotopic composition can be assumed. If the Moon was accreted from a collection of heterogeneous planetesimals, then it was homogenized following accretion. The evidence for probable whole-moon melting (Chapters 5, 6, 7) has made this idea untestable.

However, it is from a plot of $\delta^{17}O$ versus $\delta^{18}O$ that most insight has been gained (Fig. 8.2). The Earth, Moon and the differentiated meteorites all fall close to a single fractionation line. The H, L, LL and C2 and C3 chondrites are displaced from this line and so can be excluded as representing major primary planetary compositions. Various mixtures of these components can be made to match, for example, the terrestrial oxygen isotope composition [12]. The oxygen isotope data might thus constitute evidence for heterogeneous accretion hypotheses for planetary origin, providing that homogenization occurred following accretion. Our present understanding indicates that the compositions of planets and satellites depart substantially from the basic Type 1 carbonaceous chondrite composition thought to be representative of the solar nebula. The meteoritic and lunar abundance data suggest that the elements are partitioned among these bodies in groups, based on differences in volatility, and in siderophile and chalcophile properties [13]. These differences in elemental properties appear to be fundamental to an understanding of bulk planetary compositions. They are distinguished from the elemental properties which are important in crystal-liquid fractionation processes (see for example Sections 5.9, 5.11.3, 6.3 and 6.5) and many of the elements which behave similarly in magmatic crystallization processes (e.g., K, U) are widely separated when fractionation depends on relative volatility.

The refractory elements (see Appendix VII) as usually defined comprise those elements which condense from a gas of solar nebula composition at pressures of 10^{-4} atm and temperatures above $1400°$ C [14]. This classification, based on theoretical nebula condensation temperatures, should not be taken to imply either an initially hot nebula, or a simple condensation sequence (see Section 9.2). The group of refractory elements includes Ca, Al and Ti among the major elements and 34 other elements including Ir, Ba, U, Th, V (see Appendix VII for a complete list). This group thus includes many important and significant elements, of which the lithophile members (excluding the

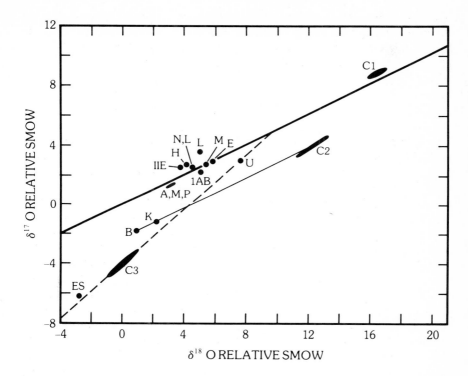

8.2 Three-isotope plot showing variations in oxygen isotope abundances. (See Chapter 6, reference [95] for notation.) The diagonal line beginning at –2 is a mass-fractionation line defined by terrestrial samples. Note that values for the Moon and differentiated meteorites fall on or close to this line. Note that values for C1 carbonaceous chondrites fall off this line, as do those for the ordinary chondrites (H, L) [9, 10]. C1, C2, C3 = carbonaceous chondrites. En = enstatite chondrites and aubrites. A, M, P = eucrites, howardites, diogenites, mesosiderites and pallasites. N, L = Nakhla and Lafayette. U = ureilites. IAB, IIE = irons. B = Bencubbin, Weatherford. ES = Eagle Station, Itzawisis. M = Moon. K = Kakangari. (Courtesy R. N. Clayton.)

siderophiles) tend to form a coherent set of elements (e.g., Ca/Al/U ratios [14]).

A second group of moderately refractory elements includes Cr, Mg, Si and Li. The important cosmochemical consequence is that Mg and Si may become decoupled from the other major elements Ca and Al.

The volatile elements include the alkali elements Na, K, Rb, as well as Cu, Ge, Ga. Elements classified as highly volatile include a large number, most of which are chalcophile (Sb, Se, Te, Zn, In, Cd, Bi, Tl, for example). The fact that chalcophile elements are also volatile has made it difficult to disentangle the cosmochemical consequences of these two properties. In contrast, the siderophile elements cover a wide range in volatility, from Os to Ag. The group includes Fe, Ni, Co, Os, Ir, Re, Pd, Au, (Ge) and Ag.

8.2   The Type 1 Carbonaceous Chondrites (C1)

A widely held assumption is that the elemental abundances (excluding H, the noble gases, and N) in carbonaceous chondrites approximate those in the primitive solar nebula. The principal philosophical support for this assumption is the close correspondence between the abundances observed in the solar photosphere [15] and the abundances in Type 1 carbonaceous chondrites (Fig. 8.1)[16]. Generally, the abundances of well-determined elements in the sun, such as Si, Mg, Al, S, Ba, Ca and Na, match the C1 abundances better than those of the other carbonaceous chondrites and better than those of other chondrite meteorite groups [17]. D, Li and Be [18] are destroyed in nuclear reactions within the sun, so that the solar Li abundance is only 1% of the meteoritic values. This deficiency, of course, constitutes evidence against theories which derive planetary material from the sun after its formation [19].

The evidence for broad homogeneity in the primitive nebula is not precluded by the isotopic evidence for local heterogeneity (e.g., [11]) which indicates separate ^{16}O reservoirs for the different meteoritic classes. Additional evidence, either for initial heterogeneity or fractionation during accretion, comes from the variation observed, for example, in the abundances of refractory lithophile elements (e.g., Mg, Si, Cr) among different meteorite classes [20–25]. The variations among volatile elements (e.g., K, Rb) and especially the highly volatile elements (e.g., Bi, Tl) with respect to the refractory elements (e.g., Ca, Al, U) among meteorite classes, and apparently among the inner planets [26], most probably indicate element fractionations based on volatility during accretion, possibly overprinting initial heterogeneities. Larimer [15] notes that two fractionation trends, denoted by variations in Mg/Si and Mg/Cr ratios (one involving ordinary and enstatite chondrites, and the other C1, Earth, Moon and the eucrite parent body), intersect at the C1 composition. Kerridge [23] reaches a similar conclusion from a study of the relative abundances of Mg, Si, Ca and Al. These observations reinforce the evidence from the high volatile element concentrations, their correlation with the solar abundances, and the other observations (e.g., [27]) that support the identification of C1 as the closest sample yet available of the initial nebula material. This conclusion seems reasonable, despite the evidence that Orgueil is not a direct nebula condensate, but is of secondary origin and has undergone some aqueous alteration on a parent body, although within a brief (10 million year) time frame [28, 29]. C1 material appears to be widespread in the solar system, comprising possibly more than three-fourths of the asteroid population [30], Phobos and Deimos [31], and probably Amalthea and rings of Uranus.

Accordingly, it is appropriate to select a set of C1 abundance data (Table 8.1, see Appendix VI for data sources). In general, the latest apparently reliable values have been selected. Such subjective judgments seem impossible to avoid in the field of trace element analysis, where the same methods of analysis may yield variable results in the hands of differing analysts. Table 8.2 gives the data for major elements expressed as oxides. Because of the high content of carbon, water and sulfur, the data in Tables 8.1 and 8.2 are expressed as "volatile-free." For the major elements, this involves multiplying the original values by 1.479. A factor of 1.5 has been used for the trace element data, the difference being within the precision limits of the data. The rationale for removing very volatile components such as water is to enable comparisons to be made on a uniform basis. The elemental abundances in the enstatite chondrites (for example, in Abee) resemble those in the carbonaceous chondrites more closely than do other classes of meteorites and they may represent another variant of the composition of the solar nebula, possibly forming closer to the sun in regions with higher C/O ratios of about 1 in comparison with C/O ratios of about 0.6 for the carbonaceous chondrites [12].

8.3   The Earth

Although the Earth is only considered briefly in this book, for reasons stated in Chapter 1, it is useful to compare the overall composition of the planet with that of other bodies in the solar system. It is fair to say, however, that the composition of the Earth is known only within the same limits of accuracy as that of the Moon, at best.

The compositions of the core and mantle are listed separately. The relationships between the core and mantle are very imperfectly understood; they are assumed here to have accreted as separate metal and silicate phases (Section 9.8). For example, disagreement exists over the nature of the light constituent in the core; oxygen [32], sulfur [33], and silicon [34] are all candidates. Some possible core compositions are given in Table 8.3, with either oxygen or sulfur as the light alloying elements.

8.3.1   Primitive Mantle (Mantle Plus Crust)

The composition of the primitive earth mantle (i.e., present mantle plus crust) is discussed in this section. The composition of the present-day mantle, although of intense interest to geochemists, is of secondary importance to cosmochemists. Deciphering the composition and structure of the present-day mantle, the relationship between the upper and lower mantles, and the

Table 8.1 Elemental abundances.†

	CI Volatile Free	Ref.	Bulk Moon	Lunar Highland Crust	Primitive Earth Mantle	Ref.	Present Earth Oceanic Crust	Continental Crust				
								Present Earth Crust (Ref. 25)	Percent in Crust	Present Upper Crust (25)	Present Lower Crust (25)	Archean Upper Crust (26–29)
Li, ppm	2.4	2	0.83	—	0.78	24	—	10	5.5	—	—	—
Be, ppm	0.072	4	0.18	—	0.10	15	—	1.5	6.6	—	—	—
B, ppm	0.4	3	0.54	—	0.8	15	—	—	—	—	—	—
Na, wt.%	0.79	1	0.06	0.33	0.25	21	2.08	2.60	5.0	2.82	2.52	2.30
Mg, wt.%	14.1	1	19.3	4.10	24.0	14	4.64	2.11	0.04	1.39	2.47	3.14
Al, wt.%	1.29	1	3.17	13.0	1.75	14	8.47	9.5	2.3	8.47	10.0	8.25
Si, wt.%	15.6	1	20.3	21.0	21.0	14	23.1	27.1	0.6	30.8	25.2	26.8
K, ppm	885	2	83	600	180	19	1250	12,500	35.7	27,400	5000	7500
Ca, wt.%	1.39	1	3.22	11.3	1.89	14	8.08	5.36	1.22	2.50	6.79	5.22
Sc, ppm	7.8	2	19	—	10.6	15	38	30	1.22	10	40	25
Ti, ppm	660	5	1800	3360	900	14	9000	4800	2.3	3600	5400	5400
V, ppm	62	2	150	21	84	17	250	175	0.90	60	230	150
Cr, ppm	3500	1	4200	680	3000	14	270	55	0.007	35	65	140
Mn, wt.%	0.27	1	0.12	—	0.10	18	0.10	0.11	0.47	0.06	0.14	0.13
Fe, wt.%	27.2	1	10.6	5.1	6.2	14	8.16	5.83	0.41	3.50	7.00	7.38
Co, ppm	765	6	—	—	100	16	47	25	0.12	10	33	30
Ni, ppm	15,100	1	—	—	2000	14	135	30	0.005	20	35	90
Cu, ppm	160	2	—	—	28	30	86	60	0.92	25	78	80
Zn, ppm	455	2	—	—	50	30	85	—	—	52	—	100
Ga, ppm	14	6	—	—	3	30	17	18	2.6	—	—	—
Ge, ppm	47	2	—	—	—	—	—	—	—	—	—	—
As, ppm	2.4	2	—	—	—	—	—	—	—	—	—	—
Se, ppm	29	2	—	—	—	—	—	—	—	—	—	—

Table 8.1 Elemental abundances (Continued).

| | CI Volatile Free | Ref. | Bulk Moon | Lunar Highland Crust | Primitive Earth Mantle | Ref. | Present Earth Oceanic Crust | Present Earth Crust | Continental Crust | | | |
									Percent in Crust	Present Upper Crust	Present Lower Crust	Archean Upper Crust
Rb, ppm	3.45	2	0.28	1.7	0.48	20	1.5	42	38	110	8	25
Sr, ppm	11.4	2	30	120	15.5	15	130	400	11.0	350	425	300
Y, ppm	2.1	2	5.1	13.4	2.9	15	32	22	3.4	22	22	15
Zr, ppm	5.7	5	14	63	7.8	15	80	100	5.5	240	30	100
Nb, ppm	0.45	7	1.1	4.5	0.60	15	2.2	11	7.9	25	4	5
Mo, ppm	2.1	8	—	—	—	—	—	—	—	—	—	—
Ru, ppb	1160	13	—	—	—	—	—	—	—	—	—	—
Rh, ppb	375	13	—	—	—	—	—	—	—	—	—	—
Pd, ppb	855	13	—	—	—	—	—	—	—	—	—	—
Ag, ppb	270	8	—	—	—	—	—	—	—	—	—	—
Cd, ppb	960	8	—	—	—	—	—	—	—	—	—	—
In, ppb	120	8	—	—	—	—	—	—	—	—	—	—
Sn, ppm	2.46	2	—	—	—	—	—	—	—	—	—	—
Sb, ppm	0.21	8	—	—	—	—	—	—	—	—	—	—
Te, ppm	4.56	8	—	—	—	—	—	—	—	—	—	—
Cs, ppm	0.29	8	0.012	0.07	<.016	22	0.015	1.7	>45	3.7	0.7	—
Ba, ppm	3.6	2	8.8	66	4.9	15	15	350	30	700	175	240
La, ppm	0.367	9	0.90	5.3	0.50	15	3.7	19	15.4	30	14	12.6
Ce, ppm	0.957	9	2.34	13	1.30	15	11.5	38	12.6	64	25	26.8
Pr, ppm	0.137	10	0.34	1.8	0.19	15	1.8	4.3	9.8	7.1	2.9	3.1
Nd, ppm	0.711	9	1.74	7.4	0.96	15	10.0	16	7.1	26	11	13.0
Sm, ppm	0.231	9	0.57	2.0	0.31	15	3.3	3.7	5.1	4.5	3.3	2.78
Eu, ppm	0.087	9	0.21	1.0	0.12	15	1.3	1.1	4.0	0.88	1.2	0.90

Table 8.1 Elemental abundances *(Concluded).*

	CI Volatile Free	Ref.	Bulk Moon	Lunar Highland Crust	Primitive Earth Mantle	Ref.	Present Earth Oceanic Crust	Continental Crust				
								Present Earth Crust	Percent in Crust	Present Upper Crust	Present Lower Crust	Archean Upper Crust
Gd, ppm	0.306	9	0.75	2.3	0.42	15	4.6	3.6	3.7	3.8	3.5	2.85
Tb, ppm	0.058	10	0.14	0.35	0.079	15	0.87	0.64	3.5	0.64	0.64	0.48
Dy, ppm	0.381	9	0.93	2.3	0.52	15	5.7	3.7	3.0	3.5	3.8	2.93
Ho, ppm	0.085	9	0.21	0.53	0.12	15	1.3	0.82	2.9	0.80	0.83	0.63
Er, ppm	0.249	9	0.61	1.51	0.34	15	3.7	2.3	2.9	2.3	2.3	1.81
Tm, ppm	0.036	10	0.088	0.22	0.048	15	0.54	0.32	2.9	0.33	0.32	0.26
Yb, ppm	0.248	9	0.61	1.4	0.34	15	5.1	2.2	2.8	2.2	2.2	1.8
Lu, ppm	0.038	9	0.093	0.21	0.052	15	0.56	0.30	2.4	0.32	0.29	0.28
Hf, ppm	0.17	11	0.42	1.4	0.23	15	2.5	3.0	5.6	5.8	1.6	3
W, ppm	0.30	2	0.74	—	—	—	—	—	—	—	—	—
Re, ppb	60	13	—	—	—	—	—	—	—	—	—	—
Os, ppb	945	13	—	—	—	—	—	—	—	—	—	—
Ir, ppb	975	13	—	—	—	—	—	—	—	—	—	—
Pt, ppb	1520	13	—	—	—	—	—	—	—	—	—	—
Au, ppb	255	13	—	—	—	—	—	—	—	—	—	—
Hg, ppb	720	2	—	—	—	—	—	—	—	—	—	—
Tl, ppb	220	8	—	—	—	—	—	—	—	—	—	—
Pb, ppm	3.6	2	—	—	—	—	—	10	—	15	7.5	10
Bi, ppb	165	8	—	—	—	—	—	—	—	—	—	—
Th, ppm	0.051	12	0.125	0.9	0.070	15	0.3	4.8	30	10.5	1.95	2.9
U, ppm	0.014	8	0.033	0.24	0.018	15	0.6	1.25	30	2.5	0.63	0.75

†Adapted from Taylor, S. R. (1980) *PLC 11:* 333 and Taylor, S. R., and McLennan, S. M. (1981) *Phil Trans. Roy. Soc.* A301: 381.
References given in Appendix V.

Table 8.2 Chemical compositions expressed as weight percent oxides.[†]

	CI Volatile Free	Bulk Moon	Lunar Highland Crust	Primitive Earth Mantle	Present Earth Oceanic Crust	Present Earth Crust	Continental Crust		
							Present Upper Crust	Present Lower Crust	Archean Upper Crust
SiO$_2$	33.3	43.4	45	45	49.5	58.0	66.0	54.0	57.4
TiO$_2$	0.11	0.3	0.56	0.15	1.4	0.8	0.6	0.9	0.9
Al$_2$O$_3$	2.43	6.0	24.6	3.3	16.0	18.0	16.0	19.0	15.6
FeO	34.0	13.0*	6.6	8.0	10.5	7.5	4.5	9.0	9.5
MgO	23.4	32.0	6.8	40	7.7	3.5	2.3	4.1	5.2
CaO	1.94	4.5	15.8	2.65	11.3	7.5	3.5	9.5	7.3
Na$_2$O	1.05	0.09	0.45	0.34	2.8	3.5	3.8	3.4	3.1
K$_2$O	0.11	0.01	0.075	0.02	0.15	1.5	3.3	0.6	0.9
NiO	1.92	—	—	0.25	—	—	—	—	—
Cr$_2$O$_3$	0.52	0.60	0.10	0.44	0.04	—	—	—	—
MnO	0.34	0.15	—	0.13	0.13	—	—	—	—
Σ	100.12	100.05	99.99	100.1	99.5	100.3	100.00	100.4	99.9

[†]Data sources are given in Appendix V.
*2.3% Fe or FeS is allotted to a lunar core and 10.7% FeO to the lunar mantle.

Table 8.3 Two proposed compositions of the Earth's core.[†]

	Model 1	Model 2
Fe	85.5	86.2
Ni	5.5	4.8
S	9.0	1.0
O	—	8.0
Σ	100.0	100.0

[†]Wt. %, with sulfur (Model 1) and oxygen (Model 2) as the light alloying element [adapted from *Basaltic Volcanism* (1981) Table 4.3.2c].

question of the relative amounts of depleted and undepleted mantle are all active topics of research [35].

The refractory elements (Ca, Al, REE, Th, U, Sr, etc.) apparently stay together in cosmochemical processes, forming, according to the condensation hypothesis, part of an early refractory-element-rich condensate, with relatively constant inter-element ratios [14, 36]. Regardless of the truth of this scenario, it is an observable fact that the major cosmochemical fractionations within the inner solar system involve the three main groupings of refractory, volatile and siderophile elements (e.g., [20]). This is shown most simply by variations in K/U ratios and Fe contents. Thus, the initial assumption may be made that the relative ratios of the refractory elements (Ca, Al, REE, Th, U, Sr) are constant for the bulk earth, and are the same as for the chondritic meteorites. Accordingly, if we can arrive at Ca and Al abundances for the Earth, then estimates for the trace elements, including uranium, can be made.

The philosophy of deriving mantle abundances for major elements is to assess their abundances from the observable ultramafic nodule data [37, 38], from deductions based on the source regions of lavas [39, 40], or from experimental petrology [41]. There is reasonably good agreement between the results based on several different approaches. Selection of the most "primitive" composition has been attempted by several workers [37, 39, 42], but is very dependent on models for depletion or enrichment of the upper mantle. This approach may be valid for the major elements, but the relative mobility of the incompatible elements (e.g., K, Rb, Th, Ba, REE) adds additional uncertainty in attempting to assess trace element compositions [42].

A major unknown is the composition of the lower mantle (below the 650 km discontinuity); it may be the same [41] or it may be more silica-rich than the upper mantle [43]. Since both K and U are "incompatible" elements on account of size and valency, they would be expected to be concentrated in the upper mantle if vertical chemical heterogeneity exists. Hence, the upper mantle abundances for these elements could be overestimated for the bulk earth. Major element compositions based on nodule data show Ca/Al ratios greater than the canonical value of 1.08 [44]. If the lower mantle is dominated by perovskite phases with compositions close to (Mg, Fe, Al) (Si, Al)O_3 [43], then Al may be concentrated in the deep mantle relative to Ca.

The high concentration of the alkali elements (5% Na, 35% K) in the crust [45] make the nodule data very unreliable for these elements, and for most of the incompatible trace elements. Cr and Ni, however, are sufficiently compatible to be estimated. If the abundances of the refractory elements Ca, Al and Ti can be calculated for the mantle, then the concentrations of the other refractory trace elements may be obtained. A critical but reasonable assumption is that there is no relative fractionation among these elements (Be, B, Sc, V, Sr, Y, Zr, Hf, Nb, Ba, REE, Th, U) during accretion. This is supported by the coherent behavior of these elements in the Group I and III Allende inclusions [46], in lunar samples [47], and in the similarity in ratios between, for example, Ca, Al, Ti, La and Sc in most classes of chondritic meteorites [15, 25]. Evaluation of these data show that the ratio of the concentrations of Al, Ca and Ti between the terrestrial mantle and the volatile-free C1 abundances is 1.36 (or 2.0 to the uncorrected C1 values). Based on the arguments given above, the assumption is now made that no relative fractionation occurred among the refractory elements during accretion of the Earth. The volatile-free C1 abundances for the other refractory elements are accordingly multiplied by this value to yield the mantle values. Many of these elements are highly incompatible, and have been strongly extracted into the continental crust during geological time (e.g., 30% Th, U) [45], so that present-day mantle samples are most likely to be compromised for these elements.

In contrast, the overall abundance of the major elements in the mantle is scarcely affected by extraction of the terrestrial crust in contrast to the situation for the Moon. Only 2.3% Al, 1.2% Ca, 0.4% Fe, 0.6% Si and 0.04% Mg are concentrated in the continental crust [45]. Therefore, the present overall mantle value for Al has decreased from 1.75% to 1.71%; for Ca from 1.89% to 1.8%; for Fe from 6.2% to 6.18%; and for Si from 21% to 20.9%, by the extraction of the continental crust. These differences are sufficiently small to be neglected. They certainly lie well within the tolerances of estimating mantle composition.

The K value for the mantle (180 ppm) was established from K-Ar and K/U systematics. Further evidence for a low value of K in comparison with the solar nebula abundances [48] comes from the K/Rb ratio (typically 300) and the Rb/Sr terrestrial ratio of 0.031. From the Sr isotopic systematics, the Earth has an order of magnitude lower Rb/Sr ratio than chondrites (Rb/Sr = 0.25) [49]. Strontium is an involatile element, so that an upper limit on the terrestrial Sr value comes from Ca/Sr ratios [36]. On this basis, the terrestrial value for Sr is 17 ppm. From the terrestrial Rb/Sr ratio (0.031) and K/Rb = 300, the K value from this calculation is 170 ppm K. Hence, the limits of 75–150 ppm K for continuous degassing models for argon appear reasonable [50]. These provide, via K/U = 10^4, terrestrial uranium abundances of 7.5–15 ppb. Thus, the K/Rb value for the mantle is 375, compared with C1 value of 256. Rb is more volatile than K and on the basis of these values, the Earth has accreted less Rb than K. The upper crustal K/Rb ratio of about 250 [45], which fortuitously resembles the chondritic value, indicates that Rb is partitioned into the crust more readily than K, in accordance with its greater ionic radius. A similar situation exists for Cs, which provides the extreme example of an incompatible lithophile element. The upper crustal Rb/Cs ratio is 30 and mantle ratios must be higher, since Cs is preferentially partitioned into the crust during partial melting episodes. The C1 value for the Rb/Cs ratio is 12, so that the Earth has accreted more Rb than Cs. This is strong evidence for selective depletion in volatile elements of material which accreted to form the Earth. The relative concentrations of Rb, K and, by extrapolation, Na in the mantle appear to be governed by relative volatility. An estimate of 0.25% Na can be made on this basis [48].

If a steady-state relationship exists between heat production from radioactive sources and heat flow, then the uranium content of the bulk earth is 30 ppb, or 40 ppb in the mantle, assuming that no uranium enters the core. From $K/U = 10^4$, this gives a mantle value of 400 ppm K, which is in serious disagreement with other estimates. However, if the relationship between radioactive heat generation and heat flow is not steady-state, that is, if heat loss from the Earth has a long time constant, then the presently observed heat flow is not directly related to the abundances of the radioactive elements [2, 3]. For example, a significant contribution to the heat-flow may come from cooling from an initially molten state. On this basis, the uranium content of the Earth becomes less than 20 ppb, consistent with the other estimates.

From these interlocking sets of constraints, a uranium value within the range 15–20 ppb appears to be a reasonable estimate of the terrestrial abundance [21, 51]. From the K/U ratio of 10^4, the potassium abundance becomes 150–200 ppm consistent with the K-Ar and Rb-Sr systematics. Rb lies within the range 0.50–0.66 ppm. The terrestrial abundance of Sr, from Rb/Sr = 0.031, lies in the range 16–21 ppm, consistent with the 17 ppm derived from the Ca/Sr systematics. Thorium lies within the range 57–76 ppb derived from the canonical Th/U ratio of 3.8.

A further consistency check may be made by comparing the values for lanthanum obtained from the terrestrial K/La crustal ratio of 400 with those obtained from the chondritic Ca/La and Al/La ratios. The respective ranges are 0.38–0.50 ppm La and 0.47–0.60 ppm La. These values would indicate abundances about 200 ppm K. The calculation is sensitive to the K/La ratio, which is probably biased in favor of high values, since K is enriched in the continental crust relative to La by factors of up to two [45], so that the bulk earth K/La ratio is probably lower than crustal values [52]. The use of correlated elements to establish planetary compositions should always be accompanied by an understanding of the geochemical reasons for the correlations, and of the factors which may change them during fractionation (see Section 5.9.1). The REE patterns for the mantle are assumed here to be parallel to those of the C1 meteorites, since they are all refractory elements. The two least refractory elements, Eu and Yb show no anomalies consistent with selective evaporation or condensation.

It is assumed that the upper and lower mantle compositions are equivalent. The presently available geochemical and geophysical data are not adequate to resolve this interesting question, although there are some indications from shock-wave and diamond anvil experiments that a change toward a higher density composition with increasing depth may be required [43, 53]. The Mg/Al ratio (13.7) for the terrestrial mantle is significantly higher than the C1 value of 10.9. This requires either overall enrichment of Mg in the Earth, or an enrichment in the upper mantle (from which the geochemical data are derived) relative to the lower mantle. The depletion in Al in the mantle due to its extraction into the continental crust is too small to produce this 26% shift in the ratio. Although a plausible case can be made for differing upper and lower mantle compositions, such calculations are at present too model-dependent to provide useful geochemical constraints, and uniformity is assumed.

The siderophile element component in the mantle has recently been established from nodule studies to have a carbonaceous chondrite (C1) signature [54]. This is not what would be expected if the upper mantle had been in equilibrium with a metal phase during core formation. Accordingly, this evidence supports the view that these elements were possibly accreted at a later stage following core separation. Metal separation during core formation would have changed the pattern markedly due to wide differences among the distribution coefficients. Accordingly, the siderophile elements in the upper mantle do not define a unique terrestrial signature.

8.3.2 The Crust

Abundance data for the present-day crust are given in Table 8.1 [45]. Separate values are given for the upper, lower and total continental crusts. An estimate is also given of the Archean crust. The oceanic crust of the Earth is difficult to model precisely, on account of seawater alteration, a varying content of sedimentary material and other factors. The important point in the context of planetary evolution is the composition of the portion derived from the mantle. The estimate given in Table 8.1 is heavily biased toward the abundances in Mid-Ocean Ridge basalts, interpreted as the volumetrically most important contributor. Smaller amounts of alkali basalt and sediment may be added to taste.

The percentage of each element now residing in the continental crust, relative to that present in the primitive mantle, is given in Table 8.1, assuming that the crust is 0.43% of the mass of the mantle. The degree of enrichment of an element in the crust depends on the difference between its ionic radius and valency and those of the elements forming the major mantle minerals (Mg, Fe). This indicates that crystal-liquid fractionation is the principal factor responsible for the derivation of the crust from the mantle. The very large percentage of elements, Cs, Rb, K, Ba, U, Th, La and Ce in the continental crust indicates the processing of at least 30% of the mantle to produce the crust [45]. Some of the more volatile constituents are mostly concentrated in the crust, for example, up to 75% of the Cl and Br budget [55].

8.4 The Moon

The composition of the bulk moon is understood at least as well as that of the Earth, and the view may be expressed that it is even better understood since the composition of the volumetrically important lower terrestrial mantle is in dispute [56]. The philosophy of calculating the overall composition of the Moon differs from the procedures used in the case of the Earth. On Earth the continental crustal composition is subordinate to estimates of mantle composition for the major elements (although not for the incompatible trace elements). Thus, the amount of Si, Al, Fe, Mg or Ca in the crust can be neglected in a first approximation of Earth mantle compositions. In contrast, the abundances of the major elements Al and Ca in the lunar highland crust, are critical for an evaluation of whole moon abundances. The crustal abundances of K, U, Th, Sr and Eu likewise form a sizeable percentage of the total moon budget, analogous to the K, U and Th contents of the terrestrial continental crust. Thus, a first step is to establish the thickness and composition of the highland crust (see Sections 5.1 and 5.9).

8.4.1 The Highland Crust

The crustal thickness, density, and Al_2O_3 content are all interrelated with the overall lunar density, moment of inertia and the center of figure–center of mass offset (2 km) of the Moon. Thus, for a density of 2.95 g/cm^3, and a composition of 24.6% Al_2O_3, the average crustal thickness is 73 km. If the crust is more aluminous (29% Al_2O_3), which appears unlikely, the average thickness drops to 70 km. If the crust is less aluminous, (20% Al_2O_3), the average thickness increases to 86 km. There is some uncertainty in the near-side thickness of the crust. If it is 50 km instead of 64 km, then the average thickness becomes 65 km for 24.6% Al_2O_3. We assume a 24.6% Al_2O_3 crust of average thickness 73 km, which contributes 2.63% Al_2O_3 to the bulk moon composition.

The question of the change in the composition of the highland crust with depth is in fact somewhat secondary, since a lower alumina crust must be thicker to meet the geophysical constraints (see Section 5.1.1). Depending on the various assumptions, the crustal contribution to the total Al_2O_3 budget ranges from 2.4–2.85% [57–59]. This is a critical value for lunar compositions, for a moon of C1 volatile-free composition would contain 2.43% Al_2O_3. Moons of a chondritic composition accordingly may be ruled out [60]. For such a composition, all the Al_2O_3 would be in the highland crust even for the lowest possible alumina content proposed (20% Al_2O_3). The highland crustal data are thus a primary piece of evidence for a moon enriched in refractory elements relative to primitive solar nebula abundances. The highland crustal composition is given in Table 5.5.

Earlier workers [61, 62] have suggested that the highland material was directly accreted and represents the last stage of accumulation of a heterogeneous moon. However, three features of the abundance patterns disprove that hypothesis [63]. The first is the reciprocal relationship of highland and mare basalt abundances, which are derived from the deep interior (Fig. 6.40); the second is the similarity in volatile/involatile element ratios; and the third is the geochemical evidence for crystal-liquid fractionation processes in the evolution of the highland crust.

The involatile elements in the highlands have been fractionated to a degree that depends on the differences in ionic radii and/or valency from those of Fe^{2+} and Mg^{2+}. This constitutes evidence for fractionation based on crystal-liquid equilibria, and hence implies large-scale partial melting in the outer layers of the Moon. The type and levels of enrichment are analogous to those observed in terrestrial crustal rocks and show no dependence on volatility (e.g., fractionation based on ionic radius and valency is observed within the REE group). From this evidence, it is concluded that the chemically distinct rock types in the highlands are derived by partial melting in post-accretion processes. Pre-accretion processes are responsible for loss of the volatile, and probably the siderophile, elements. Post-accretion processes of melting and fractionation are responsible for the production of the surface rocks (Fig. 8.3).

The Cr/Ni ratios in the highlands show the typical lunar characteristics, with Cr much enriched (Cr/Ni > 5) compared to primitive cosmic abundances (Cr/Ni = 0.25), indicating that the siderophile elements are depleted in the highly fractionated highland crust. This point is discussed later in a general section on siderophile element comparisons. This low siderophile element content also is evidence that the origin of the Mg-suite in the lunar highlands was not due to the influx of a chondritic component, for otherwise the Cr/Ni ratios would be reversed. If late-accreting material was still being added and was responsible in part for the high Mg-suite, then it was of whole moon composition.

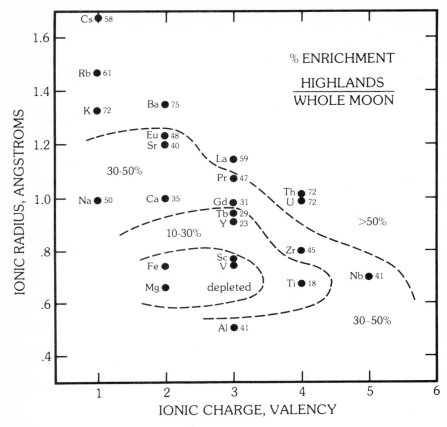

8.3 The relative enrichment of elements in the lunar highland crust, relative to whole-moon compositions. The average highland crustal thickness is conservatively taken as 60 km and uniformity of composition with depth is assumed. The elements show progressive enrichment as their ionic radius and valency depart from those of the principal mantle cations, Fe and Mg. Such patterns are consistent with the operation of crystal-liquid fractionation processes, and show no dependence on volatility. Note the extreme enrichment of the large alkali elements Ba, La, Th and U. This comprises evidence both for whole-moon melting, and for an overall enrichment of the refractory elements in the Moon. (Data from Tables 5.5 and 8.1.)

In the discussion on volatile elements (Section 5.5), it was noted that in some highland samples (the notorious rusty rock, 66095, and 61016 are extreme examples), enrichment by factors up to 10^4 are observed over average lunar volatile contents. Do these represent evidence contrary to the notion that the Moon is depleted in volatiles as attested by isotopic evidence? The view is adopted here that volatile element enrichment occurs mainly through redistribution of KREEP-rich material mainly by meteorite impacts or during multi-ring basin formation. Although the volatile content of the Moon (e.g., Tl, Bi) is low, it is not zero. The small amounts in the Moon will be concentrated into the residual melts during crystallization of the magma ocean, since all of them are highly incompatible. Accordingly, the KREEP component will be highly enriched. Redistribution by impact melting will account for the high local enrichments, which may represent a substantial proportion of the total lunar budget for these elements.

Some contribution also comes from the carbonaceous chondrite component in the regolith, but most of the volatile elements are probably of lunar origin. A cometary origin is ruled out since the water is terrestrial and the lead is lunar, not meteoritic. It is not necessary to invoke "fumaroles" with their volcanic connotations. A simpler energy source is available to redistribute the volatile elements (Section 5.5).

Data from the orbital gamma ray measurements indicate highland surface averages of 0.90 ppm Th and 500 ppm K, which are consistent with estimates from surface chemistry of 0.90 ppm Th and 600 ppm K. If the arguments for Al_2O_3 uniformity in the crust are extended to these elements as well, then the crust contributes 0.09 ppm Th and 60 ppm K to the whole-moon budget.

8.4.2   The Mare Basalts

The elemental abundances in the mare basalts have provided many constraints on lunar processes. However, they represent the end-stage product of several processes. The abundances have been altered from those in the primitive solar nebula in several steps. These include:

(a) Pre-accretion fractionation which resulted in loss of the volatile elements, and probable loss of the siderophile elements.

(b) After the accretion of the Moon, widespread melting and fractionation occurred and a zoned structure developed, in which the source regions of the mare basalts were enriched in some elements and depleted in others (e.g., Eu). This occurred shortly after formation at about 4.4 aeons.

(c) Much later, between 4.0 and 3.2 aeons, partial melting leading to element fractionation occurred in the source region of the mare basalts (due to radioactive heating), and the basaltic magma was extruded at the surface.

(d) En route to the surface, and during and following extrusion, crystallization provided the opportunity for removal of phases, and further element fractionation.

A general conclusion may be drawn with respect to abundances of the incompatible elements. During the crystallization of the magma ocean, and during processes of partial melting to generate basalts, these elements become concentrated in the residual liquids or enter early melts. There is a tendency for the ratios of these elements to remain relatively constant. Accordingly, they may reflect whole-moon abundance ratios. The use of such correlated element ratios for estimating the composition of the Moon must be treated with an appropriate comprehension of the geochemical subtleties of varying distribution coefficients, and of the parameters of ionic radii, valency, and crystal field effects which may alter the simple picture.

Accordingly, the data must be examined with these caveats in mind in using mare basalt abundances to provide information useful for calculating bulk moon compositions. The compositions only of primitive undifferentiated samples (e.g., 12009 or glasses), which are thought to have reached the surface as liquids, should be employed. It is clear from the mare basalt chemistry and isotopic systematics (Chapter 6) that the interior from which they are derived by partial melting is not unfractionated material of bulk lunar composition, but has been melted and differentiated. However, the mare basalts preserve the distinct lunar signatures of depletion of volatile and siderophile elements and enrichment in refractory elements.

It should be recalled that lunar basalts have undergone three stages of element fractionation. Thus, depending on one's views of the primitive accretionary process and its relation to primitive solar nebula abundances (see Sections 9.6–9.7), some enrichment of refractory elements occurred when the volatile elements were lost. The extent of this enrichment varies depending on whether a simple loss of volatiles and siderophiles is considered, or whether a condensation mechanism is invoked. Second, the geochemical zoning of the primitive Moon concentrates the REE, U, Th, Zr, Hf, Ba, and other elements, and produces a second enrichment in the source region of mare basalts. Third, the partial melting process again enriches the melt relative to the residue in these elements, all of which are distinguished by their inability to enter the sixfold Mg-Fe coordination sites, which are probably the main cation sites in minerals in the deeper zones of the lunar interior.

The mare basalt compositions invite comparison with those of the eucritic class of basaltic achondrites. The closest lunar analogues are the aluminous mare basalts, which show a very good match for the major elements. In particular, the chromium, sodium, and potassium concentrations, so typically lunar, match those of the eucrites. Even the trace volatile (Tl, Bi) and siderophile (Ir) elements cover about the same range in concentration, although gold is still depleted [64].

The refractory elements (Ba, Zr, Ti) are consistently higher and show their distinctive lunar enrichment. Other criteria (e.g, oxygen isotopes) are also distinctive in distinguishing the eucrites from the lunar samples, so that this class of meteorite cannot be derived from the Moon. Nevertheless, the eucrites contain evidence of many analogous processes, and indicate that the same major element chemistry is being produced elsewhere in the solar system.

8.4.3 Bulk Moon Compositions

From the information discussed throughout the text, a bulk composition may be calculated for the Moon (Table 8.1). The various chemical constraints on lunar compositions are noted below:

(a) The variations in heat flow allow for a range in uranium values of 33–46 ppb. The lower value assumes a non-steady-state solution, and is adopted here as the most conservative value, consistent with other parameters.

(b) The highland crustal values of K, U and Th abundances represent 60–70% of the whole-moon values. Accordingly, the estimates of 100 ppm K, 125 ppb Th, and 33 ppb U are conservative, and indicate a massive near-surface concentration of heat producing elements.

(c) The overall highland abundances (Table 5.5) set further limits. Assuming that U/Eu ratios are chondritic (since both are refractory), the Eu content of the bulk moon is 0.21 ppm. The highland crust averages 1.0 ppm. This value correlates with the Al_2O_3 content since both elements are contained in plagioclase, and is likely to be representative of the total crust. Thus, even if K, U and Th are concentrated near the surface, Eu is concentrated in the mineral phase, plagioclase, which is a major constituent of the highland crust. On this basis, about 50% of the Eu in the Moon is in the highland crust. This indicates that at least one-half the Moon was involved in the initial differentiation. Since some U (and Al) are retained in the interior, and since crystal-liquid fractionation and melt separation are not 100% efficient, these figures are consistent with whole moon involvement in the primary melting.

(d) Once values for Al, Ca or U are established, element ratios can be used to calculate the abundances of the other elements. The ratio for the lunar abundances to those of the volatile-free C1 abundances is 2.45. This enables data for the other refractory elements to be evaluated. Chromium is derived from the Cr/V ratio in lunar rocks of 28. It is a moderately refractory element, and is depleted in the Moon relative to V by a factor of about two. Manganese, likewise obtained from Fe/Mn ratios, is also depleted by a factor of about two. The V/Mn ratio is thus a factor of six greater than in chondrites.

(e) The geochemical data do not provide unique solutions for the Mg/Si ratio and most workers have assumed chondritic ratios in the absence of other evidence. The geophysical constraints may indicate a lower value of 0.77 rather than the chondritic value [58]. Such compositions have high SiO_2 values (48%) and depend on the Goins model for the lunar interior (Section 7.6.1), and an initial melting depth of 400 km. The seismic evidence for this discontinuity is judged to be inconclusive (see Section 7.6.1), and hence it does not provide a firm constraint on interior compositions or Mg/Si ratios. A greater depth of melting, or whole moon melting, would ease some of the problems. Removal of Al and Ca from depth makes garnet of less significance as a deep lunar phase, while a core accommodates the moment-of-inertia constraint. The Buck-Toksöz composition is, in fact, close to that adopted here. If the Mg/Si ratios for the primitive lunar mantle (Table 8.1) is changed to 0.77, then the overall compositions are in close agreement. However, such a change in the Mg/Si ratio has no justification at present. The Mg/Si ratio is of much interest in Earth-Moon comparisons and an accurate figure would be very useful.

(f) The density data require about 13% FeO or equivalent, but this provides too high an Fe/(Fe + Mg) ratio to be consistent with the petrological requirements for lunar compositions from which the highland crust and the source regions of the mare basalts can be derived [65, 66, 67]. Values of 10–11% FeO are required. These are compatible with the petrological evidence, but are too low to account for the lunar density, if the lunar mantle represented the bulk composition. The solution to this dilemma can be provided by an Fe or FeS core, which provides a sink for the excess iron. The bulk lunar composition in Table 8.1 includes 2.3% Fe (or FeS) and 10.7% FeO. The primitive mantle composition has 10.9% FeO and 32.7% MgO Mg/(Mg + Fe) = 0.84. The composition of the residual mantle, from which the mare basalts are derived, follows the extraction of the highland crust.

Petrological testing of the various models has been extensively carried out (e.g., [65, 66]), and the Taylor-Bence model [67], from which the composition given in this book are largely derived, provides an appropriate starting material from which to derive the highland crust and the source regions for the mare basalts.

The composition given in Table 8.1 is intermediate between the Moon 1 and 2 compositions listed in *Basaltic Volcanism on the Terrestrial Planets* (1981) Table 4.3.2 e. The vexing question of the siderophile abundances in the Moon is addressed in Section 8.4.4.

8.4.4   Earth-Moon Comparisons

This question is of special significance, since it enables a test of the fission hypothesis. Models of lunar origin are notoriously difficult to test, but in its simplest form, the fission hypothesis predicts that the composition of the Moon is equivalent to that of the mantle of the Earth. The similarity in density has always lent an attraction to this hypothesis.

In this comparison it is necessary to deal with bulk moon versus bulk terrestrial mantle compositions. The hazards of comparing, for example basalts, are well known. No lunar basalts can be identified as coming from primitive unfractionated material of bulk lunar composition. Also, we cannot readily identify pristine unfractionated mantle in the Earth at the present stage of mantle evolution. Hence, direct matching of terrestrial and lunar basalt data is beset with ambiguities, since they are derived from source regions with differing evolutionary histories. This means that interplanetary comparisons are best carried out at a more sophisticated level than by simple matching of surface rock compositions.

In comparing moon and earth rock compositions, we can immediately exclude many terrestrial rock types on grounds of gross incompatibility in element abundances. These include the granites and granodiorites. It should be noted that tiny amounts of K-rich granitic glasses, presumably residual late stage differentiates, formed during crystallization of mare basalts but are not found as separate rock types. Volcanic equivalents such as rhyolites and ignimbrites are not found. Neither are there representatives of the great calc-alkaline suite of andesites, dacites, or rhyolites, so typical of the orogenic areas on the Earth. Their origin on this planet is intimately related to the subduction zones, where down-going slabs of the lithosphere slide into the Earth's mantle. The absence of plate tectonics effectively rules out the mechanism for the production of these rocks on the Moon and most other planets. Even the fairly common differentiated sequences known from basaltic mag-

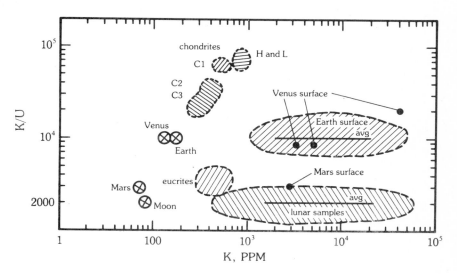

8.4 K/U ratios and potassium abundances showing the separation of chondritic ratios from those of the planets. The Venus and Earth data overlap, and are separated from those for the eucrites, the Moon and Mars. The proposed bulk compositions for Earth, Venus, Mars and the Moon are shown on the left side of the diagram. (Data from Tables 8.1 and 8.5.)

mas on the Earth are not found on the Moon. There is an absence of mugearites, hawaiites, trachytes, and other assiduously studied species, which indicates the very limited nature of near-surface fractionation in lunar mare basalts (Section 6.6.2). The general similarity of refractory element ratios and the fact that Sm-Nd systematics indicate the whole earth REE patterns are chondritic lead to the conclusion that the Earth possesses close to the primordial content of the refractory elements. Substantial loss of volatiles has occurred, as attested from Rb-Sr systematics, with the Earth having only about one quarter of the primitive abundances of K and Rb.

When we compare the bulk earth and the bulk moon compositions, K/U ratios are clearly different by a factor of four. However, the K abundances in the Earth are less than twice those of the Moon. Accordingly, the Moon *must*

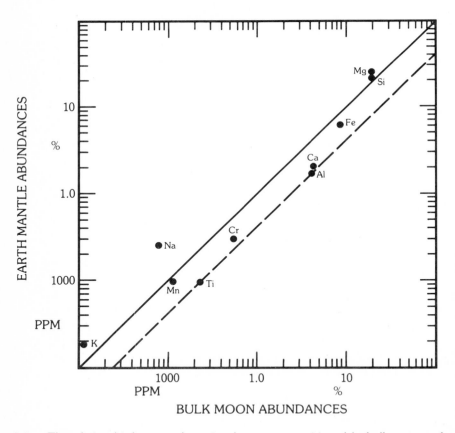

8.5a   The relationship between the major element composition of the bulk-moon and that of the primitive terrestrial mantle. Note the enrichment of Ca, Al and Ti and the depletion of Na and K in the Moon. (Data from Table 8.1.)

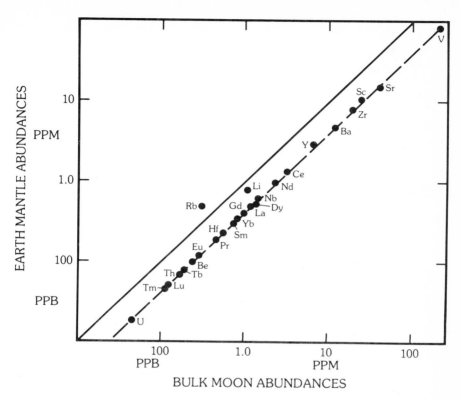

8.5b Comparison of trace element abundances for refractory and moderately volatile elements in the Moon and the primitive terrestrial mantle. (This includes the present Earth mantle and crust.) Note the uniform enrichment in refractory elements. (Data from Table 8.1.)

be enriched in U relative to the Earth (Fig. 8.4). This condition also applies to the amounts of the other refractory elements, such as Ca and Al, assuming no selective loss or enrichment during planetary formation. On this basis, the concentration of the refractory elements in the Moon is about two or three times that of the Earth (Fig. 8.5a and b), which is in agreement with much geochemical and petrological evidence and provides a high concentration of Ca, Al, Eu and Sr for the thick anorthositic crust.

The chalcophile and siderophile trace element abundances are of special interest because of the evidence which they may provide for pre-accretionary loss, core formation, and inter-planetary comparisons. A very large literature and some intense controversy has arisen over this problem (e.g.,[68, 69]). As noted earlier, the problems of attempting inter-planetary comparisons on the basis of surface samples are non-trivial. In particular, attempts to estimate indigenous lunar abundances of the siderophile elements in highland rocks seems especially hazardous. The history of the lunar highlands, as discussed in Chapter 2, 3 and 5, is replete with meteoritic bombardment and igneous fractionation. Even the pristine samples may not have escaped contamination (Section 5.7). No clear-cut solutions have emerged, except that any primitive siderophile element component is exceedingly low. This may be illustrated by reference to the lunar dunites (72415, 72417) which have Ni concentrations much lower than their terrestrial counterparts.

The situation is clearer for the mare basalts (Section 6.3) where meteoritic contamination is minimal. These data show the extreme depletion in nickel and the trace siderophile elements in lunar rocks, a fact that was noted at the

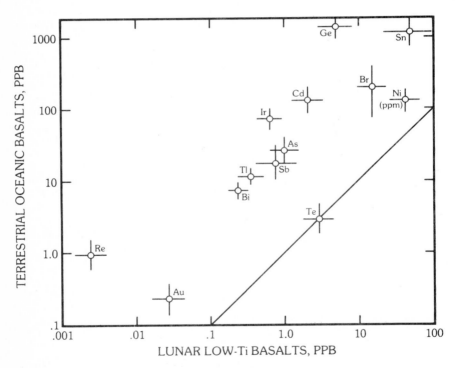

8.5c   Comparison of trace volatile and siderophile element abundances in terrestrial oceanic basalts and low-Ti lunar basalts. Many problems exist in establishing whole-earth and whole-moon abundances for these elements. Accordingly, this less satisfactory comparison of lunar and terrestrial "primitive" basalts is used, despite the hazards inherent in this approach (see text). The general depletion of these elements in the lunar samples is nevertheless taken as indicative of a whole-moon depletion in volatile and siderophile elements relative to the Earth. (Data from [69, 71, 72].)

time of the initial examination of the Apollo 11 samples. An additional problem with the trace siderophile elements lies in the problems of accurate analysis at ppb levels, and of mobility during alteration of terrestrial basalts [70]. A straightforward comparison of lunar and terrestrial basalts [69, 71, 72] shows that the lunar basalts are depleted relative to terrestrial basalts by large factors (Fig. 8.5c). The simplest interpretation of these data is that the source regions of the mare basalts are depleted in siderophile elements, relative to those of the terrestrial basalts. (The depletion of the volatile chalcophile elements follows that of the other volatile elements in the Moon, and is considered to be an intrinsic feature of bulk lunar composition.) There is no indication from the mare basalt chemistry that the source regions contained metallic phases. The Fe metal observed in some basalts crystallizes late in the paragenetic sequence and is a near-surface effect.

The depletion in nickel could arise as a consequence of prior crystallization and removal in olivine and orthopyroxene, operating during crystallization of the Moon (Section 6.5.2), and undoubtedly this mechanism operates. Such processes will not deplete the source regions of the mare basalts in trace siderophile elements (e.g., Ir, Os, Pt). Removal of these elements into a lunar core is possible, providing that equilibrium with a molten metal phase was reached. The small amount of the postulated metal phase (about two percent—Table 8.4—which may also be present as FeS) may make this unlikely, although the distribution factors are in favor of such a stripping of

Table 8.4 Chemical compositions expressed as weight percent oxides.[†]

	Bulk Moon	Primitive Mantle (mantle + crust)	Highland Crust (10%)	Mantle Following Crust Extraction
SiO_2	43.4	44.4	45.0	44.3
TiO_2	0.3	0.31	0.56	0.28
Al_2O_3	6.0	6.14	24.6	4.1
FeO	10.7	10.9	6.6	11.4
MgO	32.0	32.7	6.8	35.6
CaO	4.5	4.6	15.8	3.36
Na_2O	0.09	0.092	0.45	0.052
K_2O	0.01	0.01	0.075	0.003
Cr_2O_3	0.60	0.61	0.06	0.67
MnO	0.15	0.15	—	—
(Fe, FeS) (core)	2.3	—	—	—
Σ	100.05	99.9	99.7	99.9

[†]A lunar Fe or FeS core comprising 2.3% of the Moon is assumed to be present.

the lunar interior. All the processes make the question of the siderophile element composition of the bulk moon very difficult to estimate. The overall low iron content of the Moon indicates that the Moon accreted less than half its share of the primordial iron abundance. The scenario adopted here is that most of the lunar iron was accreted as oxide or silicate, with only 15% of the element accreting as metal although this value depends on the Fe/Mg ratio of the lunar mantle. The trace siderophiles will be mainly in the latter phase, and are now present in the lunar core.

The comparison with the Earth is complicated by a further factor. The trace siderophile elements in the upper mantle as revealed by analyses of ultramafic nodules [54, 55] have an abundance pattern parallel to that of C1 chondrites. Equilibrium with a metal phase during core formation would have distorted this pattern, on account of differing melt-metal distribution coefficients. It appears that the terrestrial upper mantle complement of these elements may be due to a late meteoritic addition [54, 55]. Accordingly, the abundances of the trace siderophile elements in the Earth possess no genetic significance nor do they provide a unique terrestrial signature for use in inter-planetary comparisons.

The following constraints are placed on Earth-Moon relationships:

(a) The Moon has an overall enrichment of refractory elements (Ca, Al, U, Th, Sr, REE) relative to the Earth by a factor of between two and three.

(b) The Moon has lower absolute abundance levels of the volatile elements, K and Rb, compared to the Earth, and very much lower abundances of the very volatile elements.

(c) Hypotheses which derive the Moon from the mantle of the Earth must account for (1) volatile loss and (2) substantial refractory element enrichment during this process.

These constraints are considered further in Section 9.7. Although complex scenarios may be invoked, they should always meet the scientific criterion of being testable.

8.5 The Planets

In considering the bulk composition of solar system bodies other than the Earth or the Moon, it is well to realize the many gaps in our knowledge. We are now approaching a consensus on the composition of the Earth and the Moon after many years of work and much data gathering and evaluation.

Nevertheless, although the Apollo missions were brief, the sample return has enabled us to have as good or better an understanding of bulk lunar composition than we do for that of the Earth. The contrast with the pre-Apollo understanding of lunar composition is complete. No model survived the encounter with the Apollo sample data. This situation is now being repeated, for example, with the rare gas atmospheric data from Venus and Mars (Section 4.13). In this context, our compositional models for the other planets resemble those for the Moon in pre-Apollo time. One advantage is our dearly bought experience with the Moon. This enables us to approach the problem of other planetary compositions with wisdom and some hindsight. These questions have been addressed most recently in *Basaltic Volcanism on the Terrestrial Planets*, and the discussions and compositions given here are principally derived from that source. It cannot be stressed too strongly that these are estimates only, with about the resolution of pre-Apollo lunar compositions, a fact which should give pause to modellers.

For Mercury, no chemical data are available except for the spectral reflectance data which indicate that the surface composition closely resembles that of the lunar highlands and that the FeO content does not exceed 5.5%. Based on density, theoretical models, and other factors, a range of compositions has been proposed for Mercury [73]. The range of reasonable models is given in Table 8.5. The composition proposed by Morgan and Anders [74] falls within this range (see Sections 2.5, 5.13, and 6.7).

Table 8.5 Proposed chemical compositions of Mercury, Venus, Mars, and the eucrite parent body.[†]

Mantle and Crust (%)	Mercury	Venus	Mars	Eucrite Parent Body
SiO_2	43.6–47.1	40.4–49.8	36.8–49.6	39.0–46.0
TiO_2	0.33	0.2–0.3	0.2–0.3	0.1–0.15
Al_2O_3	4.7–6.4	3.4–4.1	3.1–6.4	1.8–3.1
MgO	33.7–54.6	33.3–38.0	30	28.5–32.5
FeO	3.7	5.4–18.7	15.8–26.8	14.4–26.6
CaO	1.8–5.2	3.2–3.4	2.4–5.2	1.2–2.5
Na_2O	0.08	0.1–0.28	0.1–0.2	0.04–0.07
H_2O	0.016	0.22	0.001	—
K, ppm	69	130–220	59–218	40–62
U, ppb	26–34	20	17–33	13–18
Th, ppb	120	70	60–113	47–70
Core %				
Fe	93.5–94.5	79–89	64–88	82–86
Ni	5.5	4.8–5.5	8.0–8.2	12–13
S	0–0.35	1.0–5.1	3.5–9.3	0.9
O	—	8.0–9.8	0–18.7	—
Relative Masses:				
Mantle and Crust	32.0–35.2	68.0–76.4	81–82	87–92
Core	64.8–68.0	23.6–32.0	18.2–19	8–13

[†]Adapted from *Basaltic Volcanism* (1981) Table 4.3.2.

The available compositional data for Mars (Viking Lander XRF and gamma ray data, [75]) are consistent with a basaltic surface. The K/U ratios appear to be low (~3000), resembling the Moon in this parameter. Accordingly, Mars is expected to be depleted in volatile elements [76, 77] like the Moon. Based on these data and on theoretical models, a range in compositions is given in Table 8.5 [73] (see Sections 2.6, 5.13 and 6.8).

Some compositional data are available from the Venusian surface, with abundances resembling both basaltic and granitic rocks [75]. This is good evidence for differentiation and outward migration of the incompatible elements, presumably due to crystal-liquid fractionation in partial melting processes. The K/U ratios are variable (6500–18,000) which average out to about the value for the Earth [76]. The range in model compositions based on these data and on theoretical models is given in Table 8.5. The range includes the estimates of Morgan and Anders [76] (see Sections 2.7, 5.13 and 6.9).

8.6   Meteorite Parent Bodies and Satellites

This subject requires specialized book-length treatment in its own right and only brief mention can be made here. The question of the source of the meteorites appears to be in the asteroid belt, both from the observed orbits of Pribram and Lost City, and other evidence, such as the rare gas data which may forbid a cometary source [78]. Only the fractionated meteorites are considered here. The C1 and enstatite chondrites are discussed in Section 8.2.

The eucrites form a group of basaltic-type samples possessing many similarities in chemistry to lunar basalts. They have Cr, Ni, and Na contents notably similar to low-Ti mare basalts, although they lack the evidence of depletion of the light REE, which often appears in lunar basalts. Eu depletions are known only for Stannern and Nuevo Laredo, while Serra de Magé and Moore County have positive Eu anomalies. Most of the others have REE patterns parallel to chondrites. The other related meteorites (diogenites and howardites) are apparently simply related, the diogenites being dominantly composed of pyroxene, while the howardites are mixtures of eucrites and diogenites. The compositions of the eucrites could be derived by partial melting in small planetary objects (≤200 km radius), the generally flat REE patterns indicating derivation from a chondritic-type source [79]. The experimental petrological data are consistent with formation by varying degrees of partial melting from an olivine-pyroxene-plagioclase-spinel and metal source, with occasional formation of cumulates, and fractionation of plagioclase and pyroxene, all at low pressures [80]. A depletion in alkali elements similar to that of the Moon is consistent with the low alkali contents of the eucrites. The W-La correlations [81] and oxygen isotope ratios are similar to those for the Moon and the Earth. All these parameters point to a parent body with similar characteristics to the Moon (Table 8.5). The eucrites thus are apparently derived by single-stage melting of volatile and possibly siderophile-depleted, but otherwise chondritic, parent bodies [79]. The ancient radiometric ages $(4.6 \times 10^9$ yr.) apparently preclude two-stage processes while the trace element and petrological evidence is consistent with single-stage melting and crystal fractionation at low pressures. There is no inherent Eu anomaly, for example, in the source regions. Thus, there are intriguing similarities and differences between the Earth, Moon and the eucrite parent bodies. All have suffered volatile element depletion relative to the chondrites. The Moon and possibly the eucrite parent bodies are depleted in siderophile elements. It is not clear whether the depletion in Ni and other siderophiles is due to pre-accretion loss of metal, or to core formation in the parent body, although the former alternative appears more likely. The asteroid Vesta has been suggested as the parent body of the eucrites by various workers (e.g., [79]) but there are severe dynamical problems in deriving meteorites from that asteroid.

Among the more significant conclusions is that internal partial melting in the eucrite parent bodies has produced basaltic compositions, with ages very close to the condensation age of the solar nebula.

The shergottites, nakhlites and chassignites are unique in having well-determined young crystallization ages (620 million years for Shergotty, 1.3 billion years for Nakhla) [82]. The generation of young basalts in the solar system thus raises the interesting question of their location. Shergotty is of particular interest since it has trace element abundances very similar to terrestrial basalts. The relative abundances of the volatile and siderophile elements are close to those for terrestrial basalts [83]. The evidence that the abundances of siderophile elements in the terrestrial upper mantle may be due to a pre-4-billion-year meteoritic bombardment [54, 55] means that similar abundance patterns will be common throughout the inner solar system. A Martian source for these young basaltic meteorites has also been suggested [84].

The available data for the Jovian and Saturnian satellites are summarized in Table 7.1. The recent Voyager missions have resulted in refined estimates of radii, masses and densities, so that the rock-ice ratios are more firmly established. If the ice content of the Galilean satellites is removed, then they are all of size and density similar to the Moon [85], a fact of probable significance for the planetesimal hypothesis.

References and Notes

1. Morgan, J. W., et al. (1981) *Tectonophys.* 75: 47.
2. Daly, S. F., and Richter, F. M. (1978) LPS IX: 213.
3. Sharpe, H. N., and Peltier, W. R. (1979) *Geophys. J.* 59: 171; Cook, F. A., and Turcotte, D. L. (1981) *Tectonophys.* 75: 1.
4. Schubert, G., et al. (1979) *Icarus.* 38: 192.

5. Hsui, A. T. (1979) *PLC 10*: 2393.
6. Keihm, S. J., and Langseth, M. G. (1977) *PLC 8*: 499.
7. Duba, A., et al. (1976) *PLC 7*: 3173; Lambeck, K., and Pullan, S. (1980) *PEPI.* 22: 12.
8. Clayton, R. N., and Mayeda, T. K. (1975) *PLC 6*: 1761; Grossman, L., et al. (1975) *PLC 6*: 1799.
9. Clayton, R. N. (1979) PCE 11: 121.
10. Clayton, R. N. (1978) *Ann. Rev. Nucl. Part. Sci.* 28: 501.
11. Clayton, R. N., et al. (1973) *PLC 4*: 1535.
12. Smith, J. V. (1981) LPS XII: 1002, notes that a mixture of 47.5% H, 47.5% E and 5% of C2 chondrites matches the oxygen isotope signature of the Earth.
13. The terms siderophile, chalcophile, and lithophile elements were introduced by Goldschmidt to describe the distribution of elements among metallic, sulfide, and silicate phases; the classification stems from the observed distribution of elements among metallic, troilite (FeS), and silicate phases in meteorites. The behavior of an element is dependent on local conditions of temperature, pressure, oxidizing or reducing conditions, and so on, but the classification remains a useful first approximation to indicate the geochemical behavior of the elements.
14. Grossman, L., and Larimer, J. W. (1974) *Rev. Geophys. Space Phys.* 12: 71; Larimer, J. W. (1979) *Icarus.* 40: 446.
15. Ross, J. E., and Aller, L. H. (1976) *Science.* 191: 1223.
16. A summary of the element abundance data for C1 and most other meteorite types is given by Mason, B. H. (1979) USGS. Prof. Paper 440 B-1. The individual C1 values are listed in Table 84. This work revises the earlier compilation [Mason, B. H. (1971) *Handbook of Elemental Abundances in Meteorites*, Gordon and Breach, London], which is still a useful mine of information.
17. Becker, U., et al. (1980) *GCA.* 44: 2145.
18. Recent determinations of boron in chondrites indicate substantial agreement with solar abundances. Previously measured higher values in meteorites have probably been caused by terrestrial contamination [Curtis, D. B., et al. (1980) *GCA.* 44: 1945].
19. Gough, D. (1980) *Ancient Sun*, p. 533.
20. Smith, J. V. (1979) *Mineral Mag.* 43: 1, provides much useful information on meteorite chemistry.
21. Smith, J. V. (1982) *J. Geol.* (in press) provides much additional information and an appropriate perspective on the interrelationships of meteorites.
22. Kallemeyn, G. W., and Wasson, J. T. (1979) *Nature.* 282: 828.
23. Kerridge, J. F. (1979) *PLC 10*: 989.
24. Rambaldi, E. R., et al. (1979) *PLC 10*: 997.
25. Wasson, J. T. (1978) in *Photostars and Planets* (ed., T. Gehrels), Univ. Ariz. Press, p. 488.
26. Morgan, J. R., and Anders, E. (1980) *Proc. U.S. Nat. Acad. Sci.* 77: 6973.
27. Anders, E. (1971) *GCA.* 35: 516.
28. Wilkening, L. (1981) LPS XII: 1185.
29. Kerridge, J. F., and Macdougall, J. D. (1976) *EPSL.* 29: 341.
30. Morrison, D. (1977) *Icarus.* 31: 185; see however Gaffey, M. J., and McCord, T. B. (1979) in *Asteroids* (ed., T. Gehrels), Univ. Ariz. Press, p. 688 for a discussion on the difficulties of matching meteoritic and asteroidal spectra.
31. Noland, M., and Veverka, J. (1977) *Icarus.* 30: 200.
32. Ringwood, A. E. (1979) *The Origin of the Earth and Moon*, Springer.
33. Ahrens, T. J. (1979) *JGR.* 84: 985.
34. MacDonald, G. J. F., and Knopoff, L. (1958) *Geophys. J.* 1: 284.

35. E.g., see papers in Zartmann, R. E., and Taylor, S. R. (1981) *Tectonophys.* 75: 1.
36. Ahrens, L. H., and Von Michaelis, H. (1969) *EPSL.* 5: 395.
37. Jagoutz, E., et al. (1979) *PLC 10*: 2031; Palme, H., et al. (1978) *PLC 9*: 25.
38. Maaløe, S., and Aoki, K. (1977) *Contrib. Min. Pet.* 63: 161.
39. Bickle, M. J., et al. (1976) *Nature.* 263: 577.
40. Sun, S. -S. (1981) *GCA.* In press.
41. Ringwood, A. E. (1975) *Composition and Petrology of the Earth's Mantle*, McGraw-Hill.
42. *Basaltic Volcanism* (1981) Table 4.5.2.
43. Liu, L. G. (1979) *EPSL.* 42: 202.
44. Ahrens, L. H. (1970) *EPSL.* 10: 1.
45. Taylor, S. R., and McLennan, S. M. (1981) *Phil. Trans. Roy. Soc.* A301: 375.
46. Grossman, L. (1980) *Ann. Rev. Earth Planet. Sci.* 8: 559.
47. E.g., Taylor, S. R. (1975) *Lunar Science: A Post-Apollo View*, Pergamon.
48. Taylor, S. R. (1979) *PLC 10*: 2017; (1980) *PLC 11*: 333.
49. O'Nions, R. K., et al. (1979) *Ann. Rev. Earth Planet. Sci.* 7: 11.
50. Hart, R., and Hogan, L. (1978) in *Terrestrial Rare Gases* (ed., E. C. Alexander and M. Ozima), Japan Sci. Soc. Press.
51. Smith, J. V. (1977) *PLC 8*: 333.
52. Dreibus, G. (1977) *PLC 8*: 211.
53. Ahrens, T. J. (1980) *Science.* 207: 1035.
54. Chou, C. -L. (1978) *PLC 9*: 219; Morgan, J. W., et al. (1981) *Tectonophys.* 75: 47.
55. Smith, J. V. (1981) *Nature.* 289: 762.
56. See, for example, the range in values for Earth mantle compositions given in *Basaltic Volcanism* (1981), Table 4.3.2c.
57. Haines, E. L., and Metzger, A. L. (1980) *PLC 11*: 689.
58. Buck, W. R., and Toksöz, M. N. (1980) *PLC 11*: 2043.
59. *Basaltic Volcanism* (1981) Table 4.5.5.
60. Warren, P. H., and Wasson, J. T. (1979) *PLC 10*: 2051.
61. Gast, P. W. (1972) *Moon.* 5: 121.
62. See Ringwood, A. E. (1974) *GCA.* 38: 983 for comments on the position.
63. E.g., Brett, P. R. (1973) *GCA.* 37: 2697.
64. See *Basaltic Volcanism* (1981) Section 1.3.1 for detailed comparisons.
65. Longhi, J. (1977) *PLC 8*: 601.
66. Longhi, J. (1980) *PLC 10*: 289.
67. Taylor, S. R., and Bence, A. E. (1975) *PLC 6*: 1121.
68. Delano, J. W., and Ringwood, A. E. (1978) *PLC 9*: 111.
69. Wolf, R., et al. (1979) *PLC 10*: 2108. This paper and the one preceding provide good examples of the controversy over the siderophile elements.
70. E.g., Crocket, J. (1979) *Can. Mineral.* 17: 391.
71. Wolf, R., et al. (1980) *GCA.* 44: 2111.
72. Hertogen, J., et al. (1980) *GCA.* 44: 2125.
73. *Basaltic Volcanism* (1981) Section 4.5.
74. Morgan, J. W., and Anders, E. (1980) *Proc. U.S. Nat. Acad. Sci.* 77: 6973.
75. Surkov, Y. A. (1980) *PLC 11*: 664, 669.
76. Morgan, J. W., and Anders, E. (1979) *GCA.* 43: 1601.
77. Anders, E., and Owen, T. (1977) *Science.* 198: 453.
78. Anders, E. (1975) *Icarus.* 24: 363.
79. Consolmagno, G. J., and Drake, M. J. (1977) *GCA.* 41: 1271.
80. Stolper, E. (1977) *GCA.* 41: 587.
81. Rammensee, W., and Wänke, H. (1977) *PLC 8*: 399.

82. Nyquist, L. E., et al. (1979) *GCA*. 43: 1057.
83. Stolper, E. (1979) *EPSL*. 42: 239; Smith, J. V., and Hervig, R. L. (1979) *Meteoritics*. 14: 121.
84. Walker, D., et al. (1979) *PLC 10*: 1995; Wood, C. A., and Ashwal, L. D. (1981) LPS XII: 1197.
85. Russell, C. T. (1980) *Rev. Geophys. Space Phys.* 18: 95.

Chapter 9

THE ORIGIN AND EVOLUTION OF THE MOON AND PLANETS

9.1 In the Beginning

The age of the universe, as it is perceived at present, is perhaps as old as 20 aeons so that the formation of the solar system at about 4.5 aeons is a comparatively youthful event in this stupendous extent of time [1]. The visible portion of the universe consists principally of about 10^{11} galaxies, which show local marked irregularities in distribution (e.g., Virgo cluster). The expansion rate (Hubble Constant) has values currently estimated to lie between 50 and 75 km/sec/MPC [2]. The reciprocal of the constant gives ages ranging between 13 and 20 aeons but there is uncertainty due to the local perturbing effects of the Virgo cluster of galaxies on our measurement of the Hubble parameter [1]. The age of the solar system (4.56 aeons), the ages of old star clusters ($>10^{10}$ years) and the production rates of elements all suggest ages for the galaxy and the observable universe well in excess of 10^{10} years (10 aeons). The origin of the presently observable universe is usually ascribed, in current cosmologies, to a "big bang," and the 3°K background radiation is often accepted as proof of the correctness of the hypothesis. Nevertheless, there are some discrepancies. The observed ratio of hydrogen to helium which should be about 0.25 in a "big-bang" scenario may be too high. The distribution of galaxies shows much clumping, which would indicate an initial chaotic state [3], and there are variations in the spectrum of the 3°K radiation which are not predicted by the theory. Thus, the background radiation is surprisingly anisotropic and the question must be raised whether it is a relic of the big bang or the result of the superposition of many point sources (e.g., radiation from warm interstellar dust) [4–6].

There is some observational evidence that galaxies at the observational limits of about 10^{10} light years do not show the expected less-evolved characteristics, but appear to be as mature as nearby galaxies. Thus, galaxies 3C13 and 3C 427-1, at a distance of 10–11 billion light years do not contain the expected higher population of blue stars, which would be indicative of comparative youth [7].

It seems possible that we are still looking at only a small fraction of the universe and that the "big bang" hypothesis, if correct, is a local phenomenon, superimposed in a larger universe. The local group of galaxies (17 members including the Andromeda galaxy, M31) extend over 3 million light years. The expansion of the universe begins outside this group. Nevertheless, the local group is moving toward the Virgo cluster and this relative movement is superimposed on the general Hubble expansion.

The chemical composition of the universe is dominated by hydrogen and helium. The heavier elements are produced in stars and supernovae and returned to the interstellar medium, both by supernovae explosions and by mass loss at the red giant stage. A slow enrichment of the heavy elements occurs, so that younger stars (Population I) contain higher concentrations of elements heavier than helium than the older "metal-poor" stars of Population II. The galaxies show abundance gradients so that the central areas contain higher concentrations of heavy elements than the outer regions with the globular clusters in the galactic haloes having the lowest concentrations. This seems consistent with an evolutionary sequence of collapse from spherical to disk form with time [8]. Some progress is being made in understanding the origin of the beautiful spiral structures in many galaxies, including our own [8]. The spiral structure exists only in galaxies containing both gas and dust in addition to stars (typically, there are about 10^{11} stars in a galaxy), and is ascribed in current hypotheses to the passage of "density wave." The resultant perturbation of the gravitational field causes a shock wave to impinge on the gas and dust clouds, which collapse to form stars. The spiral arms are outlined by the production of massive young stars which soon evolve to a supernova stage (of which more later).

Typical interstellar clouds have densities ranging from 10 to 1000 hydrogen atoms/cm^3 and temperatures of $100°$ K [9]. Such gas clouds with masses from a few hundred to a few thousand solar masses will begin to collapse when the cloud enters the increased pressure regime of a galactic spiral arm. Thus, many stars will form in rather restricted regions. The collapse of the diffuse clouds into ones with densities typically 10^4 times that of the diffuse clouds leads to conditions favorable for the production of complex molecules. They are accordingly often labelled "molecular clouds" [10]. About 50 molecular species have been identified, with up to 11 atoms. Species as diverse as ethyl alcohol (CH_3CH_2OH), acetaldehyde (CH_3CHO), and propionitrile (CH_3CH_2CN) exist, as well as the simpler molecules of cyanogen (CN), hydroxyl (OH), ammonia (NH_3), water (H_2O), carbon monoxide (CO), SiO, SO, SO_2 and SiS [11]. Dust grains typically comprise 1% of the dense clouds. The molecular/atomic hydrogen ratio is high, so that ionization is low and the clouds do not behave as a plasma. The effects of magnetic fields on element abundance and distribution may therefore be low. Masses in dense clouds range from 20 to 10^5 solar masses and temperatures from 10 to 70°K. The cores of some dense clouds may have strong infrared sources, indicating an early stage of stellar formation.

The classic example of recent stellar formation is the Orion Nebula where star clusters are less than 10^6 yr. old and where star formation as revealed by infrared sources is probably occurring at present. A significant point is that star formation is not an isolated phenomenon. The neighborhood of a star of solar mass will also be populated with a large number of stars of larger mass (about 10–30 times solar mass). These massive stars will evolve on time scales of 10^6–10^8 years, finally exploding as Type II supernovae with an energy release of the order of 10^{51} ergs [12]. The supernovae produce two effects. Shock waves impinging on nearby dust and gas clouds (Fig. 9.1) will assist or trigger collapse into stellar and perhaps planetary systems. A second effect is that the supernovae will contribute newly formed (particularly r-process) isotopes, which may be incompletely mixed into the relatively homogeneous dust and gas clouds, at least on a short time scale.

The diffuse and the molecular clouds are expected to be broadly homogeneous for most elements and the isotopic heterogeneities which existed in the solar nebula are considered to result from supernovae occurring just before condensation of the solar system. Mixing times are uncertain, but are probably on the order of 10^6 years [13, 14].

9.2 Initial Conditions in the Solar Nebula

The concept of a well mixed and homogeneous solar nebula dominated cosmochemical thinking until about 1974. Models of the composition of the nebula at that time generally specified that it had an overall composition equivalent (except for H, He and the rare gases) to that of C1 carbonaceous chondrites. The basis for that assumption was the equivalence in composition between the solar and C1 abundances (Section 8.2, Fig. 8.1). Most models also considered the solar nebula in isolation, containing enough material from which to form the sun and the planets, but otherwise separate from, and not interacting with, other stellar systems. From the preceding section, it is clear that star formation is not a solitary event, but occurs in association with many other such events. The problem of segregating the local solar nebula may be due to localized shock waves from nearby supernovae.

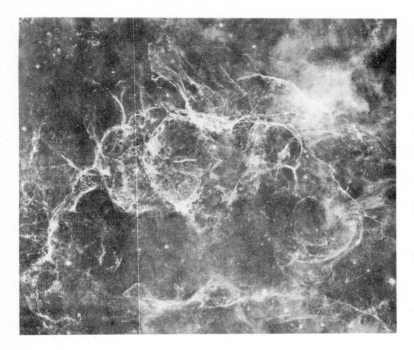

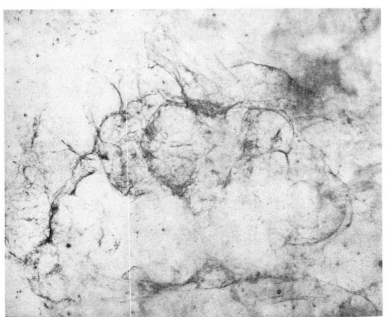

9.1   The Vela nebula. Shock waves from a supernova 30,000 years ago interact with gas and dust clouds in the interstellar medium. Compression of this material, perhaps a molecular cloud, may lead to star formation. Anomalous isotopic abundances produced in the supernova may be mixed into the interstellar gas and dust. (Courtesy Siding Spring Observatory.)

The clearly established evidence that the solar nebula was not isotopically homogeneous dates principally from the discovery of the ^{16}O anomaly [15]. This and other isotopic anomalies led to the concept that these were supplied by a nearby supernova, which acted as a trigger for condensation of the nebula and formation of the solar system. There is no shortage of such supernovae, but rather an oversupply in young stellar systems. A typical OB stellar association has 20 supernovae occurring within a period of 10^7 years or one each 10^5 years. Supernovae remnants fade into the interstellar background at about 30 parsecs ($\sim$100 light years) [16].

The concept of a supernova trigger to initiate condensation can supply the observed isotopic anomalies for ^{16}O and ^{26}Mg (derived from ^{26}Al with a half-life of 7.2×10^5 years) [17, 18]. In addition, the presence of excess ^{107}Ag due to ^{107}Pd (half-life of 6.5×10^6 years) in iron meteorites indicates that ^{107}Pd was available for incorporation in a metal phase within a few million years of formation, placing stringent new time limits on meteorite formation [19]. One problem with the supernova trigger model is that the evidence for r-process [20] isotopes produced during the supernova explosion (except for ^{26}Al and ^{16}O) appears to be lacking [21, 22].

For this reason, the alternative model has been proposed in which the isotopic anomalies were contained in "pre-solar grains" produced in expanding supernova shells, unrelated to the condensation of the solar system or to the local solar nebula [23, 24]. A more generally accepted model favors two supernovae. It has been known for a long time that the ^{129}Xe excess in primitive meteorites is due to the existence of ^{129}I (half-life of 16×10^6 years) [25]. Likewise, ^{244}Pu (half-life of 82×10^6 years) produces detectable fission Xe [26]. Along with other longer-lived radionuclides (^{238}U, ^{235}U, ^{232}Th, ^{187}Re), these are produced by the r-process. ^{26}Al is not so produced. Accordingly, models involving two supernovae events have become popular. The first, $\sim 10^8$ years before condensation, adds ^{244}Pu, ^{129}I and a spike of other r-process nuclides. A second supernova, which injects ^{26}Al, ^{107}Ag and ^{16}O, triggers the condensation 10^8 years later. The time interval for the rotation of the galactic density wave is about 10^8 years, which has led to models in which the solar nebula was successively in two spiral arms [27]. All these models are consistent with the observed conditions in the Orion nebula and other similar situations (e.g., Canis Major)[28].

The next question to be addressed concerns the thermal state of the nebula. Was the early nebula hot (1800° K) with the dust grains completely vaporized? Did it then condense, providing the sequence of condensation products observed in some meteorites (e.g., Allende [29]). Alternatively, was it never much hotter than perhaps 300° K? This important question is fundamental to our interpretation of the chemical evidence from the meteorites and has been the cause of much controversy [30–32].

The early assumptions of cosmochemists that the nebula was homogeneous led naturally to the assumption that an overall high temperature stage was needed to account for the presence of high temperature minerals. Mixing in a gas phase will smooth out isotopic heterogeneities in gas clouds on a time scale of 10^6 years and the heterogeneities which we observe could not be preserved for 10^8 years (the rotation period for the galaxy) [13]. If solid phases condensed from supernova ejecta on short time scales (1 year) such isotopic heterogeneities could be preserved and would not be modified by equilibration with the interstellar gas.

The general evidence from the meteorites is that the solar nebula was not well mixed, as attested by the survival of the isotopic anomalies. Accordingly, the whole nebula could not have been at high temperatures, a verdict in accordance with the astrophysical evidence. The most primitive meteorites, the C1 carbonaceous chondrites, with solar elemental abundances, contain an agglomeration of particles which have not been heated since they were aggregated. The evidence for high temperature phases in meteorites is attributed to melting or metamorphism within parent bodies [32]. The likely heat source was ^{26}Al. The presence of ^{107}Ag anomalies in iron meteorites indicates that metal phases capable of incorporating ^{107}Pd were present on the extremely short time scale of 10^6 years following the supernova production of Pd. High temperatures capable of vaporizing the whole nebula would have equilibrated the ^{16}O anomaly. Oxygen has the property that it can exist both within mineral lattice sites and as a gas in the nebula. A gas phase might be expected to equilibrate quickly, so that the continued presence of ^{16}O anomalies favors their addition to the nebula in solid phases.

In this context, the Neon-E (^{22}Ne) anomaly may be mentioned [33, 36]. Meteoritic Neon contains three components:

(1) The primordial or "planetary" component from the solar nebula (see Section 4.13)

(2) A solar wind implanted component

(3) A component produced by interaction with cosmic rays (= cosmogenic).

Primitive low temperature meteorites also contain a component (E) with an excess of ^{22}Ne, possibly derived from ^{22}Na (half-life of 2.6 years). Possibly it is derived from the triggering supernova, or is carried within pre-solar dust grains from previous events. Its presence attests to incomplete mixing or vaporization in the nebula, and probably indicates the survival of at least some interstellar grains [34, 35]. It should be noted that the supernova additions of "anomalous" isotopes do not have a discernible effect on the bulk composition of the nebula, but provide evidence in favor or a low temperature, heterogeneous, not well-mixed nebula, with an important short term heat source in the form of ^{26}Al.

In summary, the initial state of the nebula is assumed to be a cold cloud of dust and gas, injected with debris from at least two supernovae events, 10^8

years apart. The latter did not produce r-process elements, but triggered the condensation of the nebula. Such models do not allow a major role for condensation sequences. The evidence for high temperature is attributed to local evaporation and condensation in supernova shock waves, with the high temperature grains injected into the nebula. A subsequent complex history involves melting in condensed bodies (e.g., [37]).

Our present understanding is well summarized in the statement by Wetherill: "Many of the isotopic anomalies in the early solar system . . . probably have their origin in supernova ejecta, heterogeneously mixed into the solar system during its formation, probably including material from a supernova which triggered the collapse of the interstellar molecular cloud that evolved into the sun and other stars" [38].

9.3   Primary Accretion Models and "Condensation" in the Nebula

The origin of the solar system is one of the basic scientific questions, and most societies, mythologies, religions and philosophical systems have addressed this problem. Historical treatment of the various hypotheses is well beyond the scope of this book (see [39] for a recent review). In common with other areas of human knowledge, this field was dominated by deductive methods of inquiry until the Renaissance, and this style, so common in medieval thinking, occasionally still appears in modern hypotheses. Most progress is made, however, by inductive methods of inquiry and the appearance of unanticipated new data, of which the oxygen isotope evidence for heterogeneity in the nebula and the atmospheric rare gas data for Mars and Venus provide recent examples.

Two basic models are currently proposed to derive the final form of the solar system from the initial dispersed solar nebula. These may be referred to respectively as the protoplanet and the planetesimal hypotheses. The first suggests that collapse into large massive protoplanets is caused by gravitational instabilities in the nebula (e.g., [40–42]). The second hypothesis begins with dust grains, which form larger aggregations, growing in size from millimeter to meter, to kilometer sized objects, to asteroids, and finally accumulating into lunar and planetary-sized bodies (e.g., [43–45]). The mechanics of neither process are well understood, but the weight of the evidence as assessed in this chapter appears to favor the planetesimal hypothesis for the formation of the inner planets. Some portions of the giant gaseous protoplanet hypothesis may be applicable to the formation of Jupiter and Saturn, while the planetesimal hypothesis accounts for the smaller planets and satellites. Among other evidence in favor of the latter hypothesis, large-scale impacts (Chapter 3) are common throughout the solar system at least out to Saturn (e.g., Mimas, Dione) and probably account for the well-known obliquities of the planets. The extreme case is Uranus, inclined, along with its equatorial ring and satellite system, at 90° to the plane of the ecliptic.

Various attempts have been made to simulate the mechanism by which grains stick together in the initial stages of accretion (e.g., [46]). It is now generally supposed that weak electrostatic forces (Van der Waal forces) are probably adequate to accomplish the initial accretion of dust grains into centimeter-sized objects. The evidence from meteorites is particularly critical at this stage [47]. The meteorites do not appear to have formed by aggregation of equilibrium condensation products, but display much evidence of non-equilibrium processes. Even the most primitive of the C1 chondrites contain over 20 minerals in disequilibrium relationship. The bulk composition of the C1 chondrites reflects the overall uniformity of the nebula for major elements. The complex mixtures observed must be due to repeated recycling of material, possibly during impact processes on asteroidal surfaces. The uniformity of composition of the primitive meteorites on scales of a centimeter or so rules out direct formation of planets by accumulation of material, such as the Allende inclusions, into planetary-sized bodies. Instead, a complex hierarchy of growth, disruption, and reaggregation appears to be needed [45].

Many collisional episodes also appear necessary, as was emphasized by Urey [48]. Primary aggregation into small bodies (up to 1 km) seems an adequate process to account for the evidence. Rather than favoring condensation from a gaseous nebula, the evidence is consistent with an agglomeration of material with various histories. Most of the information has been destroyed as the small bodies are aggregated into larger objects. This primary stage of accretion into objects of about one kilometer in size probably proceeds quickly. No chemical fractionation occurs, and objects resemble the C1 meteorites in composition: a complex aggregate. Accretion to meter and kilometer-sized objects may not need a special sticking mechanism, but may be a natural consequence of low velocity collisions [49].

Before considering the second and subsequent stages of the accretionary process, it is necessary to revisit the chemical evidence for fractionation within the solar system. There is abundant evidence in the meteorite record of a wide variety of conditions which include condensation from high temperatures, gas-solid equilibria, and variations in oxidation and reduction. These effects have been discussed for many years (e.g., [50]). Difficulties with the concept of simple equilibrium cooling of a gas from temperatures greater than 1750° K and pressures of 10^{-3} to 10^{-6} atm have frequently been noted [50–52]. The observed heterogeneity of the nebula in ^{16}O which would rapidly equilibrate at high temperatures has already been commented upon. If ^{16}O excess had been added as a gas, it would have equilibrated. Accordingly, it was probably added as a solid condensate. The observed condensation sequences in meteorites and the isotopic anomalies are consistent with a complex scenario involving the injection of supernova debris, the passage of a shock front, localized high temperatures, and selective evaporation and condensation, both of grains with ^{16}O excess and other dust grains originally present in the nebula. The complex relationships among the REE [53] record several episodes of condensation and separation of grains and gas. This indicates a chaotic state, rather than a simple equilibrium cooling. Such turbulence might be expected to occur, with the development of convection, during the passage of successive shock waves during the supernova event and the dissipation of 10^{51} ergs. No other process appears capable of providing adequate energy sources, particularly at distances of a few astronomical units from the developing sun. This model calls for very early (pre-solar) formation of the "condensation" sequences observed in meteorites. Whether the evaporation and recondensation occurs outside the present solar system, or in the nebula (induced by the passage of the supernova shock waves), is not clear.

Complex evaporation and recondensation scenarios of this type probably account for the selective depletion of volatile elements, addressed in the next section, which are typical of meteorites and planets in the inner solar system. If the volatile elements failed to re-condense following evaporation, such material might be able to be swept out of the inner solar system along with H, He and the rare gases during the T Tauri phase of solar evolution, and thus not be available to be incorporated in the Earth, Moon, Venus or Mars.

The scenario described in this section can account for the primitive meteorites, the ordinary and enstatite chondrites. The differentiated meteorites form a little later.

9.4   Variations Among Refractory, Volatile, Chalcophile and Siderophile Elements

The principal differences among the planetary and satellite compositions noted in Chapter 8 lie among variations in these groups of elements. Properties based on relative volatility and siderophile character dominate. The meteorite compositions show similar differences. The diversity among asteroid compositions is considerable and indicates the complexity of processes which have operated in a local portion of the solar nebula [54], a region thought to be uniform in most early theories. Many distinct parent bodies of differing composition are required by the meteorite evidence. Clearly, substantial element fractionation has occurred at some stage between the primary accretion and the final state of the solar system. The ordinary chondrites show depletion in volatile elements and variations in siderophile element content. Such differences among meteorites could be due to small scale phenomena, but the bulk earth is depleted in K and Rb by a factor of four compared to the chondritic abundances, while the Moon shows both an enrichment in refractory elements and a depletion in volatile and siderophile elements. None of these elements can be lost once the planet is assembled so that the relative refractory/volatile element ratios (e.g., K/U) are fixed for the bulk planetary composition. Other planets show differences in bulk composition which are compatible with this scenario. The differences cannot be reconciled with distance from the sun or initial condensation compositions although much effort has been expended in such studies [55]. If the nebula was never hot enough for total vaporization, mixing and homogenization to occur, then such sequences, although important perhaps locally, cannot be the primary cause of the major element fractionation. The variations are most readily demonstrated by K/U and similar ratios (Fig. 8.4). The variations discussed in Chapter 8 involved fractionation in small bodies, selective breakup and reaccumulation. No adequate mechanism to heat the solar nebula to temperatures above 1000°C appears to exist, on the scale necessary. The major difference in refractory to volatile element ratios between the Earth and the Moon, despite the similarity in oxygen isotopic abundances, seems to require that local conditions dominated over those based on heliocentric distance. This is also apparent from the planetary atmospheric rare gas data (Section 4.13). The variations in siderophile element abundances are not explicable by volatility differences, although they are mainly responsible for the density differences. This indicates that metal and silicate fractions were already present at an early stage in nebula history and were accreted in varying proportions.

9.5   Early Solar Conditions

At some stage during the accretion process, the sun forms and begins to radiate. It does not exist earlier than disk and planetary formation since it clearly did not sweep up or accrete all the material in the original nebula. Solar material, in turn, is not subsequently returned (except in minor amounts in the solar wind) to the planets and meteorites, which contain, for example, Li and Be abundances in excess of those in the sun (Fig. 8.1).

Most mass loss from stars occurs late (post-main sequence) in stellar evolution during the red giant stage, or later in supernovae events [56]. The most important effect in the context of this review is the heating and dissipation of the solar nebula out to a few astronomical units. Jupiter and Saturn contain near solar proportions of hydrogen and helium, and have satellites with large proportions (~50%) of ice, so that temperatures in that portion of the nebula never rose above perhaps 300° K. Closer into the sun, H, He and the rare gases have effectively vanished, consistent with removal before they could be trapped in terrestrial or moon-sized objects. The most important process is usually ascribed to the so-called T Tauri stage of stellar evolution, in which a strong solar wind sweeps at least the inner portion of the solar nebula clear of H, He and the rare gases, and sometimes removes early planetary atmospheres as well. Such scenarios imply that most of planetary formation has occurred before the T Tauri stage of stellar evolution had been reached. T Tauri stars are pre-main sequence objects, typically 10^6 years old [57] with strong infrared radiation, assumed to indicate a warm gas and dust disk ($\sim700°$ K) surrounding the star. In these objects it is probable that planetary formation has already occurred since the nebula is thin enough to permit stellar radiation to be observed. "If the planets are largely formed before the T Tauri stage, then they must do so within the first 10^5 or so years of their lifetimes" ([57], p. 705). The T Tauri stage might thus assist in the post-planetary cleanup of the nebula, but many current models require much longer times for planetary formation (e.g., 10^8 years) [44]. A possible reconciliation is that primary accretion into kilometer-sized objects, resistant to sweep-out, occurs before the T Tauri stage, but that accretion into larger objects does not occur until after clearing of the solar nebula. Volatile elements present in finely dispersed phases might be so lost.

There is some question among astrophysicists of the efficiency or indeed the ability of a T Tauri enhanced solar wind to sweep away the nebula (Fig. 9.2) but observational evidence appears to support the concept that strong stellar winds can remove interstellar gas [58].

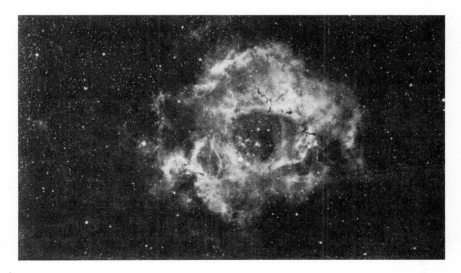

9.2   The Rosette Nebula. The cavity in the center has a diameter of a few light years and may be due to strong stellar winds from the OB stars in the center. This should not be confused with the effects due to supernova shock waves [58]. The Cerro-Tololo Inter-American Observatory.

9.6   Secondary Accretion Process

In this stage, the gravitational effects cause the small (kilometer-sized) objects to accrete into bodies of a few hundred kilometers radius. Melting due to the presence of ^{26}Al causes some of these bodies to differentiate into silicate mantles and metallic cores, while repeated breakup and reaccumulation occurs. The length of this process, which finally results in the accretion of the inner planets, may be as long as 10^8 years [44]. Selective melting and evaporation during impacts, followed by recondensation, may mimic theoretical condensation sequences from an initially hot nebula. The material at the primary accretion stage is likely to be highly reduced, in the presence of excess hydrogen. As the nebula clears, oxidizing conditions will predominate. Possibly to this stage belongs the formation of the magnetite platelets, which, together with the phyllosilicates, comprise a large proportion of the primitive carbonaceous chondrites.

At the secondary accretion stage, the differentiated meteorites are produced in asteroid-sized bodies. These include the basaltic achondrites (eucrites, diogenites, howardites, mesosiderites, ureilites, pallasites, shergottites, etc.) [45, 47]. Thus, the sequence for meteorite formation is enstatite chondrites (reduced) in the inner solar system, carbonaceous chondrites (oxidized), followed by the ordinary chondrites. Chondrule formation is not primary, but secondary in this scenario. Formation of chondrules as primary nebula condensates has been generally abandoned as a hypothesis. Current models propose an origin for these enigmatic objects by melting of pre-existing dust [59]. Accretion into meter- and kilometer-sized objects occurs at this stage. Clearing of the solar nebula also occurs, causing depletion of the volatile elements. The secondary accretion to objects about 1000 km in diameter occurs from varying mixtures of free metal and silicate depleted to varying degrees in volatile elements.

The growth of large planetary objects from this material in the inner solar system has been addressed by many workers (Fig. 9.3). We see abundant evidence for the final stages of the accretionary process in the form of large impact craters and basins [60]. Accretion of large objects in this manner explains many observations. Two criteria appear to apply, based on the lunar example. First, intense early cratering episodes were endemic in the solar system (Chapter 3), and second, initial melting and differentiation was a common feature of the accretional process.

Clearly, the bombardment extended from Mercury to the Saturnian satellites and probably to Uranus. The obliquities of the planets, which form a classical small statistical population (all planets and satellites differ in size, density, composition, inclination to the plane of the ecliptic, and subsequent evolution), are readily explained by massive late collisions [43]. This involves heterogeneous accretion since the planetesimals are not uniform in chemistry

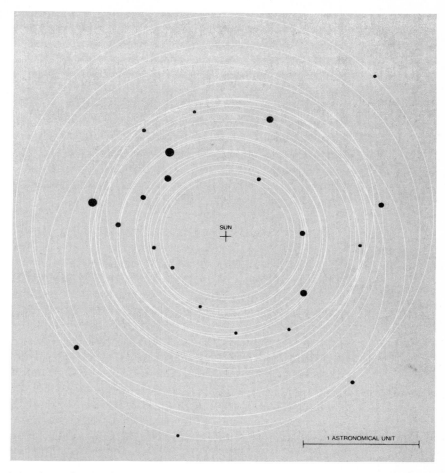

9.3   A simulation of an intermediate stage in the formation of the inner solar system, according to the planetesimal hypothesis. From an initial starting condition of 100 planetesimals (with mass totalling that of the inner planets), this diagram represents conditions after 30 million years, when accretion has resulted in the growth of 22 bodies. Total sweep-up to form the four planets takes about 10^8 years. From "The Formation of the Earth from Planetesimals" by George Wetherill. Copyright © 1981 by Scientific American, Inc. All rights reserved.

or isotopic signatures. Numerous distinct parent bodies are required to account for the presently observed meteorites [18]. These come mainly from the asteroid belt. No stony meteorite has an exposure age greater than 100 million years, indicating that a continuing supply of objects was available. The recent discovery of a probable H group chondrite in Ordovician limestone

[61]indicates that ordinary chondrites have been in Earth-crossing orbits for at least 500 million years.

The overall evidence thus favors a brick-by-brick buildup of the planets. The physical details of this process cannot be addressed here, but certain chemical consequences can be outlined. In the succeeding sections, the initial states of the Moon, Earth and other planets are considered, as well as their subsequent evolution. The concept that the planets are built from objects which have already undergone chemical fractionation (so that varying amounts of siderophile, refractory and volatile elements are accreted), is a central theme for explaining the varying compositions of the planets.

Heterogeneous accretion models in the general sense are required by the data, but within individual planets or satellites, this evidence may be largely lost if melting subsequently occurs. The curiously colored surfaces of some of the Saturnian satellites may be a relic of heterogeneous accretion. On the Moon, the chemical evidence indicates homogeneous accretion, or homogenization following this event, due to moon-wide melting, so that primary evidence for heterogeneous accretion is missing.

The presence of the giant planets, Jupiter and Saturn, comprised mainly of hydrogen and helium, accreted on to rocky cores, may indicate that they were once giant gaseous protoplanets, but they can also fit the accretion model. Once a planet growing by planetesimal accretion reaches a certain size (10–50 Earth masses), it will be able to accrete H and He from the nebula and grow to the size of Jupiter and Saturn. This scenario requires that H and He be present at 5 astronomical units during this stage [62, 63]. The outer planets, Uranus and Neptune, either did not grow large enough to accrete H and He, or else the nebula had lost much of its H and He by the time they did so. Clearly, time constraints are quite critical for the large planets with respect to accretion and clearing of the nebula. Possibly the early sun and the T Tauri stage were effective in removing gases out to about the asteroidal belt, and proto-Jupiter received a massive amount of H and He from the region of the inner planets. Condensation and accretion of the sun in relation to this sequence of events are dimly understood (except that it cannot form too early), but may have contributed toward the clearing of the inner solar system. It is clear that both Jupiter and Saturn grew in the presence of a gas phase. If this had been present during the accretion of the Earth, then it is reasonable to suppose that the Earth would contain somewhat larger amounts of trapped rare gases (e.g., Ne, Xe). Accordingly, the giant planets probably formed somewhat earlier than the inner planets. Such a scenario is consistent with the commonly held view that both the small size of Mars and the absence of a planet at the location of the asteroid belt are due to preferential accretion of material in this region by the growing Jupiter.

The low abundance of material in the asteroid belt is thus probably due to the effect of Jupiter in sweeping up material. The entire mass of the present asteroids is only about 0.3 lunar mass. It is sometimes suggested that collisions in the belt have destroyed larger objects. However, planetesimals larger than 1000 km diameter would still survive, if originally present [64].

9.7 Formation of the Moon

This book emphasizes lunar research, and its impact on our understanding of the inner solar system. For this reason, the question of the formation of the Moon rather than that of the Earth is treated first, followed by consideration of the Earth and the other bodies in the solar system. This displacement of the Earth from its usual central position is deliberate. Our understanding of the evolution of the Moon is better than that of the Earth, and has led to new insights into Earth history. The bulk composition of the Moon is probably somewhat better understood than that of the bulk composition of the Earth, where some major problems remain. The composition of the lower mantle, for example, is not yet well constrained, while that of the lower crust is also model-dependent. The role of the refractory, volatile and siderophile elements in determining planetary and meteoritic compositions was properly appreciated only after studying the lunar samples. Accordingly, the constraints on lunar origin provide useful information about the other planets. It is sometimes asserted that the Moon is an anomaly in the solar system. This view received some credence when the Moon could not be fitted into systems which equated planetary compositions with distance from the sun. When the best understood object fails to fit into the theoretical sequence, then some re-examination of the basic assumptions in the hypothesis is called for.

Comprehensive accounts of the many ideas about planetary origins have appeared recently [39, 65, 66] and will not be repeated here in detail.

The equilibrium condensation theory, which starts with an initial hot nebula, and which forms the planets at successive distances from the sun, encounters many objections. Although there is a superficial and facile attraction, the lack of evidence of an initially hot nebula is a fatal flaw. No hydrated minerals should be present in the Earth, Venus or Mars. The Moon is an anomaly in this scenario.

Variations on this theme, which accrete the Moon from mixtures of high and low temperature condensates, suffer from other difficulties. The high temperature condensates are enriched in the refractory siderophile elements such as Re, Os, and Ir. To dispose of these, it is necessary to accrete iron (at a lower temperature!) so that they may be removed by a core. A second incompatible situation arises from the oxygen isotope data. A substantial amount of low temperature condensate must be added (with a high component of volatile elements) to satisfy the oxygen values. Although the REE are refractory, both Eu and Yb are relatively volatile and the Allende inclusions

' enrichments or depletions of these two elements. Although ...scribed to crystal-liquid fractionation, are ubiquitous on the anomalies are not observed, thus constraining the amount of high ...crature condensates.

The heterogeneous accretion model sweeps up each component as it is condensed. This model also requires total vaporization and recondensation. The Moon, although comprised of refractory components, lacks the evidence of high iridium concentrations as required by the model. Neither theory accounts for the isotopic anomalies.

The reduction-during-accretion model faces the dilemma of removing the H_2O and CO_2 produced, if carbon is the reducing agent. An alternative process involving hydrogen requires special timing which seems unlikely. It was noted earlier (Section 9.6) that Jupiter and Saturn most likely formed in a gas-rich nebula before the inner planets, and certainly before Mars and the asteroids, although the time interval may be short. For this reason it seems unlikely that the inner planets accreted in a hydrogen-rich atmosphere [44]. Accordingly, models which sweep up a heterogeneous collection of differentiated planetesimals seem best to account both for the varying properties and chemistry of the planets and for the meteorite evidence.

The classical attempts to explain the origin of the Moon fell traditionally into three types of theory, which were labelled as capture, fission, or double planet. It has been realized more recently that such an approach is an oversimplification, and that probably elements of all three theories may contribute. However, some constraints on theories may be applied from the chemical evidence and some hypotheses may be ruled out as major contributors to lunar origin. Capture of an already formed Moon has been deemed "horrendously improbable" and has a *deus ex machina* flavor [67]. Apart from the dynamical difficulties, the observed heterogeneity of the nebula could be expected to contribute distinctive isotopic or chemical signatures. The similarity in oxygen isotope characteristics among the Earth, Moon, enstatite chondrites and aubrites [18] indicates that these objects accreted from a similar reservoir. An unknown aspect is that large objects in the solar system might all contain similar oxygen isotope signatures, having accreted a broad enough range of material to fall on the terrestrial line (Section 8.1). However, the carbonaceous chondrites are displaced, despite their resemblance to solar compositions, so that large-scale (~ 1-2 A. U.) differences may be present in addition to the smaller scale features preserved in the various meteorite classes.

The fission hypothesis has been extensively debated, since in principle, it is the most readily tested of the competing hypotheses. In the simplest version, the bulk composition of the Moon should resemble that of the terrestrial mantle. We have seen in Chapter 8 that there are substantial differences in composition, not only for the volatile elements, but also in major elements.

A number of critical objections appear to rule out the fission hypothesis for the origin of the Moon from the Earth. Those include:

(a) The Mg/Si ratio for the bulk moon appears to be outside the range of possible Mg/Si values for the mantle of the Earth.

(b) There is 4.6 aeon lead on the Moon, indicating that the Moon formed from material which had not participated in the core-forming event on the Earth.

(c) The Al, U, Ti, Ca and other refractory elements are enriched by a factor of two in the Moon relative to the Earth.

(d) The volatile elements and highly volatile elements are depleted.

(e) Vaporization of the terrestrial mantle could have caused fractionation of the oxygen isotopes, with enrichment of ^{16}O. This is not observed.

(f) The Ni and Ni/Fe ratios in the Moon are too low to be derived from the terrestrial mantle. The complex history of the mantle with respect to the distribution and abundance of the siderophile elements probably makes comparisons invalid for the trace elements. Thus, there is good evidence that the abundances of Os, Ir, and Pd in the Earth's upper mantle have chondritic ratios. These are probably due to a late addition of material from meteorites. Accordingly, there is not a unique terrestrial siderophile element signature.

(g) The volatile elements (Au, Sb, Ge) are depleted by factors of about 100 in the Moon, while the very volatile elements Bi, Tl and Sn are depleted rather uniformly by factors of about 50 in the Moon relative to the Earth. Chromium, a relatively volatile element is enriched on the Moon relative to the Earth. The alkali elements K, Rb and Cs are depleted by a factor of about two between the Moon and the Earth (Section 8.5.4). All these differences call for such a complex scenario as to make the hypothesis untestable. If the Moon was derived from the terrestrial mantle by fission, then the chemical evidence for such an event has been destroyed.

Some limits may be placed on pre-accretion temperatures in terms of bulk lunar compositions. Lunar Th/U ratios are 3.8, equivalent to chondritic and terrestrial values with no selective loss of more volatile U, as is seen, for example, in Group II Allende inclusions [29]. Lunar REE patterns are chondritic with variations which are explained by crystal-liquid fractionation (e.g., Eu anomalies, La/Yb ratios). The general tendency for some depletion in the heavy REE is explicable, in the cumulate model, by some preferential removal of the heavy REE by early orthopyroxene crystallization (see Section 6.5.2).

The absence of Yb anomalies indicates no selective loss of Eu and Yb due to selective volatilization, as is commonly observed in Allende inclusions [29, 68]. Palladium, more volatile than the other Pt group elements, does appear to be relatively depleted, but the data are extremely sparse. Silver and Au are likewise depleted, relative to the other siderophile elements. Conversely, Re,

refractory siderophile elements, is also depleted. These
indicate that the pre-lunar fractionation occurred at lower
..tures than those experienced by the Allende inclusions.

The timing of the initial lunar differentiation is of interest since it provides a lower age limit for the accretion of the Moon. The most precise time
comes from the lead isotopic systematics which indicate an age of 4.47 aeons
[69]. Thus, there are tight time constraints on fission models, which have to
extract the Moon from the Earth, following core formation. The number of
objections to testable versions of the fission hypothesis make it unprofitable to
consider this theory further at present.

Direct accretion from the solar nebula of a refractory component is
forbidden by the depletion in refractory siderophile elements, and indirectly
by the ^{16}O evidence and the other isotopic evidence for heterogeneity.

A number of events have to be fitted in between the collapse of the solar
nebula, and the formation of the Moon. These include:

(a) separation of siderophile elements,
(b) loss of volatile elements,
(c) accretion of the Moon,
(d) melting most probably of the whole Moon,
(e) formation of the highlands crust and the source regions of the mare
    basalts by 4.4 aeons or possibly by 4.47 aeons.

All these events have to be accomplished within 100–200 million years of
the collapse of the nebula. The evidence reviewed in this book suggests that the
inner planets accreted from a mixture of metal and silicate, which has previously undergone varying degrees of volatile element loss. This takes place in
the primary accretion stage (Section 9.3).

In this scenario, the accretion of the Moon occurs simultaneously with
that of the Earth, rather than sequentially. This conclusion follows from the
lack of time to carry out all the steps outlined above. If the planets take about
10^8 years to accrete, then there is no time to allow sequential formation of the
Earth and the Moon. Accordingly, since capture is effectively ruled out, a
simultaneous growth of the Earth and the Moon in association is required.
This conclusion is in accord with other features of the solar system. Capture of
small satellites appears to be a reasonably frequent event (e.g., Phobos,
Deimos, and Jupiter 6–12) in the early solar system, and indeed follows from
the planetesimal theory that there should be large numbers of 100–1000 km
diameter objects formed. The larger satellites of Jupiter, Saturn, Uranus, and
probably Triton, formed in equatorial orbits in a similar manner to that
envisaged for the Moon. In this model, aspects of both fission and capture will
play a role, although only a minor amount of the material necessary to form
the satellites is knocked off the primary body.

The scenario adopted here is one in which material is captured into orbit
around the growing planet; it mainly comes from nearby and with only a

small contribution from distant parts of the solar system (Fig. 9.4). Occasionally it is added to by debris ejected from the growing planet with escape velocity by extra-large impacts. A consequence of this hypothesis is that the planetary accretion results in single or multi-ring formation around growing planets. The absence of a Venusian moon may be accounted for if the proto-satellite forming in orbit around Venus crashed into the planet. If the retrograde motion of Venus, and its inclination to the ecliptic, was caused by such an event, the debris from the collision did not form a ring, or a new satellite, but was re-accreted. Mercury and Mars were too small to acquire a

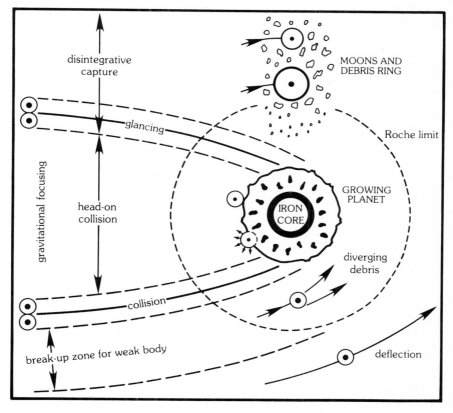

9.4 Processes in and near a growing planet. An iron-rich core is growing from blobs sinking through a silicate mantle. The approach regions are tapered because of gravitational attraction. A circular region for head-on collision is surrounded by two annular regions for glancing collision and break-up of a weak body; the latter collectively form the zone in which debris can go into orbit by the collective processes of disintegrative capture. Moons and a debris ring increase the collision probability. (Adapted from Smith, J. V. [45].)

debris ring. Satellite formation about the larger planets was common. The Earth-Moon system is thus close to a double planet system. If Jupiter has a core of 25 Earth masses, then the ratio of primary silicate and metal mass to satellite mass in the Earth-Moon system is not unusual. The difference in density, and the lack of iron and other siderophile elements in the Moon, on this model, is thus due to several causes. The Earth accreted a greater share of the available iron possibly through a greater gravitational attraction, and possibly through disintegrative capture mechanisms of the type advocated by Smith [45].

Vaporization of material during massive impacts on the Earth, followed by condensation, is an unlikely scenario since the lunar composition, although refractory, does not contain Eu and Yb anomalies, indicating that temperatures did not exceed about 1400° K. Vaporization of mantle material would have produced Eu and Yb depletion as seen in the Allende Group II inclusions [39, 68].

Once the Moon has formed, the energy of accretion, possibly assisted by ^{26}Al heating, melts at least half the body. The view is taken here that the whole Moon melted, for reasons discussed in Chapters 5, 6 and 8. Formation of a lunar core removes about 2% Fe (FeS) and some of the siderophile elements; crustal formation (Chapter 5) and the formation of a zoned interior occurs, with all phases being below the solidus by 4.4 aeons.

"In general, the Sm-Nd isotopic data demonstrate that the mare basalts were derived from light REE-depleted sources, which formed close to the origin of the Moon and that the highlands crust represents the complementary light REE-enriched material. However, the only precise age for early lunar differentiation is 4.47 aeons and comes from the upper intersection of the U-Pb cataclysm isochron" [69,70]. The meteoritic bombardment continues down to 3.8 aeons, but no late spike in the cratering record appears to be required.

There is no sign of a late veneer of volatile-rich C1 type material in the lunar chemistry (except for the later additions to the regolith, Section 4.5.3), so that models which propose such an event for the Earth must explain its absence on the Moon. Possibly the Moon is too small to retain impact debris from high-speed late collisions with material coming from the region of Jupiter [45]. The continuing addition of material from the last stages of accretion during crystallization of the magma ocean does not appear to have resulted in any discernible chemical trends (except for the late meteorite component). Note that ^{26}Al heating is not significant 10 million years after the supernova event, so that heating from this source is no longer available by the time that the final planetary accretion occurred.

The clear evidence from the Moon for early differentiation involving most of that body raises questions about the extent of early global melting on the other planets [71]. Several sources are available even if ^{26}Al has become extinct. Another early source of superluminosity from the sun is restricted to periods of 10^5 years after contraction [72]. Heating can occur in a gas-rich nebula by adiabatic compression [71], and thus is available for gaseous protoplanets, but probably not for the mechanism of planetary growth by accretion of planetesimals. Tidal heating is important only for the Moon and Earth (and Io).

Accretion energy has long been a popular source of heating, but it depends very strongly on the growth conditions [44, 73–75]. The general consensus appears to be that melting of planets definitely occurred if they accreted on short time scales (10^6 years), and probably occurred even on larger (10^8 years) time scales.

9.8   The Formation of the Earth

It is assumed that most of the core elements Fe, Ni, Co were accreted as a separate metallic phase. During melting and segregation of the core material, the evidence for equilibration with the mantle silicates is nebulous. This is shown by the large abundances of Ni and trace siderophile elements in the present upper mantle. These elements may have been added from post-core-formation meteorite bombardment. The refractory elements are not expected to be fractionated from one another during the accretion of the Earth. Accordingly, the ratio of these elements to those in the primitive solar nebula can be used to arrive at values for the refractory trace elements.

It is assumed here that these abundances are typical of the whole mantle and that the lower mantle is not chemically different from the upper mantle but this is uncertain. No K, U and Th are assumed to have entered the core. The Earth is depleted by a factor of about four in potassium relative to the composition of the solar nebula. This depletion of the Earth in alkalis relative to refractory elements, and the slight enrichment of the latter relative to the solar nebula abundances, comprises a primary piece of evidence for the assembly of the Earth from a non-equilibrium mixture of components. These include metals and silicates depleted to varying degrees in volatile elements. The presence of free metal makes the reduced enstatite chondrites more attractive than the oxidized carbonaceous chondrites as possible representatives of the primary accreting material. The enstatite chondrites (E4) contain about the same abundances of the volatile elements as do the C1 meteorites, so that a more volatile-depleted source is needed. The presence of trace siderophile elements in the upper mantle in C1 proportions has been frequently addressed, and it is generally concluded that the mantle was not in equilibrium with the core. These questions are readily understood in a non-equilibrium accretion model, where metallic iron sinks to form a core at temperatures below that required to equilibrate with silicates. If non-equilibrium is assumed, then it may be possible to accrete enough FeS for S to be the light element component in the core. The reduced enstatite chondrites contain the same sulfur concentration as the oxidized carbonaceous chondrites (as well as similar alkali contents).

The question whether the Earth was melted and hence formed a magma ocean for which so much evidence exists on the Moon has been frequently debated, without any resolution of the problem. There is no geochemical evidence for an early crust, and a differing scenario to that of the Moon is required. Clearly the rate of accretion is important, so that the core might be insulated from later additions, explaining the high Ni content of the mantle.

Any crust existing prior to 3.8 aeons would be massively altered by large scale meteoritic bombardment. The infall of the final one or two percent of the accreting projectiles will completely destroy any preexisting planetary surface [76]. This has occurred throughout the inner solar system, from Mercury to Mars, while the surfaces of Ganymede, Callisto, Mimas, Rhea, Tethys and Dione bear witness to a probable extension to the neighborhoods of Jupiter and Saturn. The termination of the bombardment is dated only from the Moon, but the widespread morphological evidence for similar cratering and estimates based on crater counting and meteorite flux estimates encourages the view that such events were common in the period up to 3.8 aeons. There is no reason to suppose that the Earth escaped the bombardment, a view consistent with the absence of identifiable crust older than 3.8 aeons.

Is there any geochemical evidence available to cast light on these early events? A primitive anorthositic crust, from analogy with the lunar highlands, appears improbable. There is no evidence for a large Eu or Sr spike, nor for the reservoir of very primitive $^{87}Sr/^{86}Sr$ ratios, which would result from the Rb/Sr fractionation inherent in such an event. A plagioclase-rich crust will float over an anhydrous lunar magma ocean, but will sink in the wet terrestrial environment. Isotopic constraints provide the major argument against primeval sialic crusts, negative Eu spikes (as yet unobserved) might also be generated. Early basaltic crusts are less easy to disprove, but the Sm-Nd isotopic systematics appears to rule out any major early mantle differentiation [77].

The overall evidence for the Earth, in contrast to the Moon, is of continuing formation of the continental crust, slow in the early Archean, very rapid between about 3 and 2.5 aeons, and proceeding since that time at a more sedate pace [78]. The degree of enrichment of an element in the crust depends on the difference between its ionic radius and valency, and those of the elements forming the major mantle minerals. This indicates that crystal-liquid fractionation is the principal factor responsible for the derivation of the crust from the mantle. The large percentage of elements such as Cs, Rb, K, Ba, U,

Th, La and Ce in the continental crust indicates the processing of at least a 30% fraction of the mantle to produce the crust.

The question of the relative positions of the Moon and the Earth and the evolution of the lunar orbit cannot be addressed here [79]. It may be noted from the geological record that the oldest unmetamorphosed sedimentary rocks, the Archean Moodies Group in South Africa (older than 3.3 aeons), show clearly documented evidence of tidal effects, similar to those observed for example in the present estuary of the Elbe River [80]. Tidal ranges appear to be about the same, constituting evidence that the Moon has been in Earth orbit, at about the present distance, throughout most of geological history [81].

The amount of evidence needed to construct reasonable models for the evolution of the Earth and the Moon indicates that it is premature to construct similar scenarios for Mercury, Mars, and Venus, except in the most general terms. Detailed attempts must be postponed until appropriate chemical and isotopic evidence is available.

References and Notes

1. Tinsley, B. M. (1980) *Nature*. 286: 911; van den Bergh, S. (1981) *Science*. 213: 825.
2. MPC = megaparsec. One parsec is the distance at which the radius of the Earth's orbit subtends an angle of one second of arc. It is equivalent to 3.262 light years or 206,265 astronomical units (1 A.U. = Earth-Sun mean distance).
3. Turner, M. S. (1979) *Nature*. 281: 549.
4. Woody, D. P., and Richards, P. L. (1979) *Phys. Rev. Lett*. 42: 925.
5. Chang, E. S., et al. (1979) *Ap. J. Lett*. 232: L.139.
6. Rubin, V. C. (1980) *Science*. 209: 64.
7. Spinrad, H., et al. (1981) *Ap. J*. 244: 382. More distant objects are known. The most distant quasar (e.g., OQ 172) is 16 billion light years away, for a Hubble universe 18 billion years old.
8. Edmunds, M. G. (1979) *Nature*. 102: 281. For a recent discussion, see McCrea, W. H., and Rees, M. J. (1980) *Phil. Trans. Roy. Soc*. A296: 269. The question of galactic evolution unfortunately lies outside the scope of this book.
9. Cameron, A. G. W., and Truran, J. W. (1977) *Icarus*. 30: 447.
10. Evans, N. J. (1978) *Protostars and Planets*, p. 153.
11. Herbst, E. (1978) *Protostars and Planets*, p. 88.
12. Gough, D. (1980) *Ancient Sun*, p. 533; Trimble, V. (1981) *Nature*. 293: 186.
13. Clayton, R. N. (1978) *Ann. Rev. Nucl. Sci*. 28: 501.
14. Additional information may be found in Trimble, V. (1975) *Rev. Mod. Phys*. 47: 877 on the origin of the elements. See also Wannier, P. G. (1980) *Ann. Rev. Astron. Astrophys*. 18: 399, who provides a review of elemental abundances in the interstellar medium. See Savage, B. D., and Mathis, J. S. (1973) *Ann. Rev. Astron. Astrophys*. 17: 73 for a review of the properties of interstellar dust, and Martin, P. G. (1979) *Cosmic Dust*, Oxford Univ. Press, 266 pp.
15. Clayton, R. N., et al. (1973) *Science*. 182: 485.

16. Reeves, H. (1978) *Protostars and Planets*, p. 399.
17. Lee, T., et al. (1976) *GRL.* 3: 109.
18. Excellent reviews of isotopic anomalies are given by Clayton, R. N. (1978) *Ann. Rev. Nucl. Sci.* 28: 501; Podosek, F. A. (1978) *Ann. Rev. Astron. Astrophys.* 16: 293; Clayton, R. N. (1981) *Phil. Trans. Roy. Soc.* In press.
19. Kelly, W. R., and Wasserburg, G. J. (1978) *GRL.* 5: 1079.
20. r-process isotopes are formed by neutron capture in the intense neutron fluxes of supernovae. Elements produced include gold, uranium and thorium.
21. Cameron, A. G. W., and Truran, J. W. (1977) *Icarus.* 30: 447.
22. Schramm, D. N. (1978) *Protostars and Planets*, p. 384.
23. Clayton, D. D. (1977) *Icarus.* 32: 255.
24. Clayton, D. D. (1977) *EPSL.* 35: 398; 36: 381.
25. Jeffrey, P. M., and Reynolds, J. H. (1961) *JGR.* 66: 3582.
26. Alexander, E. C., et al. (1971) *Science.* 172: 837.
27. Reeves, H. (1972) *Astron. Astrophys.* 19: 215.
28. Herbst, W., and Assousa, G. E. (1977) *Ap. J.* 217: 473; (1978) *Protostars and Planets*, p. 368.
29. See Grossman, L. (1980) *Ann. Rev. Earth Planet. Sci.* 8: 559 for a comprehensive review of the Allende inclusions.
30. Clayton, D. D. (1980) *Ap. J. Lett.* 239; (1980) *EPSL.* 47: 199.
31. Cameron, A. G. W. (1978) *Moon and Planets.* 18: 5.
32. Wasson, J. T. (1978) *Protostars and Planets*, p. 488.
33. Black, D. C. (1972) *GCA.* 36: 377.
34. Clayton, R. N. (1981) *Phil. Trans. Roy. Soc.* In press.
35. Lewis, R. S., et al. (1979) *Ap. J. Lett.* 234: L.165.
36. Anders, E. (1981) *Proc. Roy. Soc.* A374: 207.
37. Kurat, G., et al. (1975) *EPSL.* 26: 144.
38. Wetherill, G. W. (1980) *Ann. Report Dept. Terr. Mag.* 444.
39. A brief historical survey is given by Brush, S. G. (1981) in *Space Science Comes of Age* (ed., P. A. Hanle), Smithsonian Institution Press.
40. McCrea, W. H. (1960) *Proc. Roy. Soc.* A256: 245.
41. Cameron, A. G. W. (1978) *Protostars and Planets*, p. 453.
42. Cameron, A. G. W. (1979) *Moon and Planets.* 21: 173.
43. Safronov, V. S. (1972) Evolution of the Protoplanetary Cloud and the Formation of the Earth and Planets. NASA TTF 677; (1975) NASA SP 370, 797; (1978) *Icarus.* 33: 3.
44. Wetherill, G. W. (1978) *Protostars and Planets*, p. 565; (1980) *Ann. Rev. Astron. Astrophys.* 18: 77.
45. Smith, J. V. (1982) *J. Geol.* In press; (1979) *Mineral Mag.* 43: 1.
46. Wasson, P. S. (1981) *Moon and Planets.* 24: 339; Arrhenius, G., and Alfvren, H. (1971) *EPSL.* 10: 253.
47. The proper treatment of the meteorite evidence requires a full book in its own right. See Wasson, J. T. (1974) *Meteorites*, Springer-Verlag; Dodd, R. T. (1981) *Meteorites*, Cambridge Univ. Press and the reviews by Smith [45]. The earlier books by Wood, J. A. (1968) *Meteorites and the Origin of the Solar System*, McGraw-Hill, and Mason, B. H. (1962) *Meteorites*, Wiley, contain valuable summaries.
48. E.g., Urey, H. C. (1956) *Ap. J.* 124: 623.
49. Hartmann, W. K. (1978) *Icarus.* 33: 50.
50. A partial list of references includes the following:
    Boynton, W. V. (1975) *GCA.* 39: 569;
    Davis, A. M., and Grossman, L. (1979) *GCA.* 43: 1611;

Grossman, L. (1972) *GCA.* 36: 597;

Grossman, L. (1977) NASA SP-370, 787;

Grossman, L., and Larimer, J. W. (1974) *Rev. Geophys. Space Phys.* 12: 71;

Larimer, J. W. (1979) *Icarus.* 40: 446;

Larimer, J. W., and Anders, E. (1970) *GCA.* 34: 367;

Morgan, J. W. (1977) *J. Radionanal. Chem.* 37: 79;

Smith, J. V. (1982) *J. Geol.* In press;

Wasson, J. T. (1978) *Protostars and Planets*, p. 488.

51. E.g., Kerridge, J. F. (1979) *PLC 10*: 989.

52. Kurat, G., et al. (1975) *EPSL.* 26: 144.

53. Boynton, W. V. (1978) *Protostars and Planets*, p. 427.

54. Gehrels, T. (1979) *Asteroids*, Univ. Ariz. Press, 1181 pp.

55. E.g., Lewis, J. S. (1974) *Science.* 186: 440 (see also [50]).

56. E.g., Dearborn, D. S. P., et al. (1976) *Nature.* 261: 210; Barbaro, G., et al. (1968) *Mass Loss from Stars*, Reidel.

57. E.g., Imhoff, C. A. (1978) *Protostars and Planets*, p. 699.

58. Handbury, M. J., and Williams, I. P. (1976) *Observatory.* 96: 140, raise doubts of this nature, but Conti, P. S., and McCray, R. (1980, *Science.* 208: 9) provide evidence for removal of gas by strong interstellar winds (see Fig. 9.2).

59. Gooding, J. L., et al. (1980) *EPSL.* 50: 171; Gooding, J. L., and Keil, K. (1981) *Meteoritics.* 16: 17. These two references provide an excellent source of information on chondrule abundances and textures. See also Nagahara, H. (1981) *Nature.* 292: 135 for another statement on the secondary origin of chondrules.

60. E.g., Hartmann, W. K. (1980) *Lunar Highlands Crusts*, p. 155; Wetherill, G. W. (1975) *PLC 6*: 1539; (1976) *PLC 7*: 3245; Chapter 3, and references therein; Kaula, W. M. (1979) *JGR.* 84: 999.

61. Thorslund, P., and Wickman, F. E. (1981) *Nature.* 289: 285.

62. Pollack, J. B., et al. (1976) *Icarus.* 29: 35.

63. Soderblom, L. A., and Johnson, T. V. (1981) LPS XIIA: 12.

64. Chapman, C. R., and Davis, D. R. (1975) *Science.* 190: 553.

65. E.g., *Basaltic Volcanism* (1981) Sections 4.3, 4.4.

66. Wood, J. A. (1979) *The Solar System*, Prentice Hall, p. 174.

67. Kaula, W. M. (1971) *Rev. Geophys. Space Phys.* 9: 217.

68. E.g., Mason, B., and Martin, P. M. (1977) *Smithsonian Contrib. Earth Sci.* 19: 84; Mason, B. H., and Taylor, S. R. (1982) Ibid. In press.

69. Oberli, F., et al. (1978) LPS IX: 832.

70. Jacobsen, S. B., and Wasserburg, G. J. (1980) *EPSL.* 50: 139.

71. Hostetler, C. J., and Drake, M. J. (1980) *PLC 11*: 1915.

72. Hayashi, C. (1966) *Ann. Rev. Astron. Astrophys.* 4: 171.

73. Wetherill, G. W. (1976) *PLC 7*: 3245.

74. Greenberg, R. (1979) *Icarus.* 39: 141.

75. Ransford, G. A. (1979) *PLC 10*: 1867.

76. Smith, J. V. (1981) *Phil. Trans. Roy. Soc.* A301: 217.

77. Moorbath, S. (1975) *Nature.* 254: 395; O'Nions, R. K., et al. (1979) *JGR.* 84: 6091; Jacobsen, S. B., and Wasserburg, G. J. (1979) *JGR.* 84: 7429.

78. Taylor, S. R., and McLennan, S. M. (1981) *Phil. Trans. Roy. Soc.* A301: 381.

79. See Lambeck, K. (1980) *The Earth's Variable Rotation*, Cambridge for a detailed review.

80. Ericson, K. A. (1979) *Precambrian Res.* 8: 153. An excellent account of the tidal flats of the Elbe estuary appears in Childers, E. (1903) *The Riddle of the Sands*, Dover Publications, N.Y., 284 pp.

81.   The existence of cross-bedding in sandstones on a gigantic scale [e.g., in the Navajo (Jurassic) sandstones], sometimes used as evidence for a close approach of the Moon in later geological epochs, is due to aeolian conditions of deposition in desert-type environments, and has no significance for Earth-Moon evolution.

Chapter 10

THE SIGNIFICANCE OF LUNAR AND PLANETARY EXPLORATION

10.1   The Lunar Evidence

Having accomplished a preliminary exploration of the solar system, we now possess a general understanding of the problems of and constraints on planetary origin and evolution. Much of our information has been derived from the study of the Moon. We have been able to establish the following sequence of events from the lunar evidence:

Differentiated planetesimals accreted to form the Moon within 100 million years of the isolation of the solar nebula. The Moon became partially to completely molten and large-scale crystal-liquid differentiation produced a feldspathic crust and a complex mineralogically zoned mantle, from which some 20 differing types of mare basalts were subsequently erupted. Whether the initial melting produced a magma ocean or several smaller-scale melted regions is partly a semantic problem, since the bulk of the Moon went through this process on a timescale of 100–200 million years.

By 4.4 aeons, the interior had crystallized, as shown by the isotopic data, and the feldspathic crust was in place. For the next 500 million years, the crust was subjected to the full fury of bombardment by planetesimals and lesser objects. The resulting chaos in the upper crust explains much of the petrological and chemical complexities of the highland samples. Possibly the crust did not become stable against the raining impacts until 4.2 aeons. The declining stages of the bombardment are preserved in the majestic ringed basins and large craters of the highlands. The basin-forming events were over by 3.8 aeons, with the formation of the Imbrium and Orientale multi-ring basins. A "terminal cataclysm" does not appear to be required.

This date (3.8 aeons) provides a younger limit for heavily cratered surfaces, which are common throughout the solar system, and is perhaps the single most important result from the Apollo mission. The similar age of the oldest terrestrial crust, in Isua, Greenland, has an as yet undetermined relationship to the great bombardment, which must also have struck the Earth.

The slow generation of heat by radioactive decay produced basaltic lavas by partial melting from the deep lunar interior from about 4 to 3 aeons. This stage, involving a secondary differentiation and transport of incompatible elements to the outer regions, is still continuing on Earth and apparently persisted to young periods on Mars. Such basaltic volcanism provides probes of planetary interiors and information about the constitution of the mantles from which they are derived. The surface rocks exposed to cosmic radiation provide not only detailed evidence of their exposure history, but also provide information on the history of the sun.

The bulk composition of the Moon is sufficiently different from that of the mantle of the Earth to provide first-order evidence for accretion of planets from a heterogeneous assemblage of planetesimals. The varying compositions of the planets, satellites and meteorites demand differing proportions of metal and silicates with varying enrichment and depletions of refractory and volatile elements. The simple pre-Apollo scenarios of capture and fission have been replaced by complex models of double-planet formation. The sophisticated understanding we have acquired of the closest, most accessible object in the solar system must serve to remind us that our pre-Apollo thinking about the composition and history of the Moon was effectively unconstrained [1].

The ultimate benefit from lunar exploration is that we can, in effect, stand upon the lunar surface, understand its complexity in detail, and use it as a platform from which to contemplate the other planets and satellites from a superior vantage point.

The knowledge we have gained from the lunar samples has both promoted our understanding and created a new interdisciplinary and integrated scientific community, capable at present of dealing decisively with any solar system observation or material. The development of expertise in this field was not really predictable in advance [2]. The planetary discoveries even impinge on cosmology, for the existence of fossil surfaces 4 aeons in age places stringent limits on the variation of the gravitational constant G with time.

10.2   Future Missions

The political aspects of the space missions cannot be addressed here [3]. The current question of the value of the space program echoes the complaints levelled against most early explorers who failed to return enough gold. The lesson of history is that unexpected but generally beneficial results follow such voyages. What we are looking for is not always what we find. The Beagle expedition was organized by the premier naval power of the age, interested in better navigational charts of the South American coast. It was not forseen that the observations by Darwin would result in a fundamental change in our perception of the relationship between man and nature.

Astronomical research, well funded since the days of the Babylonian and Egyptian observers and the builders of Stonehenge, has concentrated mainly on the search for answers to fundamental questions. Many questions from the present stage of solar system exploration could be answered quickly by a series of unmanned space probes. Landers or orbiters equipped with gamma-ray detectors can readily obtain the surface distribution of the radioactive elements K, U and Th. These elements are often strongly concentrated by differentiation processes into the outer regions of planets. Their abundance and surficial distribution provide first-order geochemical information about both the geochemical evolution and the bulk composition of the planet. This information can be coupled with the relative ages of the planetary surfaces established by photogeological crater counting methods or (in the case of Venus) radar imagery. It is therefore possible to begin to establish the pattern of planetary and crustal evolution for each planet or satellite once we have data on the radioactive elements. These missions first need to be calibrated against the lunar surface by polar orbiter type missions, which would provide the full surface coverage that we still lack for the Moon.

Techniques for establishing relative ages of planetary surfaces become less secure as we depart from the terrestrial neighborhood, and the estimates of past meteorite flux rates become less certain. The precise ages of the surfaces need to be established by radiometric dating techniques. Surface ages for Mars and Mercury would provide the next quantum jump in our understanding of the history of the solar system.

How can this task be accomplished? Is it possible to make sophisticated age determinations in remote automated laboratories? This question is worth pursuing at length. Should we have obtained further remote data from the Moon, in 1969, in preference to a sample return? The Surveyor V and VI analyses confirmed the basaltic nature of the maria, but the Surveyor VII analysis on the ejecta blanket of Tycho did not lead us to the concept of an anorthositic highland crust. This was immediately deduced from the presence of a few feldspathic grains in the Apollo 11 sample return, 50 km from the nearest highlands. By the time of the Apollo 12 sample return, the isotopic analyses revealed much of lunar history. The analyses provided decisive evidence of early 4.4 aeon differentiation of the basaltic source regions, along with the eruption of lava some hundreds of millions of years later.

Our experience with the Viking data has been enlightening. The analytical data, analogous to data from the lunar Surveyor landers, tells us of the basaltic nature of the surface, confirming photogeological interpretations, but does not tell us the ages, the composition, or the history of the source regions. The vital isotopic and trace element data are missing. The major element values, subject to much instrumental and geological uncertainty, were only obtained for wind-blown material, not rocks. Thus the quality of scientific information available from remote landers may often be compromised or degraded by factors remote from the scientific objectives.

The expense of sample return missions requires that the scientific information obtained in this manner must be vastly superior to the alternative. A number of factors are relevant to this argument:

(a) The instrumental techniques available in terrestrial laboratories will always be more advanced than those available in remote automated laboratories.

(b) The experiments are difficult to plan in advance. Our whole experience with the Moon illustrates the difficulties of projection in science. If the experimental parameters cannot be altered, one either gets the expected result, or an ambiguous and possibly uninterpretable result.

(c) Returned samples enable experiments not conceived at the time of the mission to answer scientific questions. Many workers can be involved and many experiments can be performed, rather than the few possible in a remote laboratory.

(d) The acquiring of reliable age information, for example, is not only a matter of sophisticated instrumental measurements but is also critically dependent on selection of the appropriate mineral or rock sample. This complex process, familiar to terrestrial workers, is exceedingly difficult to carry out under clean room conditions on Earth. The scientific skills and abilities needed require workers of the highest quality to address these problems, so that human factors are involved as well as instrumental sophistication.

Accordingly, although remote-sensing measurements are very valuable, they can never replace the information that can be obtained from the return of planetary samples to terrestrial laboratories. The latter is judged to be more cost-effective when the time and expertise involved in designing remote laboratories is weighed against the scientific return. Who would trade one mare basalt isochron for 50 Surveyor analyses?

10.3   Man's Responsibility in the Universe

The possibility of the existence of Earth-like planets elsewhere in the universe is frequently addressed. Although the uncertainties are extreme, possibly between 10^{-5} and 10^{-7} of the stars may possess satellites suitable for the origin and evolution of life [4, 5]. This number would provide between one and 100 such planets within 1000 light years of Earth.

The question of the emergence of life is reinforced by the large variety of organic molecules available in interstellar space. There is, accordingly, a good scientific rationale for continuing a search for extraterrestrial intelligence [5, 6]. The message from biochemical studies is more sobering. The course of evolution is not predictable [7]. Even on this planet, where life arose and hominid evolution began more than 4 aeons after accretion, the chances of the appearance of Homo sapiens seem remarkably small. Pollard [4] has noted that continental drift on the Earth provided three separate regions where evolution could proceed independently. When life first invaded the land in the Devonian Period, the continental crust formed a single unit (Pangaea). Subsequently, the Southern Continent (Gondwana) separated, followed by the breaking away of Antarctica and Australia and of South America from Africa. Mammalian evolution then proceeded independently in South America, Australia and Africa. The marsupials evolved in Austrialia. Although primates appeared both in South America and Africa, the New World monkeys never left the trees. "Thus continental drift has produced for us three Earth-like 'planets'. Only on one of them . . . did evolution lead to man" [4, p. 659]. This sobering conclusion reinforces the evidence that we may be unique, alone in a universe that is older than 15 or perhaps 20 aeons, and which contains at least 10^{11} galaxies, each having 10^{11} stars. If this is our position, then we alone, comprised of atoms synthesized in stellar interiors and supernovae, and the product of 4 aeons of organic evolution on this planet, have developed the skills and ability to comprehend the universe and understand its history. If this scenario is correct, we have an awesome responsibility to ensure that our species survives and explores not only the solar system, but, in due time, ventures further afield [8, 9].

References and Notes

1. The range of uncontrolled speculation available in the pre-Apollo literature is a fertile field for the historians of science.
2. A comparison of the pre-Apollo selection of Principal Investigators for the lunar program with the community, which subsequently evolved under the stress of the lunar investigations, reveals a high casualty rate. Less than 15% of the groups remain active and perhaps less than one third contributed significantly to our understanding of lunar science. This illustrates the twin difficulties of identifying both objectives and competent individuals in advance, since unexpected problems arise. The analogy with the history of warfare is close.
3. See, for example, Hallion, R. P., and Crouch, T. D., eds. (1979) *Apollo: Ten Years since Tranquillity Base*, Smithsonian Institution Press, Washington, D.C.
4. Pollard, W. G. (1979) The prevalence of Earth-like planets. *Amer. Sci.* 67: 654.
5. Owen, T., and Goldsmith, D. (1979) *The Search for Life in the Universe*, Benjamin-Cummings.
6. Morrison, P., et al. (1977) *The Search for Extraterrestrial Intelligence: SETI*, NASA SP 419.
7. Monod, J. (1974) *Chance and Necessity*, Collins.
8. Arnold, J. R. (1980) The frontier in space. *Amer. Sci.* 68: 299. A marked feature of hominids is a tendency to travel widely. Arnold comments (p. 299) that "we seem to be

descended from the migrants and innovators. So far as we know, the chimpanzees and gorillas have not left home."

9. See also Tipler, F. J. (1981) *Quart. J. Roy. Astr. Soc.* 22: 133, 279 for a discussion of the history of the search for extraterrestrial intelligence; and Rood, R. T., and Trefil, J. S. (1981) *The Possibility of Extraterrestrial Civilizations*, C. Scribner.

APPENDICES

APPENDIX I

Reference Abbreviations

 AJS: *American Journal of Science*
 Ancient Sun: *The Ancient Sun: Fossil Record in the Earth, Moon and Meteorites* (Eds. R. O. Pepin, et al.), Pergamon Press (1980) Geochim. Cosmochim. Acta Suppl. 13
 Ap. J.: *Astrophysical Journal*
 Apollo 15: *The Apollo 15 Lunar Samples,* Lunar Science Institute, Houston, Texas (1972)
 Apollo 16 Workshop: *Workshop on Apollo 16,* LPI Technical Report 81-01, Lunar and Planetary Institute, Houston (1981)
 Basaltic Volcanism: *Basaltic Volcanism on the Terrestrial Planets*, Pergamon Press (1981)
 Bull. GSA: *Bulletin of the Geological Society of America*
 EOS: *EOS, Transactions of the American Geophysical Union*
 EPSL: *Earth and Planetary Science Letters*
 GCA: *Geochimica et Cosmochimica Acta*
 GRL: *Geophysical Research Letters*
 Impact Cratering: *Impact and Explosion Cratering* (Eds. D. J. Roddy, et al.), 1301 pp., Pergamon Press (1977)
 JGR: *Journal of Geophysical Research*
 LS III: *Lunar Science III* (Lunar Science Institute) see extended abstract of Lunar Science Conferences Appendix II
 LS IV: *Lunar Science IV* (Lunar Science Institute)
 LS V: *Lunar Science V* (Lunar Science Institute)
 LS VI: *Lunar Science VI* (Lunar Science Institute)
 LS VII: *Lunar Science VII* (Lunar Science Institute)
 LS VIII: *Lunar Science VIII* (Lunar Science Institute
 LPS IX: *Lunar and Planetary Science IX* (Lunar and Planetary Institute
 LPS X: *Lunar and Planetary Science X* (Lunar and Planetary Institute)
 LPS XI: *Lunar and Planetary Science XI* (Lunar and Planetary Institute)
 LPS XII: *Lunar and Planetary Science XII* (Lunar and Planetary Institute)

Lunar Highlands Crust: *Proceedings of the Conference in the Lunar Highlands Crust*, 505 pp., Pergamon Press (1980) Geochim. Cosmochim. Acta Suppl. 12

Mare Basalts: *Origin of Mare Basalts*, Lunar Science Institute Contrib. 234 (1975)

Mare Crisium: *Mare Crisium:* The View from Luna 24, 742 pp., Pergamon Press (1978) Geochim. Cosmochim. Acta Suppl. 9

Moon: *The Moon*

Moon and Planets: *The Moon and Planets*

Multi-Ring Basins: *Proceedings, Conference on Multi-Ring Basins*, Pergamon Press (1981)

NASA SP: National Aeronautics and Space Administration Special Publication Series

PCE: *Physics and Chemistry of the Earth*

PEPI: *Physics of the Earth and Planetary Interiors*

PLC 1: *Proceedings Apollo 11 Lunar Science Conference*, Geochim. Cosmochim. Acta Suppl. 1 (1970)

PLC 2: *Proceedings 2nd Lunar Science Conference*, Geochim. Cosmochim. Acta Suppl. 2 (1971)

PLC 3: *Proceedings 3rd Lunar Science Conference*, Geochim. Cosmochim. Acta Suppl. 3 (1972)

PLC 4: *Proceedings 4th Lunar Science Conference*, Geochim. Cosmochim. Acta Suppl. 4 (1973)

PLC 5: *Proceedings 5th Lunar Science Conference*, Geochim. Cosmochim. Acta Suppl. 5 (1974)

PLC 6: *Proceedings 6th Lunar Science Conference*, Geochim. Cosmochim. Acta Suppl. 6 (1975)

PLC 7: *Proceedings 7th Lunar Science Conference*, Geochim. Cosmochim. Acta Suppl. 7 (1976)

PLC 8: *Proceedings 8th Lunar Science Conference*, Geochim. Cosmochim. Acta Suppl. 8 (1977)

PLC 9: *Proceedings 9th Lunar and Planetary Science Conference*, Geochim. Cosmochim. Acta Suppl. 10 (1978)

PLC 10: *Proceedings 10th Lunar and Planetary Science Conference*, Geochim. Cosmochim. Acta Suppl. 11 (1979)

PLC 11: *Proceedings 11th Lunar and Planetary Science Conference*, Geochim. Cosmochim. Acta Suppl. 14 (1980)

PLC 12: *Proceedings 12th Lunar and Planetary Science Conference*, Geochim. Cosmochim. Acta Suppl. 15 (1981)

Rev. Geophys. Space Phys.: *Reviews of Geophysics and Space Physics*

USGS: United States Geological Survey

APPENDIX II

Lunar Primary Data Sources

I. Lunar Sample Preliminary Examination Team (LSPET) Reports.

    Apollo 11, *Science*. 165: 211 (Sept. 19, 1969).
    Apollo 12, *Science*. 167: 1325 (March 6, 1970).
    Apollo 14, *Science*. 173: 681 (August 20, 1971).
    Apollo 15, *Science*. 175: 363 (Jan. 28, 1972).
    Apollo 16, *Science*. 179: 23 (Jan. 5, 1973).
    Apollo 17, *Science*. 182: 659 (Nov. 16, 1973).

II. Preliminary Science Reports: Superintendent of Documents, U.S. Government Printing Office, Washington, D.C. 20402. Surveyor Program Results, NASA SP 184 (1969).

    Apollo 11 NASA SP 214 (1969) 209 pp.
    Apollo 12 NASA SP 235 (1970) 235 pp.
    Apollo 14 NASA SP 272 (1971) 313 pp.
    Apollo 15 NASA SP 289 (1972) 502 pp.
    Apollo 16 NASA SP 315 (1972) 636 pp.
    Apollo 17 NASA SP 330 (1973) 710 pp.

III. USGS Apollo Geology Team Reports.

    Apollo 11 geologic setting is included in PET article, *Science*. 165: 1211 (Sept. 19, 1969).
    Apollo 12 geologic setting contained in PET article, *Science*. 167: 1325 (March 6, 1970).
    Apollo 14, *Science*. 173: 716 (Aug. 20, 1971).
    Apollo 15, *Science*. 175: 407 (Jan. 28, 1972).
    Apollo 16, *Science*. 179: 62 (Jan. 5, 1973).
    Apollo 17, *Science*. 182: 672 (Nov. 16, 1973).

IV. Lunar Sample Information Data Catalogs. NASA Johnson Space Center, Houston, Texas.

    Apollo 11 Aug. 31, 1969 (No identifying number).
    Apollo 12 MSC-S-243 (June, 1970).

Apollo 14 TM X-58062 (Sept., 1971).
Apollo 15 MSC-03209 (Nov., 1971).
Apollo 16 MSC-03210 (July, 1972).
Apollo 17 MSC-03211 (April, 1973).
A short guide to Lunar Sample Information has been published by
the Lunar Science Institute (Contribution No. 200). For updated
information, contact the Lunar Sample Curator, NASA Johnson
Space Center, Houston, Texas 77058.

V. Extended Abstracts of the Lunar Science Conference have been
published by the Lunar Science Institute, Houston, Texas for the 3rd–
8th Lunar Science Conferences.

  Lunar Science III, 1972.
  Lunar Science IV, 1973.
  Lunar Science V, 1974.
  Lunar Science VI, 1975.
  Lunar Science VII, 1976.
  Lunar Science VIII, 1977.
  These are referred to in the references as LS III through LS VIII,
  respectively. Extended abstracts of Lunar and Planetary Science
  Conferences (9th–12th) have been published by the Lunar and
  Planetary Institute, Houston, Texas.

  Lunar and Planetary Science IX, 1978.
  Lunar and Planetary Science X, 1979.
  Lunar and Planetary Science XI, 1980.
  Lunar and Planetary Science XII, 1981.
  These are referred to in the references as LPS IX through XII. In
  addition, an extended set of abstracts deals with the Apollo 15
  samples:
    The Apollo 15 Samples, J. W. Chamberlain and Carolyn
  Watkins, editors, 1972. This is cited as "Apollo 15" in the
  references.

  Extended abstracts for other conferences are available from the
  Lunar and Planetary Institute, Houston, Texas.
  Abstracts Presented to the Conference on Luna 24 (1977) LPI
  304.
  *Abstracts Presented to the Conference on the Origins of Planetary
    Magnetism* (1978) LPI 348.

Proceedings of a Topical Conference on Origins of Planetary Magnetism. In *Physics of the Earth and Planetary Interiors* **20** (1979) Elsevier. LPI 376.

Abstracts Presented to the Conference on the Ancient Sun: Fossil Record in the Earth, Moon and Meteorites (1979) LPI 390.

Abstracts Presented to the Conference on the Lunar Highlands Crust (1979) LPI 394.

VI.   Science Conference Proceedings.

Proceedings of the Apollo 11 Lunar Science Conference, 1970 (A. A. Levinson, ed.). *Geochimica et Cosmochimica Acta, Supplement 1*. Pergamon Press, Elmsford N.Y.

Proceedings of the Second Lunar Science Conference, 1971 (A. A. Levinson, ed.). *Geochimica et Cosmochimica Acta, Supplement 2*. *MIT* Press, Cambridge, Mass. (Distributed by Pergamon Press).

Proceedings of the Third Lunar Science Conference, 1972 (E. A. King, Jr., D. Heymann, and D. R. Criswell, eds.) *Geochimica et Cosmochimica Acta, Supplement 3*. *MIT* Press, Cambridge, Mass. (Distributed by Pergamon Press).

All subsequent Proceedings have been published by Pergamon Press, Elmsford, N.Y. 10523:

Proceedings of the Fourth Lunar Science Conference (W. A. Gose, ed.), 1973. GCA Supp. 4.

Proceedings of the Fifth Lunar Science Conference (W. A. Gose, ed.), 1974. GCA Supp. 5.

Proceedings of the Sixth Lunar Science Conference (R. B. Merrill, ed.), 1975. GCA Supp. 6.

Proceedings of the Seventh Lunar Science Conference (R. B. Merrill, ed.), 1976. GCA Supp. 7.

Proceedings of the Eight Lunar Science Conference (R. B. Merrill, ed.), 1977. GCA Supp. 8.

Proceedings of the Ninth Lunar and Planetary Science Conference (R. B. Merrill, ed.), 1978. GCA Supp. 10.

Proceedings of the Tenth Lunar and Planetary Science Conference (R. B. Merrill, ed.), 1979. GCA Supp. 11.

Proceedings of the Eleventh Lunar and Planetary Science Conference (R. B. Merrill, ed.), 1980. GCA Supp. 14.

Proceedings of the Twelfth Lunar and Planetary Science Conference. Proceedings of Lunar and Planetary Science, vol. 12B (R. B. Merrill and R. Ridings, eds.), 1981. GCA Supp. 15.

*Index to the Proceedings of the Lunar and Planetary Science Conferences 1970-*1978 (Compiled by A. R. Masterson; R. B. Merrill, ed.), 1979.

VII. Other Conference Proceedings:

Mare Crisium: The View from Luna 24. Proceedings of the Conference on Luna 24, Houston, Texas, December 1977. Pergamon Press, 1978. GCA Supp. 9.

Impact and Explosion Cratering. Proceedings of the Symposium on Planetary Cratering Mechanics, Flagstaff, Arizona, September 1976. Pergamon Press, 1978.

Proceedings of the Conference on the Lunar Highlands Crust. November 1979, Houston, Texas. Pergamon Press, 1980. GCA Supp. 12.

The Ancient Sun: Fossil Record in the Earth, Moon and Meteorites. Proceedings of the Conference on the Ancient Sun: Fossil Record in the Earth, Moon and Meteorites, Boulder, Colorado, October 1979. (R. O. Pepin, J. A. Eddy, and R. B. Merrill, eds.) Pergamon Press, 1981.

Proceedings of the Conference on Multi-Ring Basins: Formation and Evolution, Houston, Texas, November 1980. Proceedings of Lunar and Planetary Science, vol. 12A (P. H. Schultz and R. B. Merrill, eds.). Pergamon Press, 1981. GCA Supp. 15.

Workshop on Apollo 16 (O. B. James and F. Hörz, eds.). LPI Tech Rpt. 81-01. Lunar and Planetary Institute, Houston, Texas, 1981.

APPENDIX III

Diameter and Location of Lunar Craters Mentioned in the Text[†]

Crater	Lat.	Long.	Diameter (km)
Abulfeda	13.8° S	13.9° E	65
Aitken	16.5° S	173.1° E	131
Albategnius	11.2° S	4.1° E	136
Alphonsus	13.4° S	2.8° W	119
Alphonsus R	14.4° S	1.9° W	3
Anaxagoras	73.4° N	10.1° W	51
Antoniadi	69.8° S	172.0° W	135
Apollo	35.5° S	149.6° W	503
Apollonius	4.5° N	61.1° E	53
Arago	6.2° N	21.4° E	26
Archimedes	29.7° N	4.0° W	83
Ariadaeus	4.6° N	17.3° E	11
Aristarchus	23.7° N	47.4° W	40
Aristillus	33.9° N	1.2° E	55
Arzachel	18.2° S	1.9° W	97
Autolycus	30.7° N	1.5° E	39
Bailly	66.8° S	69.4° W	303
Beer	27.1° N	9.1° W	10
Berzelius	36.6° N	50.9° E	51
Bessel	21.8° N	17.9° E	16
Bode	6.7° N	2.4° W	19
Bonpland	8.3° S	17.4° W	60
Cauchy	9.6° N	38.6° E	12
Chaucer	3.7° N	140.0° W	45
Clavius	58.4° S	14.4° W	225
Compton	56.0° N	105.0° E	162
Copernicus	9.7° N	20.0° W	93
Copernicus H	6.9° N	18.3° W	5
Cyrillus	13.2° S	24.0° E	98
D'Alembert	51.3° N	164.6° E	225
Darwin	19.8° S	69.1° W	130
Davy	11.8° S	8.1° W	35
DeLisle	29.9° N	34.6° W	25
Descartes	11.7° S	15.7° E	48
Dionysius	2.8° N	17.3° E	18
Diophantus	27.6° N	34.3° W	18
Dollond B	7.7° S	13.8° E	37
Doppler	12.8° S	159.9° W	100
Eratosthenes	14.5° N	11.3° W	58
Euler	23.3° N	29.2° W	28

Crater	Lat.	Long.	Diameter (km)
Fauth	6.3° N	20.1° W	12
Flammarion	3.4° S	3.7° W	75
Flamsteed	4.5° S	44.3° W	21
Fra Mauro	6.0° S	17.0° W	95
Fracastorius	21.2° S	33.0° E	124
Galilaei	10.5° N	62.7° W	16
Gassendi	17.5° S	39.9° W	110
Giordano Bruno	35.9° N	102.8° E	22
Grimaldi	5.2° S	69.6° W	410
Gruithuisen	32.9° N	39.7° W	16
H.G. Wells	41.0° N	122.7° E	103
Hausen	65.5° S	88.4° W	167
Hell	32.4° S	7.8° W	33
Herigonius	13.3° S	33.9° W	15
Herodotus	23.2° N	49.7° W	35
Herschel	5.7° S	2.1° W	41
Hevelius	2.2° N	67.3° W	106
Hippalus	24.8° S	30.2° W	58
Hommel	54.6° S	33.0° E	125
Humboldt	27.2° S	80.9° E	207
Hyginus	7.8° N	6.3° E	9
Icarus	5.3° S	173.2° W	96
Inghirami	47.5° S	68.8° W	91
Jansen	13.5° N	28.7° E	24
Janssen	44.9° S	41.5° E	190
Jules Verne	34.8° S	146.9° E	134
Julius Caesar	9.0° N	15.4° E	91
Kant	10.6° S	20.1° E	33
Kepler	8.1° N	38.0° W	32
King	5.0° N	120.5° E	77
Kopff	17.4° S	89.6° W	42
Korolev	4.4° S	157.4° W	453
Kramers	53.6° N	127.6° W	62
La Hire A	28.5° N	23.4° W	5
Lalande	4.4° S	8.6° W	24
Lambert	25.8° N	21.0° W	30
Langrenus	8.9° S	60.9° E	132
Lansberg	0.3° S	26.6° W	39
Lassell	15.5° S	7.9° W	23
Lavoisier	38.2° N	81.2° W	70
Le Monnier	26.6° N	30.6° E	61
Linné	27.7° N	11.8° E	2
Littrow	21.5° N	31.4° E	31
Longomontanus	49.5° S	21.7° W	145
Lyell	13.6° N	40.6° E	32

Crater	Lat.	Long.	Diameter (km)
Macrobius	21.3° N	46.0° E	64
Marco Polo	15.4° N	2.0° W	28
Marius	11.9° N	50.8° W	41
Maskelyne	2.2° N	30.1° E	24
Maunder	14.6° S	93.8° W	55
Maurolycus	41.8° S	14.0° E	114
Mendeleev	5.6° N	141.5° E	330
Mutus	63.6° S	30.1° E	78
Nansen	81.3° N	95.3° E	122
Neper	8.8° N	84.5° E	137
Olbers	7.4° N	75.9° W	75
Omar Khayyam	58.0° N	102.1° W	70
Peirce	18.3° N	53.5° E	19
Petavius	25.3° S	60.4° E	177
Picard	14.6° N	54.7° E	23
Piccolomini	29.7° S	32.2° E	88
Planck	57.9° S	135.8° E	344
Plato	51.6° N	9.3° W	101
Playfair	23.5° S	8.4° E	48
Plutarch	24.1° N	79.0° E	68
Poincaré	56.7° S	163.6° E	319
Posidonius	31.8° N	29.9° E	95
Proclus	16.1° N	46.8° E	28
Ptolemaeus	9.2° S	1.8° W	153
Pythagoras	63.5° N	62.8° W	130
Regiomontanus	28.4° S	1.0° W	124
Regiomontanus A	28.0° S	0.6° W	6
Reiner	7.0° N	54.9° W	30
Rheita	37.1° S	47.2° E	70
Ritter	2.0° N	19.2° E	29
Ross	11.7° N	21.7° E	25
Rothmann G	28.4° S	24.3° E	92
Sabine	1.4° N	20.1° E	30
Schickard	44.4° S	54.6° W	227
Schroter	2.6° N	7.0° W	35
Sirsalis	12.5° S	60.4° W	42
Sulpicius Gallus	19.6° N	11.6° E	12
Taruntius	5.6° N	46.5° E	56
Taylor	5.3° S	16.7° E	42
Theophilus	11.4° S	26.4° E	100
Timocharis	26.7° N	13.1° W	34
Tsiolkovskiy	20.4° S	129.1° E	180
Tycho	43.3° S	11.2° W	85
Van de Graaff	27.0° S	172.0° E	234

Crater	Lat.	Long.	Diameter (km)
Vasco da Gama	13.9° N	83.8° W	96
Vitello	30.4° S	37.5° W	42
Vitruvius	17.6° N	31.3° E	30
Wargentin	49.6° S	60.2° W	84
Widmanstatten	6.1° S	85.5° E	46
Wolf	22.7° S	16.6° W	25
Xenophon	22.8° S	122.1° E	25
Zwicky	15.9° S	167.6° E	135

†Data from D. Kinsler, Lunar and Planetary Institute, Houston, Texas.

APPENDIX IV

Lunar Multi-ring Basins[†]

Basin	Diameter of topographic basin rim (km)	Age
Orientale	930	
Imbrium	1500	Imbrian
Schrödinger	320	
Sikorsky-Rittenhouse	310	
Bailly	300	
Hertsprung	570	
Serenitatis	880	Nectarian
Crisium	1060	
Humorum	820	
Humboldtianum	700 (ave.)	
Mendeleev	330	
Mendel-Rydberg	630	
Korolev	400	
Moscoviense	445	
Nectaris	860	
Apollo	505	
Grimaldi	430	
Freundlich-Sharonov	600	
Birkhoff	330	
Planck	325	
Schiller-Zucchius	325	
(Amundsen)	355	
Lorentz	360	
Smythii	840	
Coulomb-Sarton	530	
Keeler-Heaviside	800	
Poincaré	340	
Ingenii	650	
(Lomosov-Fleming)	620	Pre-Nectarian
(Nubium)	690	
(Fecunditatis)	690	
(Mutus-Vlacq)	700	
(Tranquillitatis)	775	
Australe	880	
(Al Khwarizmi-King)	590	
(Pingre-Hausen)	300	
(Werner-Airy)	500	
(Flamsteed-Billy)	570	
(Marginis)	580	
(Insularum)	600	
(Grissom-White)	600	
(Tsiolkovsky-Stark)	700	
South Pole-Aitken	2500	
Procellarum	3200	

[†]Basins listed in order of relative age. Existence not definitely established for basin names in parentheses. List adapted from Wilhelms, D. E. (1982) *The Geologic History of the Moon* (Tables 901 and 902), NASA SP. In press.

APPENDIX V

Terrestrial Impact Craters

	Name	Diameter (km)	Age (m.y.)	Coordinates
1	Barringer, Arizona, USA	1.2	*	35° N 111° W
2	Boxhole, N.T. Australia	0.2	*	23° S 135° E
3	Campo del Cielo, Argentina	0.1	*	28° S 62° W
4	Dalgaranga, W.A., Australia	0.02	*	28° S 117° E
5	Haviland, Kansas, USA	0.01	*	38° N 99° W
6	Henbury, N.T. Australia	0.15	*	25° S 133° E
7	Kaalijarvi, Estonian SSR	0.11	*	58° N 23° E
8	Odessa, Texas, USA	0.17	*	32° N 102° W
9	Sikhote Alin, Primorye Terr. Siberia, USSR	0.03	*	46° N 135° E
10	Wabar, Saudi Arabia	0.10	*	21° N 50° E
11	Wolf Creek, W.A., Australia	0.85	*	19° S 128° E
12	Aouelloul, Mauritania	0.37	3.1 ± 0.3	20° N 13° W
13	Araguainha Dome, Brazil	40	<250	17° S 53° W
14	Beyenchime Salaatin, USSR	8	< 65	72° N 123° E
15	Boltysh, Ukrainian SSR, USSR	25	100 ± 5	49° N 32° E
16	Bosumtwi, Ghana	10.5	1.3 ± .2	6° N 1° W
17	B.P. structure, Libya	2.8	<120	25° N 24° E
18	Brent, Ontaria, Canada	3.8	450 ± 30	46° N 78° W
19	Carswell, Saskatchewan, Canada	37	485 ± 50	58° N 109° W
20	Charlevoix, Quebec, Canada	46	360 ± 25	48° N 70° W
21	Clearwater L. East, Quebec, Canada	22	290 ± 20	56° N 74° W
22	Clearwater L. West, Quebec, Canada	32	290 ± 20	56° N 74° W
23	Crooked Creek, Missouri, USA	5.6	320 ± 80	38° N 91° W
24	Decaturville, Missouri, USA	6	<300	38° N 93° W
25	Deep Bay, Saskatchewan, Canada	12	100 ± 50	56° N 103° W
26	Dellen, Sweden	15	230	62° N 17° E
27	Flynn Creek, Tennessee, USA	3.8	360 ± 20	36° N 86° W
28	Gosses Bluff, N.T. Australia	22	130 ± 6	24° S 132° E
29	Gow L., Saskatchewan, Canada	5	<200	56° N 104° W
30	Haughton Dome, N.W.T. Canada	20	15	75° N 90° W
31	Holleford, Ontario, Canada	2	550 ± 100	44° N 77° W
32	Ile Rouleau, Quebec, Canada	4	<300	51° N 74° W
33	Ilintsy, USSR	4.5	495 ± 5	49° N 28° E
34	Janisjarvi, USSR	14	700	62° N 31° E
35	Kaluga, USSR	15	360 ± 10	54° N 36° E
36	Kamensk, USSR	25	65	48° N 40° E
37	Kara, USSR	50	57	69° N 65° E
38	Kentland, Indiana, USA	13	300	41° N 87° W
39	Kelley West, N.T. Australia	2.5	550	19° S 133° E
40	Kursk, USSR	5	25 ± 80	52° N 36° E
41	Lac Couture, Quebec, Canada	8	420	60° N 75° W
42	Lac La Moinerie, Quebec, Canada	8	400	57° N 67° W
43	Lappajärvi, Finland	14	<600	63° N 24° E
44	Liverpool, N.T. Australia	1.6	150 ± 70	12° S 134° E
45	Lonar, India	1.83	0.05	20° N 77° E

	Name	Diameter (km)	Age (m.y.)	Coordinates
46	Manicouagan, Quebec, Canada	70	210 ± 4	51°N 69°W
47	Manson, Iowa, USA	32	< 70	43°N 95°W
48	Mien L., Sweden	5	118 ± 2	56°N 15°E
49	Middlesboro, Kentucky, USA	6	300	37°N 84°W
50	Mishina Gora, USSR	2.5	<360	59°N 28°E
51	Mistastin, Labrador, Canada	28	38 ± 4	56°N 63°W
52	Monturaqui, Chile	0.46	1	24°S 68°W
53	New Quebec Crater, New Quebec, Canada	3.2	5	61°N 74°W
54	Nicholson L., N.W.T., Canada	12.5	<450	63°N 103°W
55	Oasis, Libya	11.5	<120	25°N 24°E
56	Pilot L., N.W.T., Canada	6	<300	60°N 111°W
57	Popigai, USSR	100	38 ± 9	71°N 111°E
58	Puchezh-Katun, USSR	80	183 ± 3	57°N 44°E
59	Ries, Germany	24	14.8 ± 7	49°N 11°E
60	Rochechouart, France	23	160 ± 5	46°N 1°E
61	Rotmistrov, USSR	2.5	70	49°N 32°E
62	Sääksjärvi, Finland	5	490	61°N 22°E
63	St. Martin, Manitoba, Canada	23	225 ± 40	52°N 99°W
64	Serpent Mound, Ohio, USA	6.4	300	39°S 83°W
65	Sierra Madera, Texas, USA	13	100	31°N 103°W
66	Siljan, Sweden	52	365 ± 7	61°N 15°E
67	Slate Island, Superior, Ontario, Canada	30	350	49°N 87°W
68	Steen River, Alberta, Canada	25	95 ± 7	60°N 118°W
69	Steinheim, Germany	3.4	14.8 ± 7	49°N 10°E
70	Strangways, N.T. Australia	24	150 ± 70	15°S 134°E
71	Sudbury, Ontario, Canada	140	1840 ± 150	47°N 81°W
72	Tenoumer, Mauritania	1.9	2.5 ± .5	23°N 10°W
73	Vredefort, South Africa	140	1970 ± 100	27°S 27°E
74	Wanapitei L., Ontario, Canada	8.5	37 ± 2	47°N 81°W
75	Wells Creek, Tennessee, USA	14	200 ± 100	36°N 88°W
76	West Hawk L., Manitoba, Canada	2.7	100 ± 50	50°N 95°W
77	Morasko, Poland	100		52°N 16°E
78	Sobolev, Siberia, USSR	51		46°N 138°E
79	Kjardla, Est. SSR	4	500 ± 50	57°N 23°E
80	Logoisk, Bel. SSR	17	100 ± 20	54°N 28°E
81	Misarai, Lith. SSR	5	500 ± 80	54°N 24°E
82	Obolon, USSR	15	160	50°N 33°E
83	Serra da Canghala, Brazil	12	<300	8°S 47°W
84	Shunak, Kaz. SSR	2.5	12	43°N 73°E
85	Vepriaj, Lith. SSR	8	160 ± 30	55°N 25°E
86	Zeleny Gai, Ukr. SSR	1.4	120 ± 20	47°N 35°E
87	Zhamanshin, Aktyubinsk, USSR	10	4.5 ± 0.5	49°N 61°E
88	Elgygtgyn, Siberia		3.5 ± 0.5	67°N 172°E
89	Cambodia	6 × 10	—	14°N 106°E
90	Darwin, Tasmania	1.2	—	42°S 145°E

* All are Quaternary in age, i.e. $< 10^6$ y. Most are $< 10^5$ y. Meteorites have been found at these locations.

Source of data: Grieve, R. A. F. and Robertson, P. B. Lunar and Planetary Institute, Houston, Texas. Contrib. No. 367.

APPENDIX VI

Appendix to Table 8.1: Sources of data

1. Mason, B. H. (1971) *Handbook of Elemental Abundances in Meteorites*, Gordon and Breach, London. Orgueil major element analysis by Wiik, Table 3, p. 8, Anal. No. 1. Data $\times$ 1.479 to allow for loss of H_2O, C, S.

2. Mason, B. H. (1979) USGS Prof. Paper 440-B-1. Data for Orgueil from Table 84, pp. B111–B112. Values $\times$ 1.5 to allow for volatile loss.

3. Weller, M. R., et al. (1978) *GCA*. 42: 999, gives a value ($\times$ 1.5) of 2.4 ppm. A more recent determination by Curtis, D. B. (1980) *Meteoritics*. 15, gives a value of 0.4 ppm.

4. From Li/Be = 33. Dreibus, G., et al. (1976) *PLC 7*: 3383.

5. Shima, M. (1979) *GCA*. 43: 353 ($\times$ 1.5).

6. Mason, B. H. (1979) Ibid., Average C1 value ($\times$ 1.5).

7. Mason, B. H. (1979) Ibid., Ivuna data ($\times$ 1.5).

8. Krahenbühl, U., et al. (1973) *GCA*. 37: 1353 ($\times$ 1.5).

9. Evensen, N. M., et al. (1978) *GCA*. 42: 1203 (Mean C1 $\times$ 1.5).

10. E $\times$ trapolated value from neighboring REE.

11. From Zr/Hf = 33.

12. From Th/U = 3.8.

13. Crocket, J. H., pers. comm., 1977.

14. Taylor, S. R. (1979) *PLC 10*, Table 1. Values based mainly on nodule data. Ratio of Al, Ca and Ti to volatile free chondritic values (Col. 1) is 1.36. This ratio is used to obtain mantle abundances for refractory elements from the chondritic data, assuming no relative fractionation.

15. Refractory trace element value from Col. 1, $\times$ 1.36 (see [14]).

16. From Ni/Co = 20 (C1 value).

17. C1 value $\times$ 1.36.

18. From FeO/MnO = 60. Dreibus, G., et al. (1977) *PLC 8*: 211.

19. K from K-Ar (Taylor, 1979, ibid.) and $K/U = 10^4$.

20. Rb from Rb/Sr = 0.031 [O'Nions, R. K., et al. (1979) *Ann. Rev. Earth Planet. Sci.* 7: 11]. This gives K/Rb = 375. This implies loss of Rb relative to K during accretion of Earth since K/Rb for C1 = 260 (see [25]).

21. From Na/K = 14 [Taylor, S. R. (1980) *PLC 11*: 333].

22. Rb/Cs in crust is $\sim$ 30 compared to 12 in chondrites. Cs will be enriched in crust relative to Rb. Thus 0.016 ppm (from Rb/Cs = 30) is an upper limit.

23. From Tl/U = .022 [Krahenbühl, U., et al. (1973) *PLC 4*: 1325].

24. From Li/Zr = 0.10 (Dreibus, G., et al., ibid., 18).

25. Total crustal data based on Taylor, S. R., and McLennan, S. M. (1981) *Phil. Trans. Roy. Soc.* A301: 381. The uranium abundance has been revised from 1.0 to 1.25 ppm using $K/U = 10^4$. Th has been raised to 4.8 ppm to conform to Th/U = 3.8. Hf has been revised to 3.0 ppm (from 2.2 ppm) to be consistent with chondritic Zr/Hf = 33. Tl from Shaw, D. M. (1976) *GCA*. 40: 73. Rb has been lowered from 50 ppm (Taylor, 1977) to 42 ppm to be

consistent with $K/Rb = 300$. This ratio implies that Rb is partitioned into the crust preferentially to K (cf. Cs [22]).

26. Major element data, Ba, Co, Cu and Zn from Goodwin (1977) *Geol. Soc. Canada Spec. Paper 16*, 205.
27. REE from Average Archean Sedimentary rock (AAS) [Taylor, S. R., and McLennan, S. M. (1980) *Precambrian Plate Tectonics* (A. Kroner, ed.), Elsevier].
28. Rb from $K/Rb = 300$, U from $K/U = 10^4$, Th from $Th/U = 3.8$, Hf from $Zr/Hf = 33$.
29. Sr, Zr, V, Ni from tholeiite/tonalite – trondhjemite mix (50/50). Tholeiite data from Sun, S. -S., and Nesbitt, R. (1978) *Contrib. Mineral. Petrol.* 65: 301 and (1977) *EPSL.* 35: 429. Tonalite and trondhjemite data from Glikson, A. (1978) *Earth Sci. Rev.* 15: 1.
30. H. Palme, pers. comm., 1980.

APPENDIX VII

Element classification, Condensation sequences

Part A Element classification based on relative volatility and siderophile character.

Refractory		Al	Ba	Be	Ca	Hf	Ir	Mo
		Nb	Os	Pt	REE	Re	Rh	Ru
		Sc	Sr	Ta	Ti	U	V	W
		Y	Zr					
Moderately refractory	Cr	Li	Mg	Si				
Volatile		Ag	Cs	Cu	F	Ga	Ge	K
1300–600° K		Mn	Na	Rb	S	Sb	Se	Sn
		Te	Zn					
Very volatile		B	Bi, Br,	C	Cd	Cl	Hg	I
(< 600° K)		In	Pb			Rare gases		Tl
Siderophile		Fe	Ni	Co	Cr	P	As	Au
		Pt	group					

Part B Sequence of condensations from an oxidized $(C/O \sim 0.6)$ gas of solar composition at 10^{-3} atm. total pressure.

	Temperature °K
Refractory noble metals and hibonite $(CaAl_{12}O_{19})$	1800–1700
Refractory oxides (Al_2O_3) Perovskite $(CaTiO_3)$ Melilite $(CaAl_2Si_2O_7)$ Spinel $(MgAl_2O_4)$ U, Th, REE in solid solution.	1700–1500
Metallic iron (Ni, Co)	1470
Diopside $(CaMgSi_2O_6)$ Forsterite (Mg_2SiO_4) Anorthite $(CaAl_2Si_2O_8)$ Enstatite $(MgSiO_3)$	1450–1300
Alkali feldspar (NaK) $AlSi_3O_8$, volatile metals, halogens	1000
Troilite (FeS) Pentlandite (Fe, Ni) S	700
Magnetite Fe_3O_4 phyllosilicate minerals (e.g., $Mg_3Si_2O_7 2H_2O$)	400
Ices	< 200

Part C Sequence of condensation from a reduced (C/O ~ 1.0) gas of solar composition at 10^{-3} atm. total pressure.

	Temperature °K
Refractory noble metals Os, W, Re, Pt	1800–1700
Reduced high-temperature condensates (TiCSiC, C, Te$_3$C, AlN Ca S*)	1900–1400
Metallic Fe (with Si)	1350–1200
Refractory oxides	1200–1100
Diopside (CaMgSi$_2$O$_6$) Forsterite (Mg$_2$SiO$_4$) Enstatite (MgSiO$_3$)	1150–1050

Below 10000° K, the sequence is the same as for an oxidized gas.

APPENDIX VIII

Chondritic Rare Earth Element Normalizing Factors.

La	0.367
Ce	0.957
Pr	0.137
Nd	0.711
Sm	0.231
Eu	0.087
Gd	0.306
Tb	0.058
Dy	0.381
Ho	0.0851
Er	0.249
Tm	0.0356
Yb	0.248
Lu	0.0381
Y	2.1

Values from Evensen, M.N., et al. (1978) *GCA. 42*: 1203, for Type 1 carbonaceous chondrites (volatile-free: 1.5 $\times$ original data).

APPENDIX IX

The Lunar Sample Numbering System

A total of 2196 samples, weighing 381.69 kg, was collected during the Apollo Missions. The basic numbering system employs a five digit number for the original samples collected on the luanr surface. A different prefix is employed for each mission:

Mission	Prefix
Apollo 11	10—
Apollo 12	12—
Apollo 14	14—
Apollo 15	15—
Apollo 16	6—
Apollo 17	7—
Luna 16	21—
Luna 20	22—
Luna 24	24—

For the Apollo 11, 12 and 14 missions, the remaining three digits have no special significance beyond designating the sampling sequence. For the Apollo 15, 16 and 17 rocks, the remaining digits serve to indicate both sample location and rock type. The Apollo 15 samples are numbered in groups more or less consecutively around the traverses. For the Apollo 16 and 17 missions, the second digit specifies the station on the traverses. The final digit specifies whether the sample is an unsieved soil (0), a sieved fraction of a soil (1 through 4) or a rock (5 through 9). Thus 64826 is a rock from station 4 on the Apollo 16 mission. The main exceptions to this system are the lowest sample numbers (e.g., 70007), which are for drill cores, drive tubes, and other special samples, regardless of location or type. Among the Apollo 15 samples, 15315–15392 and 15605–15689, including numbers ending in 0 through 4, are all rocks. Not all possible numbers were used.

Samples split from the original are identified by additional digits, separated by a comma from the original sample number. Thus 15455,20 is split number 20 from rock 15455.

A total of 69,000 such individual numbers now exists, each representing a documented piece of the Moon.

APPENDIX X

Planetary and asteroidal data arranged in order of increasing distance from the sun.[†]

		Semi-major axis (A.U.)	Inclination to ecliptic	Inclination of equator to orbit	Orbital period	Radius (km)	Mean density (g/cm³)
	Mercury	0.387	7°00'	0°	86.0 d.	2439	5.435
	Venus	0.723	3°24'	177°	224.7 d.	6051	5.245
	Earth	1.000	0°00'	24°27'	1.00 yr.	6371	5.514
	Moon	-	5°09'	1°32'	-	1738	3.344
	Mars	1.524	1°51'	25°1	1.88 yr.	3390	3.934
8	Flora	2.202	5°89'	-	-	77	-
4	Vesta	2.362	7°13'	-	-	269	2.9
1	Ceres	2.767	10°61'	-	-	510	-
2	Pallas	2.769	34°83'	-	-	269	-
10	Hygiea	3.151	3°81'	-	-	225	-
65	Cybele	3.434	3°55'	-	-	154	-
107	Camilla	3.489	9°92'	-	-	105	-
624	Hektor	5.150	18°26'	-	-	108	-
	Jupiter	5.203	1°18'	3°6'	11.86 yr.	71,600	1.314
	Saturn	9.523	2°29'	26°42'	29.46 yr.	60,000	0.704
	Chiron	13.7	-	-	51.0 yr	175	-
	Uranus	19.164	0°46'	97°	84.01 yr.	25,900	1.21
	Neptune	29.987	1°47'	29°48'	164 yr.	24,750	1.66
	Pluto	39.37	17°10'	-	247 yr.	~1300	~1.0

[†]Sources: Newburn, R. (1978) *PLC 9*: 3; Newburn, R., and Matson, D. (1978) *PLC 9*: 2; Chapman, C., and Zellner, B. (1978) *PLC 9*: 1; Gehrels, T. (ed.) (1979) *Asteroids*, Univ. Arizona Press.

APPENDIX XI

Satellite data, arranged by planet.

Planet	Satellite	Semi-major axis (10^3 km)	Orbital period (r = retrograde)	Diameter (km)	Density (g/cm^3)	Inclination to equatorial plane of planet
Earth:	Moon	384.4	27.32 d.	3476	3.344	18.5° to 28.5°
Mars:	Phobos	9.38	7.65 h.	27×21×19	1.9	1.04°
	Deimos	23.5	30.3 h.	15×12×11	—	2.79°
Jupiter:						
	1979J 3	126	7.1 h.	40	—	—
	1979J 1	128	7.1 h.	35	—	—
J V	Amalthea	182	12 h.	270×170×155	—	0.45°
	1979J 2	223	16.2 h.	75	—	—
J I	Io	422	1.8 d.	3632	3.55	0.02°
J II	Europe	671	3.6 d.	3072	2.99	0.47°
J III	Ganymede	1071	7.2 d.	5276	1.93	0.18°
J IV	Callisto	1884	16.8 d.	4820	1.83	0.25°
J XIII	Leda	11,094	239 d.	10	—	26.7°
J VI	Himalia	11,487	251 d.	170	—	27.6°
J VII	Erala	11,747	260.0 d.	80	—	24.8°
J X	Lysithea	11,861	264.0 d.	25	—	29.0°
J XII	Anake	21,250	r. 1.72 yr.	20	—	147°
J XI	Carme	22,540	r. 1.89 yr.	30	—	164°
J VIII	Parisphae	23,510	r. 2.02 yr.	35	—	145°
J IX	Sinope	23,670	r. 2.07 yr.	30	—	153°
Saturn:	S 15	136	14.3 h.	40×20×20	—	—
	S 14	138	14.6 h.	220	—	—
	S 13	141	15.0 h.	200	—	—
	S 11	151	16.7 h.	180×80	—	—
	S 10	151	16.7 h.	200×180×150	—	—
S II	Mimas	186	22.6 h.	390	1.2	1.52°
S III	Enceladus	238	1.4 d.	500	1.1	0.02°
S IV	Tethys	295	1.9 d.	1050	1.0	1.09°
	S 16	295	1.9 d.	20 ?	—	—
	S 17	295	1.9 d.	30 ?	—	—
S V	Dione	377	2.7 d.	1120	1.4	0.02°
	Dione B	377	2.7 d.	160	—	—
S VI	Rhea	527	4.5 d.	1530	1.3	0.35°
S VII	Titan	1222	15.9 d.	5144	1.9	0.33°
S VIII	Hyperion	1484	21.3 d.	290	—	—
S IX	Iapetus	3562	79.3 d.	1440	1.2	14.72°
S X	Phoebe	12,960	r. 550 d.	240	—	150°
Uranus:	Miranda	130	1.4 d.	300	—	3.4°
	Ariel	191	2.5 d.	800	—	~0°
	Umbriel	266	4.1 d.	550	—	~0°
	Titania	436	8.7 d.	1000	—	~0°
	Oberon	583	13.5 d.	900	—	~0°
Neptune:	Triton	356	r. 5.9 d.	4000	—	159.9°
	Nereid	5567	360 d.	300	—	27.7°
Pluto:	Charon	19	6.4 d.	1300	—	~105°

Sources: Newburn, R., and Matson D. (1978) *PLC 9*: 2;
Science. (1979) 204, 945;
Science. (1979) 206, 925;
Science. (1981) 212, 159.

GLOSSARY

Because of the wide-ranging nature of planetary science, it is to be expected that readers of this book may be unfamiliar with one or more of the fields covered. For such a reason this glossary has been added. It is intended as a dictionary-type aid, to preserve continuity and understanding through unfamiliar territory.

Glossaries possess both the advantages and disadvantages of dictionaries. They should preserve the virtue of simplicity and are not intended for the serious student or specialist, who is at best likely to become displeased by the definitions adopted. The latter reader is referred to the subject index, the references and notes and to the text itself for enlightenment.

Accessory Mineral—A term applied to a mineral occurring in small amounts in a rock.

Accretion—A term applied to the growth of planets from smaller fragments or dust.

Achondrite—Stony meteorite lacking chondrules.

Aeon—One billion years ($= 10^9$ years).

Agglutinate—A common particle type in lunar soils, agglutinates consist of comminuted rock, mineral and glass fragments bonded together with glass.

Albedo—The percentage of the incoming radiation that is reflected by a natural surface.

Alkali Element—A general term applied to the univalent metals Li, Na, K, Rb and Cs.

ALSEP—Apollo lunar surface experiments package.

An—Abbreviation for anorthite.

Angstrom (Å)—A unit of length, 10^{-8} cm; commonly used in crystallography and mineralogy.

Anorthite (An)—$CaAl_2Si_2O_8$, the most calcium-rich member of the plagioclase (feldspar) series of minerals.

Anorthosite—An igneous rock made up almost entirely of plagioclase feldspar.

ANT—An acronym for the suite of lunar highland rocks: anorthosite, troctolite, and norite.

BABI—Best estimate for initial $^{87}Sr/^{86}Sr$ ratio of basaltic achondrites, regarded as approximating that of the solar nebula.

Bar—The international unit of pressure (one bar $= 10^6$ dynes/cm^2).

Basalt—A fine-grained, dark colored igneous rock composed primarily of plagioclase (feldspar) and pyroxene; usually other minerals such as olivine, ilmenite, etc. are present.

Basaltic achondrite—Calcium-rich stony meteorites, lacking chondrules and nickel-iron, showing some similarities to terrestrial and lunar basalts; eucrites and howardites are types of basaltic achondrites.

Bouguer gravity—The free-air gravity corrected for the attraction of the topography, so it is only dependent on the internal density distribution.

Bow shock—A shock wave in front of a body.

Breccia—A rock consisting of angular, coarse fragments embedded in a fine-grained matrix.

Brownlee particles—Interplanetary dust particles, approximately 10μm in size, collected in the Earth's stratosphere.

Carbonaceous chondrites—The most primitive stony-meteorites, in which the abundances of the non-volatile elements are thought to approximate most closely to those of the primordial solar nebula.

Chalcophile element—An element which enters sulfide minerals preferentially.

Chondrite—The most abundant class of stony meteorite characterized by the presence of chondrules.

Chondrule—Small, rounded body in meteorites (generally less than 1 mm in diameter), commonly composed of olivine and/or orthopyroxene.

Clast—A discrete particle or fragment of rock or mineral; commonly included in a larger rock.

Clinopyroxene—Minerals of the pyroxene group, such as augite and pigeonite, which crystallize in the monoclinic system.

Cumulate—A plutonic igneous rock composed chiefly of crystals accumulated by sinking or floating from a magma.

Curie temperature—The temperature in a ferromagnetic material above which the material becomes substantially nonmagnetic.

Dunite—A peridotite that consists almost entirely of olivine and that contains accessory chromite and pyroxene.

Eclogite—A dense rock consisting of garnet and pyroxene, similar in chemical composition to basalt.

Ejecta—Materials ejected from the crater by a volcanic or meteorite impact explosion.

Epicenter—The point on a planetary surface directly above the focus of an earthquake.

Eu*—See footnote [51], Chapter 5, page 259.

Eucrite—A meteorite composed essentially of feldspar and augite.

Exsolution-unmixing—The separation of some mineral-pair solutions during slow cooling.

Exposure age—Period of time during which a sample has been at or near the lunar surface, assessed on the basis of cosmogenic rare gas contents, particle track densities, short-lived radioisotopes, or agglutinate contents in the case of soil samples.

Ferromagnetic—Possessing magnetic properties similar to those of iron (paramagnetic substances with a magnetic permeability much greater than one).

Fines—Lunar material arbitrarily defined as less than 1 cm in diameter; synonymous with "soils."

Fractional crystallization—Formation and separation of mineral phases of varying composition during crystallization of a silicate melt or magma, resulting in a continuous change of composition of the magma.

Fractionation—The separation of chemical elements from an initially homogeneous state into different phases or systems.

Free-air gravity—As applied to the Moon, it is usually synonymous with the observed gravity anomaly.

Fugacity—A thermodynamic function used instead of pressure in describing the behavior of non-ideal gases.

Gabbro—A coarse-grained, dark igneous rock made up chiefly of plagioclase (usually labradorite) and pyroxene. Other minerals often present include olivine, apatite, ilmenite, etc. A coarse grained equivalent of basalt.

Gardening—The process of turning over the lunar soil or regolith by meteorite bombardment.

Geomagnetic tail—A portion of the magnetic field of the Earth that is pulled back to form a tail by the solar wind plasma.

Granite—An igneous rock composed chiefly of quartz and alkali feldspar.

Granular—A rock texture in which the mineral grains are nearly equi-dimensional.

Groundmass—The fine-grained rock material between larger mineral grains; used interchangeably with matrix.

Half-life—The time interval during which a number of atoms of a radioactive nuclide decay to one half of that number.

Howardite—A type of basaltic achondrite.

Impact melting—The process by which country rock is melted by the impact of meteorite or comet.

Intersertal—A term used to describe the texture of igneous rocks in which a base or mesostasis of glass and small crystals fills the interstices between unoriented feldspar laths.

Ionic radius—The effective radius of ionized atoms in crystalline solids; ionic radii commonly lie between 0.4–1.5 Angstrom units.

Ionosphere—The ionized layers of a planet's atmosphere.

Isochron—A line on a diagram passing through plots of samples with the same age but differing isotope ratios.

Isostasy—Weight balancing of topography by underlying density anomalies. The distribution is such that at some uniform depth the pressure is everywhere constant and beneath this depth a state of hydrostatic equilibrium exists.

Isotopes—Atoms of a specific element which differ in number of neutrons in the nucleus; this results in different atomic weight, and very slightly differing chemical properties (for example, ^{235}U and ^{238}U).

Iron meteorite—A class of meteorite composed chiefly of iron or iron-nickel.

KREEP—An acronym for a lunar crustal component rich in potassium (K), the rare-earth elements (REE), phosphorous (P) and other incompatible elements.

Lattice site—The position occupied by an atom in a crystalline solid.

Layered igneous intrusion—A body of plutonic igneous rock which has formed layers of different minerals during solidification; it is divisible into a succession of extensive sheets lying one above the other.

LIL—Large Ion Lithophile elements. Those lithophile elements (e.g., K, Rb, Ba, REE, U, Th) which have ionic radii larger than common lunar rock-forming elements, and which usually behave as trace elements in lunar rocks and in meteorites.

Liquidus—The line or surface in a phase diagram above which the system is completely liquid.

Lithophile element—An element tending to concentrate in oxygen-containing compounds, particularly silicates.

LUNI—Best estimate for initial $^{87}Sr/^{86}Sr$ ratio of Moon, determined from lunar anorthosites.

Magmatic differentiation—The production of rocks of differing chemical composition during cooling and crystallization of a silicate melt or magma by processes such as removal of early formed mineral phases.

Mascon—Regions on the Moon of excess mass concentrations per unit area identified by positive gravity anomalies and associated with mare-filled multi-ring basins.

Maskelynite—Feldspar converted to glass by shock effects due to meteorite impact.

Matrix—The fine-grained material in which large mineral or rock fragments are embedded; often used interchangeably with groundmass.

Mesostasis—The interstitial, generally fine-grained material, between larger mineral grains in a rock; may be used synonymously with matrix and groundmass.

Microcrater (Zap Pit)—Crater produced by impact of interplanetary particles generally having masses less than 10^{-3} g.

Model age—The age of a rock sample determined from radioactive decay by *assuming* an initial isotopic composition of the daughter product.

Moment of inertia—A quantity which gives a measure of the density distribution within a planet, specifically, the tendency for an increase of density with depth. It is derived from gravity and dynamical considerations of the planet.

Noble gases—The rare gases, helium, neon, argon, krypton, xenon and radon.

Norite—A type of gabbro in which orthopyroxene is dominant over clinopyroxene.

NRM - Natural remanent magnetization—That portion of the magnetization of a rock which is permanent and usually acquired by the cooling of ferromagnetic minerals through the Curie temperature.

Nucleon—Sub-atomic nuclear particles (mainly protons and neutrons).

Nuclides—Atoms characterized by the number of protons (Z) and neutrons (N). The masss number (A) = N + Z; isotopes are nuclides with the same number of protons (Z) but differing numbers of neutrons (N); isobars have same mass number (A) but different numbers of protons (Z) and neutrons (N).

Oersted—The cgs unit of magnetic intensity.

Ophitic—A rock texture which is composed of elongated feldspar crystals embedded in pyroxene or olivine.

Orthopyroxene—An orthorhombic member of the pyroxene mineral group.

Peridotite—An igneous rock characterized by pyroxene and olivine (but no feldspar).

Phenocryst—A large, early formed crystal in igneous rocks, surrounded by a fine-grained groundmass.

Plagioclase—A sub-group (or series) of the feldspar group of minerals.

Plutonic—A term applied to igneous rocks which have crystallized at depth, usually with coarsely crystalline texture.

Poikilitic—A rock texture in which one mineral, commonly anhedral, encloses numerous other much smaller crystals, commonly euhedral.

Poikiloblastic—A metamorphic texture in which large crystals form in the solid state during recrystallization, to enclose preexisting smaller crystals.

Poise, cgs—Unit of viscosity ($=$ 1 dyne sec/cm^2).

Porphyritic—Having larger crystals set in a finer groundmass.

PPB—Parts per billion — 1 ppb $=$ 0.001 ppm.

PPM—Parts per million — 1 ppm $=$ 0.0001%.

P-wave velocity—Seismic body wave velocity associated with particle motion (alternating compression and expansion) in the direction of wave propagation.

Pyroxene—A closely related group of minerals which includes augite, pigeonite, etc.

Radiogenic lead—Lead isotopes formed by radioactive decay of uranium and thorium (^{206}Pb from ^{238}U; ^{207}Pb from ^{235}U; ^{208}Pb from ^{232}Th).

Rare-earth (RE or REE)—A collective term for elements with atomic number 57-71, which includes La, Ce, etc.

Rare gases—The noble gases, helium, neon, argon, krypton, xenon, and radon.

Regolith—Loose surface material, composed of rock fragments and soil, which overlies consolidated bedrock.

Residual liquid—The material remaining after most of a magma has crystallized; it is sometimes characterized by an abundance of volatile constituents.

Siderophile element—An element which preferentially enters the metallic phase (see Appendix VII).

Silicate—A mineral (or compound) whose crystal structure contains SiO_4, tetrahedra.

SMOW - Standard Mean Ocean Water—Used as a reference for oxygen isotope work.

Soil breccia—Polymict breccia composed of cemented or sintered lunar soil.

Solar nebula—The primitive disk-shaped cloud of dust and gas from which all bodies in the solar system originated.

Solar wind—The stream of charged particles (mainly ionized hydrogen) moving outward from the sun with velocities in the range 300–500 km/sec.

Solidus—The line or surface in a phase diagram below which the system is completely solid.

Spinel-group—General term for several minerals (e.g., chromite) with chemical, physical and structural properties similar to spinel; general formula is AB_2O_4.

Stony meteorite—A class of meteorite composed chiefly of silicate minerals such as pyroxene, olivine, etc.

Suntan—Period of time during which a sample has resided on the lunar surface without shielding; determined from solar flare-induced particle tracks.

S-wave velocity—Seismic body wave velocity associated with shearing motion perpendicular to the direction of wave propagation.

Tektites—Small glassy objects of wide geographic distribution formed by splashing of melted terrestrial country rock during meteorite, asteroid, or cometary impacts.

Texture—The arrangement, shape and size of grains composing a rock.

Trace element—An element found in very low (trace) amounts; generally less than 0.1%.

T Tauri—An early pre-main-sequence state of stellar evolution, characterized by extensive mass loss.

Vesicle (Vesicular)—Bubble-shaped, smooth-walled cavity, usually produced by expansion of vapor (gas) in a magma.

Volatile element—An element volatile at temperatures below 1300° C.

Vug (Vuggy)—Small, irregular-shaped, rough-walled cavity in a rock.

XRF—X-ray fluorescence spectroscopy.

Zap Pit—See Microcrater.

$\delta^{18}O$—See footnote [95], Chapter 6, page 341.

μ—The ratio of ^{238}U to ^{204}Pb.

INDEX

Accretion 236, 416, 429
  heterogeneous 177, 220, 228, 422
  homogeneous 228-9, 422
  secondary 420
Achondrites, basaltic 338, 420
Agglutinates 128
  elemental fractionation 130
  formation 126, 148
  melts 127
Airy hypothesis 183
Akaganéite 217
Alba Patera 52
Allende 378
Allende inclusions 416, 425, 428
Altai Scarp 77
Aluminum-26 413-4, 420, 428
Anorthosites 205, 229, 249
Anorthosites, gabbroic 217
Apennine Bench Formation 38, 78, 93, 215,
  224, 273-4
Apennine Mountains 215, 346
Apenninian series 26
Aphrodite Terra 55, 354
Apollo landing sites 11
  Apollo 14 37, 110, 227, 241
  Apollo 15 187, 273, 351
  Apollo 16 110, 117, 187, 224, 238, 240
  Apollo 17 109, 130, 167, 187, 267, 272
Apollo 11 samples 267, 300
Apollo 14 samples 187
Apollo 16 samples 201
Apollo Soil Survey 202
Apollonian metamorphism 200
Archean 253
Archean crustal composition 255
Argon
  dates, resetting 239
  Argon-40 anomaly 162
Aristarchus Plateau 224
Assimilation models 326
Asteroid belt 155, 422

Basaltic volcanism 263, 317, 436
Basalts
  classification 284
  on Mars 335
  on Vesta 338
Basalts, Apollo 17 299
Basalts, Apollo 17 VLT 308, 330

Basalts, Columbia River 269,272
Basalts, Fra Mauro 203, 217
Basalts, high-Al 300, 317, 330
Basalts, high-Ti 301, 308, 318
Basalts, highland 202, 228, 232
Basalts, KREEP 214, 229, 251, 253, 326-7,
  330, 348
Basalts, low-K Fra Mauro (LKFM) 202,
  214
Basalts, low-Ti 300, 318
Basalts, Luna 16 301, 317, 330
Basalts, mare 243
  ages 318
  chalcophile 313
  depths of melting 331
  ferromagnesian elements 311
  high valency cations 309
  large cations 301
  lead 313
  nickel 311
  origin 321
  oxidation state 292
  rock types 282
  sulfur 313
  thickness 350
  trace elements 301
  volume 325
Basalts, mid-ocean ridge (MORB) 255
Basalts, VLT 200
Basins (see also Multi-ring basins)
  morphology 91
  peak ring 91, 92
  secondary 96
Basins, Callisto—Valhalla 76
Basins, impact—ages 240
Basins, lunar
  Al Khwarizmi 107
  Antoniadi 87
  Crisium 107
  Gargantuan 87
  Grimaldi 348, 351
  Imbrium 26, 107, 209, 224, 241
  Ingenii 348
  Mendeleev 348
  Nectaris 42, 77, 107, 187, 238, 240
  Orientale 38, 107, 187, 348
  Procellarum 81, 87
  Schrödinger 82, 87
  Serenitatis 107, 241

Basins, mare 12
Basins, Martian
  Argyre 51, 54
  Hellas 51, 54
  Isidis 51, 54
Basins, Mercurian
  Caloris 45, 50, 76
Benches 277
Beta Regio 336
Big bang hypothesis 410
Breccias
  basalts in 200
  highland, magnetic properties 365-6
Breccias, "dike" 194
Breccias, dimict 192
Breccias, feldspathic fragmental 194
Breccias, granulitic 198, 228, 232
Breccias, impact melt 195
Breccias, lunar 187-201
Breccias, monomict 187
Breccias, soil 187
Bunte breccia 89

Calderas 64, 97
Callisto 257
Capture hypothesis 424
Carbon 316
  in lunar soils 159
Carbonaceous chondrites 13, 97, 313, 378,
        380, 393, 411, 414, 420, 424, 429
  C1 abundance data 151, 381
Catena Davy 94
Cayley Formation 18, 27, 36, 40, 47, 109,
        142, 187, 224
Cayley Plains 109, 240
Center of mass/center of figure offset 344
Chalcophile elements 313, 379
Chondrules 134
  formation 420
Chromium 251, 329, 331, 395, 425
  in lunar crust 391
  in mare basalts 312
Chromium/nickel ratios 391
Clinopyroxene 325
Continental drift 439
Continental growth model 256
Copernican system 39
Cordillera Formation 29, 38
Cordillera Scarp 77, 187
Cosmic rays 158
Cosmogenic nuclides 157
Cosmogenic radionuclides 122, 163
Cosmology 436

Crater counting 29, 93, 96
Cratering flux 107
Cratering rates 61-2, 107
Cratering record 104
Cratering, early 12
Cratering, secondary 28
Craters
  cavity 65
  central peaks 76
  production rate 101
Craters, dark halo 95-6, 265, 317
Craters, lunar
  Anaxagoras 75
  Archimedes 22, 27, 331
  Aristarchus 139, 224
  Camelot 119, 150, 167
  Cone 195
  Copernicus 17, 26, 75, 76, 93, 99, 352
  Cyrillus 71
  Eratosthenes 26, 39
  Gassendi 98
  Humboldt 98, 351
  Inghirami 38
  Janssen 33, 97
  Kepler 17, 26, 265
  Kopff 97
  Linné 68, 100
  North Ray 119, 195, 238, 240
  Picard 266
  Plato 27
  Proclus 63
  Peirce 265
  Regiomontanus A 75
  Shorty 97, 150, 298
  Sinus Iridum 27
  South Ray 166-7
  Spur 298
  Theophilus 71, 81, 238
  Tycho 17, 26, 93, 109, 138, 167, 227, 355,
        437
  Van de Graff 87
Craters, mare 186
  central peaks 72
  wall terraces 72
Craters, Martian 95
Craters, morphology 68
Craters, secondary 31, 94
Craters, summit 74
Craters, terrestrial 101
  Arizona Meteor 68
  Gosses Bluff 74
  Lonar 133
  Manicouagan 133

Mauna Loa 14
Prarie Flat 85
Ries 36, 82, 89, 109, 194, 233
Sierra Madera 74
Craters, transient 90
Craters, volcanic 95
Crust, continental 180, 253, 377, 387, 430
Crust, lunar
  composition 200, 390
  highland composition 201, 227
  thickness 356, 390
Crust, Martian 257
Crust, Mercurian 256
Crust, oceanic 253
Crust, terrestrial
  oceanic 389
  primitive 430
Crustal evolution 437
Crustal uniformity 181
Crystal settling 331
Crystal-liquid fractionation 378, 390-1, 430
Crystallization, fractional 333
Crystallization trends
  lunar 248
  Stillwater 248
Cumulate model 327
Curie temperature 365, 368
Curved rilles 282

Dark mantle deposits 272
Descartes Formation 34, 36, 109, 142, 186,
     203, 240
Descartes Mountains 119, 187
Differentiation 236
  crystal-liquid 435
  lunar 244
  whole-moon 358
Double-planet hypothesis 424
Dunite 205, 207, 233, 249, 311, 378, 400
Dynamic assimilation model 326

Earth
  accretion 428
  age 234
  core 381
  crust 430
  expansion 49
  formation 429
  magnetic field 369
  mantle 381, 396
  mantle, abundances 387
  oceanic crust 389
  primitive mantle 381

  upper mantle 402
Eclogite 322, 324, 351
Ejecta 68
Element correlations 228
Element fractionation 376
Enstatite chondrites 381, 417, 420, 429
Equilibrium condensation 416, 423
Equipotential surface 345
Eratosthenian system 39
Erosion
  micrometeorite bombardment 165
  rates 163
Eucrite parent body 338, 404
Eucrites 394, 404, 420
Europium 205, 243, 329
  enrichment 232
  in lunar highlands 395
Europium anomaly 205, 308, 326
Exobiology 171
Exposure ages 163, 165
Fire fountains 148, 299-300
Fission hypothesis 396, 424
Fission tracks 158
Fluidization 77
Fra Mauro Formation 26, 36, 368
Fractionation 416
  crystal-liquid 391
  melt-mineral 334
  pre-accretion 393

Galactic cosmic rays 155, 157, 165, 168
Galaxies 410
Gallilean satellites 57
Ganymede 257
Gardening 122, 165
Garnet 309, 325, 344, 396
Geochemical anomalies 226
Germanium 315
Glasses (see also Agglutinates, impact glas-
     ses, tektites) 128
  color 130
  Emerald Green 297, 330
  green, composition 299
  iron spherules 131
  morphology 130
  orange, origin 299
  selective vaporization 133
  volcanic 297
Gold 315
Granodiorite 254
Gravity 345
  anomalies 348
Gravity, lunar 345

Gravity, Martian 352
Gravity, Venusian 354
Grooved terrain 257

Hadley Rille 115, 118, 269, 279, 281
Haemus Mountains 36
Heat flow, lunar 14, 361
Heterogeneous accretion model 424
Hevelius Formation 29, 38, 77, 109
Highlands 8, 21, 180-1
  volcanism 63
Highlands, crust 13, 87, 242
  abundances 230
  ages 233, 238
  composition 201, 227, 390
  major elements 230
  thickness 180-1, 345, 390
  trace elements 231
Hot spots 115
Hubble constant 409
Hybrid liquids 326

Ice 257
Imbrian system 26, 36
Impact glass 115, 130
  color 130
  morphology 130
  terrestrial 135
Impact melts 66, 92, 187, 321
Incompatible elements 229, 330, 386, 394
Intercumulus liquids 246
Interstellar clouds 410
Involatile elements 391
Io 257, 429
  lava flows 337
Iodine-129 413
Iridium 315
Iridium/gold ratios 220
Iron 147, 365, 402
Iron/nickel ratios 311, 328
Ishtar Terra 55, 354
Island-arc model 255
Isostasy 99, 351
Isotopic anomalies 413
Isotopic heterogeneities 414
Isotopic variations 168

Janssen Formation 29, 33
Jupiter 422, 424, 428

Kant Plateau 35, 110, 187, 224, 238
KREEP (see also Super-KREEP) 208, 214,
  229, 251, 253, 317, 326
  volcanism 346

Lateral transport 181, 122, 224
Lava, basaltic 21
Lava flows 269
  on Io 337
Lava lakes 351
Lead 299, 426
Liquid immiscibility 137, 333
Lithosphere, lunar 356
  thickness 352
Lunar Cataclysm 104, 242
Lunar core 344-345, 396
Lunar granites 208, 243, 333
Lunar Sounder Experiment 356
Lunar-impact Theory 139

Magma mixing 251
Magma ocean 104, 205, 209, 216,
  236, 242-3, 245, 308, 315, 327, 345-
  6, 430, 435
  criteria 244
  depth 244
Magmas, primary 331
Magnetic anomalies 367
Magnetic dipole field 370
Magnetic fields
  Earth 369
  impact generated 368
  lunar 368
  Mercury 370
  solar wind 369
Magnetism, permanent model 369
Magnetite 365
Mantle
  discontinuity 244
  lower 386
  terrestrial 396, 402
  uplift 352
  upper 386
Mare
  Crisium 63, 224, 264-5
  Fecunditatis 301
  Humorum 78, 282
  Imbrium 19, 22, 26, 265-6, 269,
    270, 350
  Ingenii 87, 100
  Marginis 100
  Orientale 19, 76-7, 266, 351
  Procellarum 27
  Serenitatis 24, 266, 274, 350-1
  Smythii 264
  surfaces 264, 344
  Tranquilliatis 267, 272, 318
Mare-Highland contacts 122

Maria 8, 13, 32
Marius Hills 38, 274, 350
Mars (see also Basins, Martian; Plains, Martian; Volcanoes, Martian) 91, 335, 422, 427
  composition 404
  core 359
  crust 257
  gravity 352
  seismology 356
  soil 121
  surface 116
Mascons 184, 322, 325, 346, 350, 362, 363
Maunder Formation 77
Maunder Minimum 168
Maxwell Mountains 57
Megaregolith 104, 186, 246
Megaterraces 79, 82, 84
Melt rocks 201
Melt sheets 68, 93
Melting
  depths of 331
  whole-moon 328
Mercury 335, 427
  composition 403
  core 359, 371
  crust 256
  magnetic field 370
  mantle 43
Metamorphism, impact-induced 216
Meteorite flux 100, 123, 246, 430
Meteorites
  age 234
  chondritic 387
  gas-rich 166
Magnesium component 202, 229
Magnesium suite 206, 249, 251
Magnesium/silicon ratios 395
Microfossils 134
Mineral, shock effects 66
Mixing models 205
Model ages 321
Mohorovičić discontinuity 254
Molecular clouds 410-1
Moment of inertia 343-4, 358, 361
Montes Cordillera 76
Montes Rook 76-7
Moon
  accretion 426
  atmosphere 169
  core 315, 344, 358-9, 363
  core formation 428

crust composition 200
crustal thickness 356
differentiation 428
electrical conductivity 358, 361
highland crust thickness 390
interior 324
interior discontinuities 357
magnetic field 364, 368
magnetism 364
melting 328
moonquakes 354-5
regolith 3
Rima Fresnel 233
seismology 354
soil 159, 161, 167
Sulpicius Gallus 272
surface temperature 116
temperature profile 362
Multi-ring basins 12, 64, 76, 110, 186, 264, 323, 345

Nectarian system 29, 33
Neon-E 414
Neptune 422
Nested Crater model 81
Neutron fluxes 163
Nickel 131, 311, 326-7, 365, 401, 430
Nickel/cobalt ratios 222
Nitrogen 159
Norite 206, 378
Nuclear tracks
  densities 165
  production 165

Oceanus Procellarum 264-5, 333
Olivines 251
Orbital chemical data 110, 223
Orbital gamma-ray data 223
Orbital XRF data 223
Organic geochemistry 171
Orgueil 380
Orion Nebula 411, 413
Oscillating Peak model 85
Oxygen fugacity 294
Oxygen isotope data 377
Oxygen isotopes 378, 424
Oxygen-16 413-4, 416

P-wave velocities 181, 358
Paleointensities, magnetic 366
Palladium-107 414
Palus Putredinis 281, 298
Partial melting 325, 393

Peak rings 76
Perovskite 387
Pioneer Venus 55
Plagioclase 246, 248-9, 306, 335, 395, 430
Plagioclase feldspar 205
Plains, Martian 51
    Lunae Planum 52
    Syrtis Major Planitia 52
    Tempe Fossae 52
Planetary formation 419
Planetesimal hypotheses 415
Planetesimals 202, 245, 423, 435-6
Planetology 11
Plate tectonics 397
Polar orbiter 11, 437
Potassium 243, 388, 395, 399, 429
    abundances on Earth 387
Potassium/uranium ratios 310, 376, 386-7, 398, 404, 418
Pratt hypothesis 183
Pre-Imbrian system 29, 32
    pitted terrain 42
Pre-Nectarian system 29, 33, 242
Pre-solar grains 413
Primitive source model 325
Pristine rocks 221
Protoplanet 415
Pyroxene, crystallation history 334

Radioactive elements 437
Radiogenic lead 217
Rare gases
    in lunar soils 161
    in solar wind 155
Rare-earth element patterns 202
Reduction-during-accretion model 424
Refractory elements 229, 378, 380, 386, 391, 395, 398, 418, 425
Regolith 3, 186, 356
    accumulation rate 123
    chemistry 140-2
    core samples 121
    exotic components 142
    metallic Fe 142
    petrology 140-1
Reiner Gamma 100, 367-8
Remanent magnetism 364
Remote landers 438
Rhenium 315
"Rusty rock" (66095) 203

S-wave velocities 358
SCCRV 205
Sampling, lunar 3, 9
Saturn 422, 424
Saturnian satellites 57
Scandium/samarium relationships 249
Shock melting 196
Shock waves 65, 411,415, 417
Siderophile elements 389, 400, 418, 425-6
Siderophile trace elements 220, 399
Silver 315
Silver-107 413-4
Silver Spur 186
Sinuous rilles 278
Sinus Iridum 349
Skaergaard intrusion 248, 334
Soil
    breccias 127
    density 119
    maturity 128
    mechanical properties 120
    meteoritic component 151
    model ages 123
    orange 298
Soil, lunar 159, 161, 167
Soil, Martian 121
Solar flare activity 155, 161, 168
Solar history 167
Solar luminosity 169
Solar nebula 380-1, 411, 413, 415
    temperature 413, 418
Solar wind 115, 155, 161
    magnetic field 369
Spinel troctolites 207
Star formation 411
Stillwater intrusion 248
Stonewall effect 246, 248
Straight rilles 282
Strontium 249, 329
Subsatellite magnetometer 358
Subsurface reflectors 119
Suevite 90, 194-5
Sulfur 159, 313, 337
Super-KREEP 214
Superheavy elements 158
Supernovae 410, 413, 418
Surface dust 117
Surface magnetometer experiment 364
Surveyor V 437

Surveyor VI 437
Surveyor VII 227, 437
Swirls 100

T Tauri stage 418-9, 422
Target characteristics 70
Taurus-Littrow 119, 241, 272, 331, 351
Taylor-Jakeš model 312, 327
Tectonic features, lunar surface 352
Tektites 135, 227
  REE patterns 137
  ages 138
  bediasites 135
  chemistry 135
  microtektites 135-7
  moldavites 135
  water content 137
Tektites, bottle-green 137
Tektites, lunar
  impact theory 139
  volcanic hypothesis 138
Tektites, terrestrial 138
Terminal cataclysm 435
Tharsis plateau 352
Thellier method 366
Thermal neutron flux 163
Thorium 223
Thorium/uranium ratios 310, 425
Tides 431
Titanium 283,334
Titanium/samarium ratios 249
Track densities 157-8

Universe 409-10
Uranium 363, 377, 388, 395,399
Uranus 416, 422

Vallis Rheita 33
Vanadium 312
Vent widths 272
Venus 336, 427
  Aphrodite Terra 55, 354
  Beta Regio 336
  composition 404
  core 359, 372
  gravity 354
  Ishtar Terra 55, 354
  Maxwell Mountains 57
  surface 116
Vesta 338, 405
Viking landing sites 55

Viking missions 11, 116, 437
Viscosity 331
Vitruvius 241
Volatile elements 147, 151, 216, 229, 299, 300, 313, 379, 380, 393, 420. 425-6
Volatile/involatile element ratios 306, 391
Volatility 378
Volatilization 297
Volcanic processes 19
Volcanic vents 272
Volcanism, lunar 214
Volcanoes, Martian
  Alba Patera 52
  Arsia Mons 52
  Ascraeus Mons 52
  Elysium Mons 52
  Olympus Mons 14, 52, 76, 116, 274, 335, 354
  Pavonis Mons 52
Volcanoes, terrestrial
  Mauna Loa 274, 336

Wall terraces 71
Water 116, 151
Wrinkle ridges 276, 352

Zap pits 64
Zirconium/niobium ratios 310